Linear Algebra and Probability
for Computer Science Applications

Linear Algebra and Probability
for Computer Science Applications

Ernest Davis

CRC Press
Taylor & Francis Group
Boca Raton London New York

CRC Press is an imprint of the
Taylor & Francis Group, an **informa** business

AN A K PETERS BOOK

CRC Press
Taylor & Francis Group
6000 Broken Sound Parkway NW, Suite 300
Boca Raton, FL 33487-2742

© 2012 by Taylor & Francis Group, LLC
CRC Press is an imprint of Taylor & Francis Group, an Informa business

No claim to original U.S. Government works

Printed in the United States of America on acid-free paper
Version Date: 20120110

International Standard Book Number: 978-1-4665-0155-3 (Hardback)

Library of Congress Cataloging-in-Publication Data

Davis, Ernest.
 Linear algebra and probability for computer science applications / Ernest Davis.
  p. cm.
 Summary: "Taking a computer scientist's point of view, this classroom-tested text gives an introduction to linear algebra and probability theory, including some basic aspects of statistics. It discusses examples of applications from a wide range of areas of computer science, including computer graphics, computer vision, robotics, natural language processing, web search, machine learning, statistical analysis, game playing, graph theory, scientific computing, decision theory, coding, cryptography, network analysis, data compression, and signal processing. It includes an extensive discussion of MATLAB, and includes numerous MATLAB exercises and programming assignments"-- Provided by publisher.
 Includes bibliographical references and index.
 ISBN 978-1-4665-0155-3 (hardback)
  1. Computer science--Mathematics. 2. Algebras, Linear. 3. Probabilities. I. Title.

QA76.9.M35D38 2012
004.01'51--dc23   2011047969

Visit the Taylor & Francis Web site at
http://www.taylorandfrancis.com

and the CRC Press Web site at
http://www.crcpress.com

For my beloved father,
Philip J. Davis,
who has had a lifelong love of *matrices*,
a profound respect for the workings of *chance,*
and a healthy distrust of the *applications of statistical inference.*

Consider the recent flight to Mars that put a "laboratory vehicle" on that planet. …From first to last, the Mars shot would have been impossible without a tremendous underlay of mathematics built into chips and software. It would defy the most knowledgeable historian of mathematics to discover and describe all the mathematics involved.

—Philip Davis, "A Letter to Christina of Denmark,"
Newsletter of the European Mathematical Society 51 (March 2004), 21–24

Contents

Preface

Since computer science (CS) first became an academic discipline almost 50 years ago, a central question in defining the computer science curriculum has always been, "How much, and what kind of, college-level mathematics does a computer scientist need to know?" As with all curricular questions, the correct answer, of course, is that everyone should know everything about everything. Indeed, if you raise the question over lunch with computer science professors, you will soon hear the familiar lament that the students these days don't know Desargues' theorem and have never heard of a p-adic number. However, in view of the limited time available for a degree program, every additional math course that a CS student takes is essentially one fewer CS course that he/she has time to take. The conventional wisdom of the field has, for the most part, therefore converged on the decision that, beyond first-semester calculus, what CS students really need is a one-semester course in "discrete math," a pleasant smorgasbord of logic, set theory, graph theory, and combinatorics. More math than that can be left as electives.

I do not wish to depart from that conventional wisdom; I think it is accurate to say that a discrete math course indeed provides sufficient mathematical background for a computer scientist working in the "core" areas of the field, such as databases, compilers, operating systems, architecture, and networks. Many computer scientists have had very successful careers knowing little or no more math. Other mathematical issues no doubt arise even in these areas, but peripherally and unsystematically; the computer scientist can learn the math needed as it arises.

However, other areas of computer science, including artificial intelligence, graphics, machine learning, optimization, data mining, computational finance, computational biology, bioinformatics, computer vision, information retrieval, and web search, require a different mathematical background. The importance of these areas within the field of computer science has steadily increased over the last 20 years. These fields and their subfields vary widely, of course, in terms of exactly what areas of math they require and at what depth. Three mathematical subjects stand out, however, as particularly important in all or most of these fields: linear algebra, probability, and multivariable calculus.

An undergraduate degree typically involves about 32 semester courses; it may not be unreasonable to suggest or even to require that CS majors take the undergraduate courses in linear algebra, multivariable calculus, and probability given by the math department. However, the need for many CS students to learn these subjects is particularly difficult to address in a master's degree program, which typically requires only 10 or 12 courses. Many students entering a master's program have weak mathematical backgrounds, but they are eager to get started on CS courses that have a substantial mathematical prerequisite. One can hardly ask them to delay their computer science work until they have completed three or four mathematics courses. Moreover, they cannot get graduate credit for undergraduate math courses, and if the math department does offer graduate courses in these areas, the courses are almost certainly too difficult for the CS master's student.

To fill this gap, I have created a new course entitled, vaguely, "Mathematical Techniques for Computer Science Applications," for the master's program at New York University (NYU), which gives an intensive introduction to linear algebra and probability, and is particularly addressed to students with weak mathematical backgrounds. This course has been offered once a year, in the fall semester, every year starting in 2009. I wrote this textbook specifically for use in this course.

Master's courses in the computer science department at NYU meet 14 times in a semester, once a week, in two-hour sessions. In my class in 2010, Chapters 1 and 2 were covered together in a single lecture; Chapters 6 and 9 required two lectures each; Chapters 3, 4, 5, 7, 8, 10, 11, 12, and 13 were each covered in a single lecture; and Chapter 14 was omitted. About halfway through the class, I decided it was necessary to add a recitation section for an additional hour a week; I certainly recommend that.

Multivariable calculus remains a gap that I have not found any practical way of closing. Obviously, it would not be possible to squeeze all three topics into a single semester, but neither, probably, is it practical to suggest that master's students take two semesters of "mathematical techniques."

The course as I teach it involves extensive programming in MATLAB. Correspondingly, the book contains an introductory chapter on MATLAB, discussions in each chapter of the relevant MATLAB functions and features, and many MATLAB assignments.

Figure 1 illustrates the strong dependencies among chapters. (Weaker dependencies also exist, in which one section of a later chapter depends on material from an earlier chapter; these are not shown.)

There are, of course, plenty of textbooks for probability and a huge number of textbooks for linear algebra, so why write a new one? For one thing, I wanted

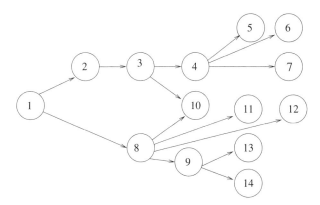

Figure 1. Chapter dependencies.

to have both subjects in a single book; the only other book I have seen that does this is *The Mathematics of Digital Images* by Stuart Hoggar (2006). More important, this book is distinguished by being addressed to the computer scientist, rather than the mathematician, physical scientist or engineer. This focus affects the *background assumed,* the *choice of topics,* the *examples,* and the *presentation.*

Background

The textbook assumes as little mathematical background as I could manage. Most of the book assumes only high-school mathematics. Complex numbers are nowhere used. In the linear algebra section of the book, calculus is entirely avoided, except in one optional short discussion of Jacobians. In the probability section of the book, this is less feasible, but I have for the most part segregated the sections of the text that do require calculus from those that do not. A basic understanding of integrals is an unavoidable prerequisite for understanding continuous probability densities, and understanding multiple integrals is a prerequisite for understanding the joint distribution of multiple continuous variables.

   The issue of mathematical proof in a course of this kind is a difficult one. The book includes the proofs of most, though not all, of the theorems that are stated, including a few somewhat lengthy proofs. In Chapter 4 on vector spaces, I have in fact split the chapter into two parts: the first contains the minimal material needed for subsequent chapters with almost no proofs, and the second presents more abstract material and more proofs. My own practice in teaching is that in lectures I present some of the proofs that I feel to be enlightening. I do try to keep in mind, however, that—whatever the mathematically trained instructor may imagine—a proof is not the same as an explanation, even for students who are mathematically inclined, and for students who are math-averse, a proof bears no relation to an explanation. This textbook includes a number of problems that ask the students to write proofs. My own practice in teaching, however, is that I do not assign problems that require proofs. My experience is that the "proofs" produced by students with weak mathematical backgrounds tend to be random sequences of sentences, only some of which are true. Unless an instructor is willing to invest substantial effort into teaching what is and is not a valid proof and how one constructs a valid proof, assigning proofs as homework or exam problems is merely frustrating and wearisome for both the student and the grader.

I have, however, assumed some familiarity with the basic concepts of computer science. I assume throughout that the student is comfortable writing programs, and I discuss issues such as computational complexity, round-off error, and programming language design (in connection with MATLAB).

Choice of Topics

In both parts of the course, the topics included are intended to be those areas that are most important to the computer scientist. These topics are somewhat different from the usual material in the corresponding introductory math classes, where the intended audience usually comprises math and science students. I have been guided here both by own impression and by discussions with colleagues.

In the linear algebra section, I have restricted the discussion to finite-dimensional vectors and matrices over the reals. Determinants are mentioned only briefly as a measure of volume change and handedness change in geometric transformations. I have almost entirely omitted discussion of eigenvalues and eigenvectors, both because they seem to be more important in physics and engineering than in CS applications, and because the theory of eigenvalues really cannot be reasonably presented without using complex eigenvalues. Instead, I have included a discussion of the singular value decomposition, which has more CS applications and involves only real values. I have also included a more extensive discussion of geometric applications and of issues of floating-point computation than is found in many standard linear algebra textbooks.

In the probability section, I have included only a minimal discussion of combinatorics, such as counting combinations and permutations. I have also omitted a number of well-known distributions such as the Poisson distribution. I have, however, included the inverse power-law "Zipf" distribution, which arises often in CS applications but is not often discussed in probability textbooks. I have also included discussions of the likelihood interpretation versus the sample space interpretation of probability, and of the basic elements of information theory as well as a very rudimentary introduction to basic techniques of statistics.

Examples and Presentation

The examples and the programming assignments focus on computer science applications. The applications discussed here are drawn from a wide range of areas of computer science, including computer graphics, computer vision, robotics, natural language processing, web search, machine learning, statistical analysis, game playing, graph theory, scientific computing, decision theory, coding, cryptography, network analysis, data compression, and signal processing. There is, no doubt, a bias toward artificial intelligence, particularly natural language processing, and toward geometric problems, partly because of my own interests, and partly because these areas lend themselves to simple programs that do interesting things. Likewise, the presentation is geared toward problems that arise in programming and computer science.

Homework problems are provided at the end of each chapter. These are divided into three categories. *Exercises* are problems that involve a single calculation; some of these can be done by hand, and some require MATLAB. Most exercises are short, but a few are quite demanding, such as Exercise 10.2, which asks students to compute the Markov model and stationary distribution for the game of Monopoly. *Programming Assignments* require the student to write a MATLAB function with parameters. These vary considerably in difficulty; a few are as short as one line of MATLAB, whereas others require some hundreds of lines. I have not included any assignments that would qualify for a semester project. *Problems* include everything else; generally, they are "thought problems," particularly proofs.

Course Website

The website for course materials is

http://www.cs.nyu.edu/faculty/davise/MathTechniques/

In particular, MATLAB code discussed in this text can be found here.

Errors, queries, and suggestions for improvements should be emailed to davise@cs.nyu.edu.

Acknowledgments

I am very grateful to my colleagues for their encouragement and suggestions in the development of this course and the writing of this book. I am especially grateful to Michael Overton, who was deeply involved in the design of the linear algebra section of the course. He read two drafts of that section of the text and made many suggestions that greatly enriched it. I have also received valuable suggestions and information from Marsha Berger, Zvi Kedem, Gregory Lawler, Dennis Shasha, Alan Siegel, and Olga Sorkine.

I owe a special debt of gratitude to the students in "Mathematical Techniques for Computer Science Applications," who suffered through early drafts of this text as well as many of the assignments, and who gave invaluable feedback. My thanks to Klaus Peters, Charlotte Henderson, Sarah Chow, and Sandy Rush for all their careful and patient work in preparing this book for publication.

During part of the writing of this book, I had support from the National Science Foundation on grant #IIS-0534809.

My own introduction to linear algebra and to probability theory came from courses that I took with Bruno Harris in spring 1974 and Gian-Carlo Rota in fall 1975. I hope that this book is a credit to their teaching. Finally, as always, thanks to Bianca for everything.

Chapter 1

MATLAB

MATLAB (short for MATrix LABoratory) is a programming language, together with a programming environment, designed to facilitate mathematical calculations and rapid prototyping of mathematical programs. It was created in the late 1970s by Cleve Moler, and it has become very popular in the mathematical, scientific, and engineering communities.

There are many fine handbooks for MATLAB, including those by Driscoll (2009) and Gilat (2008). The online documentation is also good.

A number of freeware clones of MATLAB are available, including Octave and Scilab. These should certainly be adequate for the programs discussed and assigned in this book. All the MATLAB examples in this book were generated by using MATLAB 7.8.0 (R2009a).

MATLAB creates a collection of windows. These may be "docked"—that is, all placed together within a single window—or "undocked." The most important is the Command window. The user types MATLAB commands into the Command window. The MATLAB interpreter executes the command and prints the value in the Command window. To suppress a printout (useful with a large object), the user can type a semicolon (;) at the end of the command. The user prompt in the Command window is >>. Comments begin with a percent sign (%) and continue to the end of the line.

This chapter presents some basic features of the MATLAB language. More advanced features, including operations on vectors and matrices, are discussed in the chapters where the associated math is presented.

1.1 Desk Calculator Operations

The basic arithmetic operations in MATLAB use a standard format. The command window can be used as a convenient interactive desk calculator.

```
>> format compact
>> % Comment.  format compact eliminates extra line spaces.
>> x=2+7
```

1

```
x =
     9

>> y=2*x
y =
    18

>> x+y
ans =
    27

>>% If the user types an expression, the value is assigned to variable "ans"
>> ans^(1/3)
ans =
     3

>> sqrt(2)
ans =
    1.4142

>> format long
>> ans

ans =
    1.414213562373095

>> format short

>> sin(pi/3)
ans =
    0.8660

>> format rat   % display in "rational format,'' a close rational approximation

>> (1/7)+(1/5)
ans =
      12/35

>> sqrt(2)
ans =
    1393/985
```

1.2   Booleans

MATLAB uses 1 for true and 0 for false. Strictly speaking, these are not the same as the integers 1 and 0, but MATLAB automatically casts from one to the other as needed, so the distinction only occasionally makes a difference. (We will see an example in Section 2.5.)

```
>> a=5==5
a =
     1

>> b=5==6
b =
     0

>> a&b
ans =
     0

>> a|b
ans =
     1

>> ~a
ans =
     0

>> a+a %Casting from Boolean to integer
ans =
     2

>> (2-1)&a  %Casting from integer to Boolean
ans =
     1

>> a==2-1
ans =
     1
```

1.3 Nonstandard Numbers

MATLAB conforms to the IEEE standard for floating-point arithmetic (see Overton, 2001), which mandates that a system of floating-point arithmetic support
the three nonstandard values Inf (positive infinity), -Inf (negative infinity),
and NaN (not a number). These values are considered as numbers in MATLAB.
The infinite values can generally be used numerically in any context where an
infinite value makes sense; some examples are shown next. NaN is used for
values that are completely undefined, such as 0/0 or 0*Inf. Any computation involving NaN gives NaN, and any comparison involving NaN is considered
false.

```
>> 1/0
ans =
   Inf
```

```
>> Inf+Inf
ans =
   Inf

>> Inf*-3
ans =
  -Inf

>> Inf^2
ans =
   Inf

>> 5 < Inf
ans =
     1

>> 0/0
ans =
   NaN


>> 5 < NaN
ans =
     0

>> 5 >= NaN
ans =
     0

>> sin(Inf)
ans =
   NaN

>> atan(Inf)
ans =
    1.5708
```

1.4   Loops and Conditionals

MATLAB has the usual conditional and looping constructs to build complex statements out of simple ones. Note that

- Loops and conditionals end with the key word end. Therefore, there is no need for "begin ... end" or "{ ... }" blocks.

- Atomic statements are separated by line breaks. Once a compound statement has been entered, the interpreter continues to read input until the compound statement is ended.

- One can continue an atomic statement past the line break by typing "…" at the end of the line.

- The value of each atomic statement is printed out as it is executed. This is helpful in debugging; we can trace the execution of a statement just by deleting the semicolon at the end. To suppress printout, put a semicolon at the end.

```
>> for n=1:5
s=s+n
end  % End of user input

s =
        1
s =
        3
s =
        6
s =
        10
s =
        15

>> for n=1:3
t=a; % Note suppression of printout
a=b
b=t+b
end

a =
        1
b =
        2
a =
        2
b =
        3
a =
        3
b =
        5

>> % Note: Unlike many programming languages, statements are separated by line
>> % breaks
>> x=1;
>> while x < 50
x=x+x
end
x =
        2
x =
        4
```

```
x =
        8
x =
       16
x =
       32
x =
       64

>> % The hailstone procedure
>> x=3;
x =
        3
>> while (x ~= 1)
       if (mod(x,2)==1)
              x=3*x+1
          else x=x/2
          end
      end

x =
       10
x =
        5
x =
       16
x =
        8
x =
        4
x =
        2
x =
        1
```

1.5 Script File

A script file is a plain text file with MATLAB commands. The file has the extension .m. To execute a script file that is in the working directory, just enter the name of the file (omitting the ".m") in the command window. The working directory can be changed in the "directory" window. An example is the file p35.m:

```
% p35.m computes x^35 by repeated squaring followed by multiplication

x2=x*x          %  x^2
x4=x2*x2        %  x^4
x8=x4*x4        %  x^8
x16=x8*x8       %  x^16
```

```
x32=x16*x16   % x^32
x35=x32*x2*x  % x^35
```

And this is how to execute it in the command window:

```
>> x=2
x =
        2

>> p35
x2 =
        4
x4 =
       16
x8 =
      256
x16 =
    65536
x32 =
4294967296
x35 =
34359738368

>> % Variables defined in a script are visible after the script has executed.
>> x4
x4 =
       16

>> % help prints out the comments at the head of the script file.
>> help p35
  p35.m computes x^35 by repeated squaring followed by multiplication
```

1.6   Functions

The MATLAB code for the function named foo is in the file foo.m. To call foo in the command window, just type "foo(...)"; this both loads and executes the code.

The main differences between functions and scripts is that variables in a function, including the input and output parameters, are local to the function. (It is also possible to define and access global variables.) The function declaration has one of two forms:

```
function  <output variable> =  <function name>(input variables)
```

or

```
function [<output variables>] = <function name>(input variables)
```

For instance, for the file fib.m:

```
% Recursive, very inefficient, definition of Fibonacci numbers

function x = fib(n)
  if (n==0) x=1;
  elseif (n==1) x=1;
  else x=fib(n-1)+fib(n-2);
  end
end
```

And its use in MATLAB is:

```
>> fib(5)
ans =
     8

>> help fib
  Recursive, very inefficient, definition of Fibonacci numbers
```

Functions may return multiple values. For example, for quadform.m:

```
% quadform(a,b,c) returns the two roots of a quadratic equation
% ax^{2} + bx + c = 0

function [r1,r2] = quadform(a,b,c)
  x=sqrt(b^2 - 4*a*c);
  r1=(-b-x)/(2*a);
  r2=(-b+x)/(2*a);
end


>> [p,q]=quadform(1,0,-1)
p =
    -1
q =
     1
```

Function files can have subfunctions. These are placed in the file after the main function (the one with the same name as the file). Subfunctions can be called only by functions within this same file.

Important note: i and j are used by MATLAB as predefined symbols for the square root of -1. However, MATLAB does not prevent us from using these as variables and reassigning them. If we aren't using complex numbers in our code (none of the assignments in this book require complex numbers), this does not usually lead to trouble but occasionally can cause confusion. The most common case is that if our code uses i or j without initializing them, we do not get an error, as with other variables.

```
% In an environment where the user has not initialized either 'a' or 'i'.
>> a+1
??? Undefined function or variable 'a'.

>> i+1
ans =
   1.0000 + 1.0000i
```

1.7   Variable Scope and Parameter Passing

By default a variable used in a function is local to the function call; that is, if
a function calls itself recursively, then the variables in the two instances of the
functions are different. Variables declared in the command window are local
to the command window.

If a variable is declared global in a number of functions, then it is shared
among all the calls of those functions. If it is also declared global in the command window, then it is also shared with the command window environment.

Parameters are *always* passed call-by-value. That is, the formal parameters
of a function are local variables; the value of the actual parameter is copied into
the formal parameter when the function is called, but not copied back when
the function returns.

Here is a toy example. Suppose that we have the two functions t1 and t2
defined in the file t1.m as follows:

```
% t1(x) is a toy example used to illustrate global and local variables

function y = t1(x)
global k m
    p=x;
    x=1;
    k=2;
    m=3;
    t2(x)
    y=[p,x,k,m]
end

function t2(z)
global k
    p=11;
    k=12;
    m=14;
    z=16;
end
```

The variable k is a global variable shared between t1 and t2. The variable m in t1 is a global variable that will be shared with the command

window; it is not shared with t2. We now execute the following in the command
window:

```
>> global m

>> k=100
k =
    100

>> m=200
m =
    200

>> p=300
p =
    300

>> x=400
x =
    400

>> t1(x)
y =
    400     1    12     3
% This is the execution of the last statement of t1:
    y=[p,x,k,m]
% Note that k has been changed by the execution of t2, because it was
% declared global between t1 and t2; the others are unchanged by the
% execution of t2.

% Now back to the command window
y =
    400     1    12     3
>> [p,x,k,m]
ans =
    300   400   100     3

% Note that m has been changed by the execution of t1, because it was
% declared global between the command window and t1. The others are
% unchanged .
```

The use of call-by-value in MATLAB means that if function f has parameter
y and function g executes the statement f(x), then the value of x in g must be
copied into the variable y in f before execution of f can begin. If x is a large matrix, this can involve a significant overhead. Therefore, the programming style
that is encouraged in LISP and other similar languages—using large numbers
of small functions and using recursion for implementing loops—is ill-suited to
MATLAB. Loops should generally be implemented, if possible, by using MATLAB
operators or library functions, or else by using iteration.

Copying also occurs at any assignment statement; if x is a large matrix, then executing the statement y=x involves copying x into y. If we are executing a long loop involving large arrays and are concerned with efficiency, we may need to give some thought to reducing the number of unnecessary copies.

The truth is that if our program is largely executing our own MATLAB code (rather than the built-in MATLAB functions, which are implemented efficiently), if it manipulates large quantities of data, and if efficiency is a concern, then we should probably be using some other programming language. Or at least we should write the critical sections of the program in another language; MATLAB has facilities for interfacing with code written in C or C++.

Problem

Problem 1.1. Can you write a function "swap(A,B)" in MATLAB that swaps the values of its arguments? That is, the function should have the following behavior:

```
>> i=1;
>> j=5;
>> k=10;

>> swap(i,j);
>> i
i =
     5
>> j
j =
     1

>> swap(j,k);
>> j
j =
     10
>> k
k =
     1
```

Explain your answer.

Programming Assignments

Note: We have not yet defined any data structures, so the programming assignments for this chapter are necessarily very numerical in flavor.

Assignment 1.1. A pair of *twin primes* is a pair of prime numbers that differ by two. For example, $\langle 3,5 \rangle$, $\langle 5,7 \rangle$, $\langle 11,13 \rangle$, $\langle 17,19 \rangle$, and $\langle 29,31 \rangle$ are the first five pairs of twin primes. It has been conjectured that for large N, the number of twin prime pairs less than N is approximately the function $f(N) = 1.3203 \cdot N/(\log_e N)^2$.

Write a MATLAB function `CountTwinPrimes(N)` that counts the number of twin prime pairs less than N and compares the result to the above estimate. That is, your function should return two values:

- C, the number of twin-prime pairs less than N

- $|(C - f(N))/f(N)|$, where $f(N)$ is the expression defined above.

You may use the built-in MATLAB function `isprime(X)`.

Assignment 1.2. Goldbach's conjecture asserts that every even number greater than 2 is the sum of two primes. For instance, $4 = 2+2$, $6 = 3+3$, $8 = 3+5$, $132 = 23+109$, and so on.

Write a function `Goldbach(N)` which takes as argument an even number N and returns a pair of primes P,Q such that N=P+Q.

Assignment 1.3. The *four squares* theorem states that every positive integer can be written as the sum of four square integers. For example,

$$26 = 5^2 + 1^2 + 0^2 + 0^2,$$
$$56 = 6^2 + 4^2 + 2^2 + 0^2,$$
$$71 = 7^2 + 3^2 + 3^2 + 2^2.$$

Write a MATLAB function `FourSquares(N)` that returns four integers A,B,C,D such that $N = A^2 + B^2 + C^2 + D^2$.

Assignment 1.4. Write a MATLAB function `TriangleArea(A,B,C)` that computes the area of a triangle with sides of lengths A,B,C. For instance,

`TriangleArea(3,4,5)`	should return 6;
`TriangleArea(1,1,1)`	should return 0.4330 $(= \sqrt{3}/4)$;
`TriangleArea(1,1,sqrt(2))`	should return 0.5.

Hint: Look up "Heron's formula" in Wikipedia or your favorite search engine.

Assignment 1.5. The recurrence equation $x_{n+1} = 2x_n^2 - 1$ exhibits *chaotic* behavior for x between -1 and 1. That is, if you compute the series starting from a particular starting point y_0 and then recompute the series starting from a starting point $z_0 = y_0 + \epsilon$, which is very close to y_0, the two series grow apart very quickly, and soon the values in one are entirely unrelated to the values in the other.

For example, the two series starting at $y_0 = 0.75$ and $z_0 = 0.76$ are

$y_0 = 0.75$, $y_1 = 0.125$, $y_2 = -0.9688$, $y_3 = 0.8770$, $y_4 = 0.5381$, $y_5 = -0.4209$, $y_6 = -0.6457$;
$z_0 = 0.76$, $z_1 = 0.1552$, $z_2 = -0.9518$, $z_3 = 0.8119$, $z_4 = 0.3185$, $z_5 = -0.7971$, $z_6 = 0.2707$.

(a) Write a MATLAB function `CompareChaotic(Y0,Z0)` that takes as arguments two starting values y_0 and z_0 and returns the minimum value of n for which $|y_n - z_n| > 0.5$. For instance `CompareChaotic(0.75,0.76)` should return 6, since in the above series $|y_6 - z_6| = 0.9164 > 0.5$.

(b) Experiment with a sequence of pairs that are increasingly close together, such as $\langle Y0 = .75, Z0 = 0.751 \rangle$, $\langle Y0 = 0.75, Z0 = 0.7501 \rangle$, $\langle Y0 = 0.75, Z0 = 0.75001 \rangle$, and so on. Formulate a conjecture as to how the value of `Compare-Chaotic(Y0,Z0)` increases with the value of $1/|Y0 - Z0|$.

(c) Double precision numbers are represented with about 16 digits (51 bits) of precision. Suppose that you start with a value of y_0 in double precision and compute the series $y_0, y_1, y_2, \dots$. Given your conjecture in (b), how many terms of the series can you compute before the values become completely unrelated to the true value?

I

Linear Algebra

Chapter 2

Vectors

Linear algebra is the study of vectors, discussed in this chapter, and matrices, discussed in Chapter 3.

2.1 Definition of Vectors

An *n-dimensional vector* is a *n*-tuple of numbers. (A more general definition will be discussed in Section 4.3.1.) The indices $1, \ldots, n$ are the *dimensions* of the vector. The values of the tuple are the *components* of the vector.

In this text, we use angle brackets $\langle \ldots \rangle$ to delimit vectors. We use a letter with an arrow over it, such as $\vec{v}$, to denote a vector. The *i*th component of vector $\vec{v}$ is notated $\vec{v}[i]$. (In mathematical writings, it is often notated $\vec{v}_i$, but we use the subscript exclusively for naming different vectors.) For example, $\vec{v} = \langle 0, 6, -2.5 \rangle$ is a three-dimensional vector. $\vec{v}[1] = 0$. $\vec{v}[3] = -2.5$.

A *zero vector*, denoted $\vec{0}$, is a vector whose components are all 0. The four-dimensional zero-vector is $\langle 0, 0, 0, 0 \rangle$. (The dimension associated with the notation $\vec{0}$ is determined by context.)

The *unit vector in the i dimension*, denoted $\vec{e}^{\,i}$, is the vector with a 1 in the *i*th coordinate and 0s everywhere else. For instance, in four dimensions, $\vec{e}^{\,2} = \langle 0, 1, 0, 0 \rangle$. The *n*-dimensional one vector, denoted $\vec{1}$, is the vector with 1s in each dimension; for instance, in four dimensions, $\vec{1} = \langle 1, 1, 1, 1 \rangle$.

The set of all *n*-dimensional vectors is the *n-dimensional Euclidean vector space,* denoted $\mathbb{R}^n$.

2.2 Applications of Vectors

Vectors can be used in many different kinds of applications. A few typical examples are discussed here; many more examples are encountered in the course of this book.

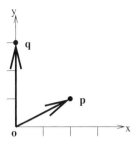

Figure 2.1. Points represented as vectors.

Application 2.1 (Geometric points). In two-dimensional geometry, fix a coordinate system $\mathscr{C}$ by specifying an origin, an x-axis, and a y-axis. A point $\mathbf{p}$ can then represented by using the vector $\vec{p} = \langle \mathbf{p}[x], \mathbf{p}[y] \rangle$, where $\mathbf{p}[x], \mathbf{p}[y]$ are the coordinates of $\mathbf{p}$ in the x and y dimensions, respectively. The vector $\vec{p}$ is called the *coordinate vector* for point $\mathbf{p}$ relative to the coordinate system $\mathscr{C}$. For instance, in Figure 2.1, $\mathbf{p}$ is associated with the vector $\langle 2, 1 \rangle$ and $\mathbf{q}$ is associated with the vector $\langle 0, 3 \rangle$.

Likewise, in three-dimensional geometry, point $\mathbf{p}$ can be associated with the vector $\langle \mathbf{p}[x], \mathbf{p}[y], \mathbf{p}[z] \rangle$ of components in the x, y, z dimensions relative to a particular coordinate system. In n-dimensional geometry, point $\mathbf{p}$ can be associated with the vector $\langle \mathbf{p}[1], \ldots, \mathbf{p}[n] \rangle$ where $\mathbf{p}[i]$ is the component along the ith coordinate axis. Geometric coordinates are discussed at length in Chapter 6.

Application 2.2 (Time series). A vector can be used to represent a sequence of numeric values of some quantity over time. For instance, the daily closing value of the S&P 500 index over a particular week might be $\langle 900.1, 905.2, 903.7, 904.8, 905.5 \rangle$. A patient's hourly body temperature in degrees Farenheit over a four-hour period might be $\langle 103.1, 102.8, 102.0, 100.9 \rangle$. Here the dimensions correspond to points in time; the components are the values of the quantity.

Application 2.3 (Almanac information). Numerical information about the 50 states could be recorded in 50-dimensional vectors, where the dimension corresponds to the states in alphabetical order: Alabama, Alaska, Arizona, Arkansas, etc. We could then have a vector $\vec{p} = \langle 4530000, 650000, 5740000, 2750000, \ldots \rangle$ representing the populations of the states; or $\vec{a} = \langle 52400, 663200, 114000, 2750000, \ldots \rangle$ representing the areas of the states in square miles; and so on.

Application 2.4 (Shopping). The dimensions correspond to a sequence of grocery products: gallon of milk, loaf of bread, apple, carrot, and so on. Each store has a *price vector*; for example, $\vec{s}[i]$ is the price of the ith product at the Stop and Shop, and $\vec{g}[i]$ is the price of the ith product at Gristedes. Each customer has a *shopping list vector*; for example, a vector $\vec{a}$, where $\vec{a}[i]$ is the number of item i on Amy's shopping list, and $\vec{b}$, where $\vec{b}[i]$ is the number of item i on Bob's shopping list.

Note that if the dimensions include all products standardly sold in grocery stores, then each shopping list vector has mostly zeroes. A vector that has mostly zeroes is called a *sparse* vector; these are common in many applications.

We may well ask, "How should we represent the price of an item that Stop and Shop does not sell?" These are known as null values. In MATLAB, we can use the special value NaN (not a number) to represent null values. In programming languages that do not support NaN, we either have to use a value that is known to be not the price of any actual groceries, such as a negative value or a very large value, or we have a second vector of 1s and 0s that indicates whether or not the object is sold. In either case, applications that use these vectors have to be aware of this representation and take it into account. In fact, in some applications it may be necessary to have more than one kind of null value; for instance, to distinguish items that Stop and Shop does not sell from items for which the price is unknown. We ignore all these issues here, however, and assume that every store has a price for every item.

Application 2.5 (Personal database). Consider a database of personal information for various kinds of applications, such as medical diagnosis, mortgage evaluation, security screening, and so on. Each dimension is the result of a numerical measurement or test. Each person has a corresponding vector; for example, the value $\vec{p}[i]$ is the ith feature of person $\vec{p}$. In medical applications, we might have age, height, weight, body temperature, various kinds of blood tests, and so on. In financial applications, we might have net worth, income, credit score, and so on. (Note that features that are critical in one application may be illegal to use in another.)

Alternatively, we can have a vector for each feature, indexed by person. For instance, $\vec{a}$ could be the vector of ages, where $\vec{a}[i]$ is the age of person i.

Application 2.6 (Document analysis). In a library of documents, one can have one dimension corresponding to each word that appears in any document in the collection and a vector for each document in the collection. Then, for document $\vec{d}$ and word w, vector $\vec{d}$ denotes a document, and $\vec{d}[w]$ is the number of times word w appears in document d.

Alternatively, we can have the dimensions correspond to documents and the vectors correspond to words. That is, for each word w in the collection there is a vector $\vec{w}$, and the value of $\vec{w}[d]$ is the number of times word w appears in document d. Document vectors and word vectors are mostly sparse in most collections.

This use of document vectors was introduced in information retrieval by Gerard Salton (1971). In actual information retrieval systems, the definition of document vector is a little more complex. It gives different weights to different words, depending on their frequency; more common words are considered less important. Specifically, let N be the number of documents in a given collection. For any word w, let m_w be the number of documents in the collection that contain w, and let $i_w = \log(N/m_w)$, called the *inverse document frequency*. Note that the fewer the documents that contain w, the larger i_w; for common words such as "the," which are found in all documents, $i_w = 0$. For any document d, let $t_{w,d}$ be the number of occurrences of word w in document d. Then we can define the document vector $\vec{d}$, indexed by words, as $\vec{d}[w] = t_{w,d}\, i_w$.

2.2.1   General Comments about Applications

Broadly speaking, applications of vectors come in three categories: geometric interpretations (Application 2.1), sequential interpretations (Application 2.2), and numeric functions of two entities or features (Applications 2.3–2.6). We will see some other categories, but these three categories include many of the applications of vectors. A couple of general points may be noted.

First, the geometric interpretations depend on an arbitrary choice of coordinate system: how a *point* is modeled as a *vector* depends the choice of the origin and x- and y-axes. An important question, therefore, concerns how different choices of coordinate system affect the association of vectors with points. Other categories generally have a natural coordinate system, and the only similar arbitrary choice is the choice of unit (e.g., feet rather than meters, US dollars rather than euros, liters instead of gallons). Even in those cases, as we see at length in Chapter 7, it is often important to think about alternative, less natural coordinate systems.

Second, in the first and third categories, the association of dimensions with numerical indices is essentially arbitrary. There is no particular reason that the states should be listed in alphabetical order; they could just as well be listed in backward alphabetical order (Wyoming, West Virginia, Washington, ...); in order of admission to the union (Delaware, Pennsylvania, New Jersey, ...); or any other order. Any question for which the answer depends on having a particular order, such as "Are there three consecutive states with populations greater than 7 million?" is almost certain to be meaningless. Likewise, in geometric applications, there is no particular reason that the x, y, and z directions are enumerated in that order; the order z, x, y is just as good.[1] For time series and other sequential vectors, by contrast, the numeric value of the index is significant; it represents the time of the measurement.

In a programming language such as Ada, which supports enumerated types and arrays that are indexed on enumerated types, the programmer can declare explicitly that, for instance, the vectors in Application 2.3 are indexed on states or that the person vectors in Application 5 are indexed on a specified list of

---

[1]There is, however, a significant distinction between left- and right-handed coordinate systems, which is related to the order of the coordinates; see Section 7.1.2.

features. However, MATLAB does not support the type structure needed for this. For the vectors of Application 2.4 indexed by product type, or the vectors of Application 2.6 indexed by document or words, where the class of indices is not predetermined, we would want arrays indexed on *open* types, such as words; I don't know of any programming language that supports this.

This distinction is not reflected in standard mathematical notations; mathematicians, by and large, are not interested in issues of types in the programming language sense.

2.3 Basic Operations on Vectors

There are two basic[2] operations on n-dimensional vectors:

- *Multiplying a n-dimensional vector by a number.* This is done component-by-component and the result is a new n-dimensional vector. That is, if $\vec{w} = r \cdot \vec{v}$ then $\vec{w}[i] = r \cdot \vec{v}[i]$. So, for example, $4 \cdot \langle 3, 1, 10 \rangle = \langle 4 \cdot 3, 4 \cdot 1, 4 \cdot 10 \rangle = \langle 12, 4, 40 \rangle$.

  In linear algebra, a number is often called a *scalar*, so multiplication by a number is called *scalar multiplication*. It is conventional to write $a \cdot \vec{v}$ rather than $\vec{v} \cdot a$. The value $(-1) \cdot \vec{v}$ is written $-\vec{v}$. As usual in mathematical notation, we may omit the multiplicative dot symbol and just write $a\vec{v}$ when that is not confusing.

- *Adding two vectors of the same dimension.* This is done component-by-component; that is, if $\vec{w} = \vec{v} + \vec{u}$ then $\vec{w}[i] = \vec{v}[i] + \vec{u}[i]$. So, for example, $\langle 3, 1, 10 \rangle + \langle 1, 4, -2 \rangle = \langle 3 + 1, \ 1 + 4, \ 10 - 2 \rangle = \langle 4, 5, 8 \rangle$.

  The difference of two vectors $\vec{v} - \vec{u}$ is defined as $\vec{v} + (-\vec{u})$. Two vectors of different dimensions may not be added or subtracted.

2.3.1 Algebraic Properties of the Operations

The following basic properties of these operators are important. They are all easily proven from the definitions. In all the following properties, $\vec{v}$, $\vec{u}$, $\vec{w}$, and

---

[2]"Says who?" you may well ask. What's so "not basic" about operations such as finding the length of a vector, adding a scalar to a vector, finding the largest element of a vector, or sorting a vector? Certainly these are important, and, of course, they are built into MATLAB. There are, however, two related reasons for emphasizing these two operations. First, as we shall see, linear algebra is essentially about linear transformations; the two operations of vector addition and scalar multiplication are linear transformations, whereas operations such as sorting are not. Second, in the more general definition of vector discussed in Section 4.3.1, these other operations may not be pertinent. For instance, the components of a vector may be entities that are not ordered in terms of "greater" and "less." In that case, sorting is not meaningful.

$\vec{0}$ are n-dimensional vectors, and a and b are numbers:

$$\vec{v} + \vec{u} = \vec{u} + \vec{v} \qquad \text{(commutative)}$$
$$(\vec{v} + \vec{u}) + \vec{w} = \vec{v} + (\vec{u} + \vec{w}) \qquad \text{(associative)}$$
$$a \cdot (\vec{v} + \vec{u}) = (a \cdot \vec{v}) + (a \cdot \vec{u}) \qquad \text{(distributive)}$$
$$(a + b) \cdot \vec{v} = (a \cdot \vec{v}) + (b \cdot \vec{v}) \qquad \text{(distributive)}$$
$$a \cdot (b \cdot \vec{v}) = (ab) \cdot \vec{v}$$
$$0 \cdot \vec{v} = \vec{0}$$
$$\vec{v} + \vec{0} = \vec{v} \qquad (\vec{0} \text{ is the additive identity})$$
$$\vec{v} - \vec{v} = \vec{0} \qquad (-\vec{v} \text{ is the additive inverse of } \vec{v})$$

2.3.2 Applications of Basic Operations

Geometric. Fix a coordinate system in space with origin **o**. Suppose that $\vec{p}$ and $\vec{q}$ are the coordinate vectors for points **p** and **q**, respectively. Let a be a number. Draw arrows from **o** to **p** and from **o** to **q**.

Now make the arrow from **o** to **p** a times as long, keeping the tail of the arrow at **o** and the direction the same. Then the coordinate vector of the head of the arrow is $a\vec{p}$.

Then copy the arrow from **o** to **q**, keeping the direction and length the same, but put its tail at **p**. Then the coordinate vector of the head of the arrow is $\vec{p} + \vec{q}$.

Figure 2.2 illustrates this geometric application with $\vec{p} = \langle 2, 1 \rangle$, $\vec{q} = \langle 0, 3 \rangle$, $a = 1.5$.

Nongeometric. The nongeometric interpretations of vector addition and scalar multiplication are mostly obvious. We multiply by a scalar when we need to multiply every component by that scalar; we add two vectors when we need to add every component. For example, consider Application 2.4 again, where the dimensions correspond to different kinds of groceries. If $\vec{a}$ is Amy's shopping list, and $\vec{b}$ is Bob's, and they decide to shop together, then $\vec{a} + \vec{b}$ is their joint shopping list. If $\vec{s}$ is the price list for Stop and Shop and $\vec{g}$ is the price

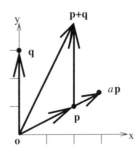

Figure 2.2. Points represented as vectors.

for Gristede's then $\vec{s} - \vec{g}$ is the amount they save at Gristede's for each item. (If $(\vec{s} - \vec{g})[i]$ is negative, then item i is more expensive at Gristede's.) If Shop and Shop announces a 20% off sale on all items in the store, then the new price vector is $0.8 \cdot \vec{s}$. To convert a price vector from dollars to euros, we multiply by the current exchange rate. Other applications work similarly.

2.4 Dot Product

The *dot product* of two *n*-dimensional vectors is computed by multiplying corresponding components, and then adding all these products. That is,

$$\vec{v} \bullet \vec{w} = \vec{v}[1] \cdot \vec{w}[1] + \ldots + \vec{v}[n] \cdot \vec{w}[n].$$

For example, $\langle 3, 1, 10 \rangle \bullet \langle -2, 0, 4 \rangle = (3 \cdot -2) + (1 \cdot 0) + (10 \cdot 4) = -6 + 0 + 40 = 34$.

The dot product is also known as the *scalar product* or the *inner product*. We indicate it by a large solid dot $\bullet$, as above; this is not standard mathematical notation, but a long sequence of small dots can be confusing.

The dot product is not defined for vectors of different dimensions.

2.4.1 Algebraic Properties of the Dot Product

The following basic properties of the dot product are important. They are all easily proven from the above definition. In the following properties, $\vec{v}$, $\vec{u}$ and $\vec{w}$ are *n*-dimensional vectors, and a is a number:

$$\vec{v} \bullet \vec{u} = \vec{u} \bullet \vec{v} \qquad \text{(commutative)}$$
$$(\vec{u} + \vec{v}) \bullet \vec{w} = (\vec{u} \bullet \vec{w}) + (\vec{v} \bullet \vec{w}) \qquad \text{(distributive)}$$
$$(a \cdot \vec{u}) \bullet \vec{v} = a \cdot (\vec{u} \bullet \vec{v}) = \vec{u} \bullet (a \cdot \vec{v})$$

Another obvious but important property is that the dot product of vector $\vec{v}$ with the *i*th unit vector $\vec{e}^{\,i}$ is equal to *i*th coordinate $\vec{v}[i]$. A generalization of this is presented in Section 4.1.3.

2.4.2 Application of the Dot Product: Weighted Sum

The simplest application of the dot product is to compute weighted sums. A few examples follow.

In the grocery shopping application (Application 2.4), $\vec{s}$ is the vector of prices at Stop and Shop, and $\vec{a}$ is Amy's shopping list. If Amy goes shopping at Stop and Shop, she will pay $\vec{a}[i] \cdot \vec{s}[i]$ for item i, and therefore her total bill will be $\sum_i \vec{a}[i] \cdot \vec{s}[i] = \vec{a} \bullet \vec{s}$.

The sum of the elements of an n-dimensional vector $\vec{v}$ is $\vec{1} \cdot \vec{v}$. The average value of a n-dimensional vector $\vec{v}$ is the sum divided by n; thus, $(1/n) \cdot \vec{1} \bullet \vec{v}$.

In the almanac application (Application 2.3), $\vec{p}$ is the population of each state. Let $\vec{q}$ be the average income in each state. Then $\vec{p}[i] \cdot \vec{q}[i]$ is the total income of all people in state i, so $\vec{p} \bullet \vec{q}$ is the total income of everyone in the country. The average income across the country is the total income divided by the population, $(\vec{p} \bullet \vec{q})/(\vec{p} \bullet \vec{1})$. Note that we are not allowed to "cancel out" the $\vec{p}$ in the numerator and denominator; dot products don't work that way.

Application 2.7 (Linear classifiers). An important category of applications involves the *classification* problem. Suppose we are given a description of an entity in terms of a vector $\vec{v}$ of some kind, and we want to know whether the entity belongs to some specific category. As examples, a bank has collected information about a loan applicant and wants to decide whether the applicant is a reasonable risk; a doctor has a collection of information (symptoms, medical history, test results, and so forth) about a patient and wants to know whether the patient suffers from a particular disease; or a spam filter has the text of an email message, encoded as a document vector (see Application 2.6) and wants to know whether the message is spam.

A *classifier* for the category is an algorithm that takes the vector of features as input and tries to calculate whether the entity is an instance of the category. One of the simplest and most widely used kinds of classifier are *linear classifiers*, which consist of a vector of weights $\vec{w}$ and a threshhold t. If the weighted sum of the feature vector, $\vec{w} \bullet \vec{v} > t$, then the classifier predicts that the entity is in the category; otherwise, the classifier predicts that it is not.

Most machine learning programs work by constructing a classifier for a category, based on a corpus of examples. One simple algorithm to do this, the *Naive Bayes* algorithm, is discussed in Section 8.11.

2.4.3   Geometric Properties of the Dot Product

Geometrical analysis yields further interesting properties of the dot product operation that can then be used in nongeometric applications. This takes a little work.

Consider a fixed two-dimensional coordinate system with origin **o**, an x-axis, and a y-axis. Let **p** and **q** be points and let $\vec{p}$ and $\vec{q}$ be the associated coordinate vectors.

First, note that, by the Pythagorean theorem, the distance from **o** to **p** (in the units of the coordinate system), which we denote as $d(\mathbf{o}, \mathbf{p})$ is

$$\sqrt{\vec{p}[x]^2 + \vec{p}[y]^2}.$$

But $\vec{p}\,[x]^2 + \vec{p}\,[y]^2 = \vec{p} \bullet \vec{p}$. So $d(\mathbf{o},\mathbf{p}) = \sqrt{\vec{p} \bullet \vec{p}}$. This quantity, $\sqrt{\vec{p} \bullet \vec{p}}$, is called the *length* of vector $\vec{p}$ and is denoted $|\vec{p}|$. Similarly,

$$d(\mathbf{p},\mathbf{q}) = \sqrt{(\vec{q}\,[x] - \vec{p}\,[x])^2 + (\vec{q}\,[y] - \vec{p}\,[y])^2}.$$

So

$$\begin{aligned}
d(\mathbf{p},\mathbf{q})^2 &= (\vec{q}\,[x] - \vec{p}\,[x])^2 + (\vec{q}\,[y] - \vec{p}\,[y])^2 \\
&= (\vec{q} - \vec{p}) \bullet (\vec{q} - \vec{p}) \\
&= \vec{q} \bullet \vec{q} - 2\vec{p} \bullet \vec{q} + \vec{p} \bullet \vec{p} \\
&= d(\mathbf{o},\mathbf{q})^2 - 2\vec{p} \bullet \vec{q} + d(\mathbf{o},\mathbf{p})^2.
\end{aligned}$$

Therefore,

$$\vec{p} \bullet \vec{q} = \frac{d(\mathbf{o},\mathbf{p})^2 + d(\mathbf{o},\mathbf{q})^2 - d(\mathbf{p},\mathbf{q})^2}{2}. \tag{2.1}$$

This proof is for two-dimensional points, but in fact the same proof works in Euclidean space of any dimension.

Proceeding from the formula for $d(\mathbf{p},\mathbf{q})$, a number of important conclusions can be deduced. First, by the triangle inequality,

$$|d(\mathbf{o},\mathbf{p}) - d(\mathbf{o},\mathbf{q})| \le d(\mathbf{p},\mathbf{q}) \le d(\mathbf{o},\mathbf{p}) + d(\mathbf{o},\mathbf{q}).$$

Since all these terms are nonnegative, we may square all parts of the inequality, giving

$$\begin{aligned}
d(\mathbf{o},\mathbf{p})^2 - 2d(\mathbf{o},\mathbf{p})d(\mathbf{o},\mathbf{q}) + d(\mathbf{o},\mathbf{q})^2 &\le d(\mathbf{p},\mathbf{q})^2 \\
&\le d(\mathbf{o},\mathbf{p})^2 + 2d(\mathbf{o},\mathbf{p})d(\mathbf{o},\mathbf{q}) + d(\mathbf{o},\mathbf{q})^2.
\end{aligned}$$

Therefore,

$$-2d(\mathbf{o},\mathbf{p})d(\mathbf{o},\mathbf{q}) \le d(\mathbf{o},\mathbf{p})^2 + d(\mathbf{o},\mathbf{q})^2 - d(\mathbf{p},\mathbf{q})^2 \le 2d(\mathbf{o},\mathbf{p})d(\mathbf{o},\mathbf{q}).$$

But, by Equation (2.1), the middle term here is just $2\vec{p} \bullet \vec{q}$. Substituting and dividing through by 2 gives

$$-d(\mathbf{o},\mathbf{p})d(\mathbf{o},\mathbf{q}) \le \vec{p} \bullet \vec{q} \le d(\mathbf{o},\mathbf{p})d(\mathbf{o},\mathbf{q}).$$

Using the facts that $d(\mathbf{o},\mathbf{p}) = |\vec{p}|$ and $d(\mathbf{o},\mathbf{q}) = |\vec{q}|$, and dividing through gives

$$-1 \le \frac{\vec{p} \bullet \vec{q}}{|\vec{p}| \cdot |\vec{q}|} \le 1. \tag{2.2}$$

Equation (2.2) is known as the *Cauchy-Schwarz inequality*.[3]

---

[3]This may seem a little suspicious. How did we derive this nontrivial algebraic inequality from this simple geometric argument? If it does seem suspicious, then you have good instincts; what we've pushed under the rug here is proving that the triangle inequality holds for Euclidean distance in n-dimensions.

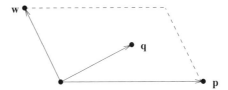

Figure 2.3. Geometric interpretation of the dot product.

Next, consider the case where $\mathbf{p}, \mathbf{o}, \mathbf{q}$ form a right angle; that is, the arrow from $\mathbf{o}$ to $\mathbf{q}$ is at right angles to the arrow from $\mathbf{o}$ to $\mathbf{p}$. Then, by the Pythagorean theorem, $d(\mathbf{p}, \mathbf{q})^2 = d(\mathbf{o}, \mathbf{p})^2 + d(\mathbf{o}, \mathbf{q})^2$. Using Equation (2.1), it follows that $\vec{p} \bullet \vec{q} = 0$. In this case, we say that $\vec{p}$ and $\vec{q}$ are *orthogonal*.

Equation (2.2) can be made more precise. Let θ be the angle between the arrow from $\mathbf{o}$ to $\mathbf{q}$ and the arrow from $\mathbf{o}$ to $\mathbf{p}$. Recalling the law of cosines from trigonometry class in the distant past and applying it to the triangle $\mathbf{opq}$ yields

$$d(\mathbf{p}, \mathbf{q})^2 = d(\mathbf{o}, \mathbf{p})^2 + d(\mathbf{o}, \mathbf{q})^2 - 2d(\mathbf{o}, \mathbf{p})d(\mathbf{o}, \mathbf{q})\cos(\theta),$$

so

$$\cos(\theta) = \frac{d(\mathbf{o}, \mathbf{p})^2 + d(\mathbf{o}, \mathbf{q})^2 - d(\mathbf{p}, \mathbf{q})^2}{2d(\mathbf{o}, \mathbf{p})d(\mathbf{o}, \mathbf{q})} = \frac{\vec{p} \bullet \vec{q}}{|\vec{p}| \cdot |\vec{q}|}.$$

Thus, the angle θ between the two vectors $\vec{p}$ and $\vec{q}$ can be calculated as $\cos^{-1}(\vec{p} \bullet \vec{q}/|\vec{p}||\vec{q}|)$.

Another way to write this formula is by using *unit vectors*. A unit vector is a vector of length 1. For any vector $\vec{v} \neq \vec{0}$, the unit vector in the same direction as $\vec{v}$ is just $\vec{v}/|\vec{v}|$; this is often written $\hat{v}$. So we can rewrite the formula for $\cos(\theta)$ as

$$\cos(\theta) = \frac{\vec{p}}{|\vec{p}|} \bullet \frac{\vec{q}}{|\vec{q}|} = \hat{p} \bullet \hat{q}. \qquad (2.3)$$

We have used this geometric argumentation to derive important properties of the dot product, but we have not said what, in general, the dot product actually *means*, geometrically. Sadly, there is no very intuitive or interesting explanation. (Equations (2.1) and (2.3) are geometrical definitions but they do not give a clear intuitive sense for the dot product.) The best, perhaps, is this: Given $\vec{p}$ and $\vec{q}$, let $\vec{w}$ be the vector that is perpendicular to $\vec{q}$, in the same plane as $\vec{p}$ and $\vec{q}$, and the same length as $\vec{q}$. Then $|\vec{p} \bullet \vec{q}|$ is the area of the parallelogram with sides $\vec{p}$ and $\vec{w}$ (Figure 2.3). This is not a very helpful definition, however; first, because the distributivity rule is not geometrically obvious; and second, because the sign of the dot product is not easy to define in this way.

2.4.4 Metacomment: How to Read Formula Manipulations

The derivation of the Cauchy-Schwarz inequality in Section 2.4.3 is a sterling example of what people dislike about math and math books: "⟨Dull long formula⟩ so ⟨series of even duller, even longer formulas⟩ so ⟨pretty dull but short formula⟩. How wonderful!"

I can't eliminate formula manipulation from this book because that is an important part of how math works. I can't always replace formula manipulation proofs with "insightful" proofs because a lot of the time there aren't any insightful proofs. Moreover, proofs that avoid formula manipulation are not necessarily more insightful. Sometimes they rely on a trick that is actually largely beside the point. Sometimes they don't generalize as well.

The one thing that I can do is to tell you how I read this kind of thing, and I suggest doing likewise. Trying to read a proof the way one would read the newspaper, or even the way one would read the earlier sections of this chapter, is (for me) a complete waste of time. (Of course, people are all different, and there may be someone somewhere who gets something out of that; but no one I've ever met.) I find that there are three possible approaches to reading a proof:

1. The best is to glance through it, get kind of an idea how it works, close the book, and try to write out the proof without looking at the book. Unless the proof is quite short, you should write it out rather than trying to think it through in your head. Once you've done this, you really understand how the proof works and what it means.

2. If you're dealing with a long proof or if you're short on time, approach (1) will probably be too difficult. In that case, the second best approach is to go through the proof *slowly*. Check the simple steps of the proof by eye, making sure that you keep track of where all the terms go and that you understand why the author brought in another formula from before at this point. If you can't follow the step by eye, then copy it down and work it out on paper. The limitation of this approach, as opposed to (1), is that you can end up understanding each of the steps but not the overall progress of the proof; you hear the notes but not the overall tune.

In either (1) or (2), if the book provides a picture or example, then you should carefully look at it and make sure you understand the connection between the picture or example and the proof. If the book doesn't provide one, then you should draw a picture if the proof is geometric, or you should work through some numeric examples if the proof is numeric. Manipulating the numbers often gives a clearer sense of what is going on than looking at the algebraic symbols.

Approaches (1) and (2) are both difficult and time-consuming, but there is no way around that if you want to learn the math—"There is no royal road to geometry" (Euclid). That leaves

3. Skip the manipulation and just learn the conclusion.

Unless you are responsible for the manipulation—you are reviewing the proof for a scientific journal, or your instructor has told you that you are required to know the proof for the exam—this is probably safe to do. I regularly read books and articles with proofs in them; I work through perhaps 5% of them, probably less. Your time is valuable, and there are other things to do in life. But even though you can do this sometimes, you can't always take approach (3) if you want to learn how to do math.

2.4.5   Application of the Dot Product: Similarity of Two Vectors

The geometric discussion of the previous section can be applied to nongeometric vectors to give two measures of how close or similar two vectors are.

The obvious measure of the similarity between two vectors $\vec{p}$ and $\vec{q}$ is just the distance between them,

$$|\vec{q} - \vec{p}| = \sqrt{(\vec{q} - \vec{p}) \bullet (\vec{q} - \vec{p})}.$$

For instance, if we have a set of pricing vectors from different stores, as in Application 2.4, and we want to determine which stores have most similar pricing policies (perhaps with a view to detecting price fixing), this might be a reasonable measure to use. Note that if we are just comparing greater and lesser distances, there is no need to extract the square root; we can just use the distance squared, which is equal to $(\vec{q} - \vec{p}) \bullet (\vec{q} - \vec{p})$.

However, it is often more pertinent to consider a *scale invariant* measure of similarity of two vectors, which concerns the difference in the direction of the two vectors independent of their magnitude. In that case, a natural measure is the angle between the two vectors, which can be calculated in terms of the expression $(\vec{p} \bullet \vec{q}) / (|\vec{p}| \cdot |\vec{q}|)$. Again, if we are just comparing greater and smaller angles, there is no need to calculate the inverse cosine function to get the actual angle; it suffices just to calculate the cosine of the angle by using this formula.

For example, suppose we want to evaluate the similarity of two documents in terms of the words they use—for example, to suggest "related documents" in a search engine. We can then use the document model described in Application 2.6. If we just use the distance between the document vectors, then long documents will tend to be close to other long documents and far from short documents, which is not at all what we want. Rather, we want to base the similarity judgment on the relative frequency of words in the documents. This can be done by using the angle cosine between the document vectors.

Application 2.8 (Recommender systems). A company wants to recommend specific products to specific users. The company has available a large database of purchases by customers.

One way to determine the recommendations, in principle,[4] is the following: For each customer in the data base, we construct a vector indexed by product. That is, corresponding to customer c, we construct an n-dimensional vector $\vec{c}$, where n is the number of different products and $\vec{c}\,[i]$ is equal to the quantity of product i that customer c has bought. Of course, $\vec{c}\,[i]$ is 0 for most i; thus, this is a sparse vector. Now, in making recommendations for customer d, we find the k customer vectors that are most similar to $\vec{d}$ in terms of the above measure and recommend products that have been popular among these customers.

There is an alternative approach: for each product p, we construct an m-dimensional vector $\vec{p}$, where m is the number of different customers and $\vec{p}\,[j]$ is the quantity of product p that customer j has bought. To find customers who might like a specific product q, we look for the k product vectors most similar to $\vec{q}$ and recommend q to customers who have bought some of these products.

Application 2.9 (Pattern comparison). Suppose we want to compare the pattern of the stock market crash of fall 1929 with the stock market crash of fall 2008. Since the Dow Jones average was around 700 at the start of September 1929 and around 12,000 at the start of September 2008, there is no point in using the distance between the corresponding vectors. The angle cosine between the two vectors might be a more reasonable measure.

Application 2.10 (Statistical correlation).[5] A course instructor wishes to determine how well the results on his final exam correspond to the results on the problem sets. Suppose that there are six students in the class; the vector of average problem set scores was $\vec{p} = \langle 9.1, 6.2, 7.2, 9.9, 8.3, 8.6 \rangle$ and the vector of final exam scores was $\vec{x} = \langle 85, 73, 68, 95, 77, 100 \rangle$.

The first step is to shift each vector so that they are both zero-centered; otherwise, the comparison will mostly just reflect the fact that students as a whole did fairly well on both. We are interested in the performances of individual students relative to the average, so we subtract the average value from each vector. The average problem set score $\bar{p} = 8.22$ and the average exam score $\bar{e} = 83$, so the shifted vectors are $\vec{p}\,' = \vec{p} - \bar{p} \cdot \vec{1} = \langle 0.8833, -2.0167, -1.0167, 1.6833, 0.0833,$ $0.3833 \rangle$ and $\vec{x}\,' = \vec{x} - \bar{e} \cdot \vec{1} = \langle 2, -10, -15, 12, -6, 17 \rangle$. The correlation is the angle cosine between these, $\vec{p}\,' \bullet \vec{x}\,' / |\vec{p}\,'||\vec{x}\,'| = 0.7536$.

In general, the correlation between two vectors $\vec{p}$ and $\vec{q}$ is defined by the algorithm

$$\vec{p}\,' \leftarrow \vec{p} - \mathrm{mean}(\vec{p}) \cdot \vec{1};$$
$$\vec{q}\,' \leftarrow \vec{q} - \mathrm{mean}(\vec{q}) \cdot \vec{1};$$
$$\text{correlation} \leftarrow \vec{p}\,' \bullet \vec{q}\,' / |\vec{p}\,'||\vec{q}\,'|.$$

---

[4]Getting this method to work efficiently for a huge database is not easy.

[5]This example is adapted from Steven Leon, *Linear Algebra with Applications*.

2.4.6   Dot Product and Linear Transformations

The fundamental significance of a dot product is that it is a linear transformation of vectors. This is stated in Theorem 2.2, which, though easily proven, is very profound. First we need a definition.

Definition 2.1. Let $f(\vec{v})$ be a function from n-dimensional vectors to numbers. Function f is a *linear transformation* if it satisfies the following two properties:

- For any vector $\vec{v}$ and scalar a, $f(a \cdot \vec{v}) = a \cdot f(\vec{v})$.

- For any vectors $\vec{v}, \vec{u}$, $f(\vec{v} + \vec{u}) = f(\vec{v}) + f(\vec{u})$.

The algebraic properties of the dot product listed in Section 2.4.1 show the following: For any vector $\vec{w}$, the function $f(\vec{v}) = \vec{w} \bullet \vec{v}$ is a linear transformation. Theorem 2.2 establishes the converse: Any linear transformation $f(\vec{v})$ corresponds to the dot product with a weight vector $\vec{w}$.

Theorem 2.2. *Let f be a linear transformation from n-dimensional vectors to numbers. Then there exists a unique vector $\vec{w}$ such that, for all $\vec{v}$, $f(\vec{v}) = \vec{w} \bullet \vec{v}$.*

For example, imagine that Stop and Shop does not post the price of individual items; the checkout clerk just tells you the price of an entire basket. Let f be the function that maps a given shopping basket to the total price of that basket. The function f is linear, assuming that Stop and Shop has no "two for the price of one" offers, "maximum of one per customer" restrictions, or other stipulations. Specifically,

1. If you multiply the number of each kind of item in a basket $\vec{b}$ by a, then the total price of the basket increases by a factor of a.

2. If you combine two baskets $\vec{b}$ and $\vec{c}$ into a single basket $\vec{b} + \vec{c}$, then the price of the combined basket is just the sum of the prices of the individual baskets.

Then Theorem 2.2 states that there is a price vector $\vec{s}$ such that $f(\vec{b}) = \vec{s} \bullet \vec{b}$ for any basket $\vec{b}$. How do we find the vector $\vec{s}$? Couldn't be simpler. Put a single unit of a single item i into a basket, and ask Stop and Shop the price of that basket. We will call that the "price of item i," and we set $\vec{s}[i]$ to be that price. By property (1), a basket that contains $\vec{b}[i]$ units of item i and nothing else costs $\vec{b}[i] \cdot \vec{s}[i]$. By property (2), a basket that contains $\vec{b}[1]$ units of item 1, $\vec{b}[2]$ units of item 2, ..., and $\vec{b}[n]$ units of item n costs $\vec{b}[1] \cdot \vec{s}[1] + \vec{b}[2] \cdot \vec{s}[2] + \ldots + \vec{b}[n] \cdot \vec{s}[n] = \vec{s} \bullet \vec{b}$.

The general proof of Theorem 2.2 is exactly the same as the argument we have given above for Stop and Shop prices.[6]

---

[6] The theorem does not hold, however, for infinite-dimensional vector spaces.

2.5 Vectors in MATLAB: Basic Operations

In this section, we illustrate the basic vector operations and functions by example. For the most part, these examples are self-explanatory; in a few cases, we provide a brief explanation.

Strictly speaking, MATLAB does not have a separate category of vectors as such; vectors are either $1 \times n$ matrices (row vectors) or $n \times 1$ matrices (column vectors). Most functions that take a vector argument, such as dot, norm, and plot, do the same thing with either kind of vector. Functions that return a vector value, such as size and randperm, have to choose, of course; these mostly return row vectors, presumably because they print out more compactly.

2.5.1 Creating a Vector and Indexing

```
>> v = [2,3,2,5,11]
v =
     2     3     2     5    11

>> a=v(4)
a =
     5

>> v(4)=8
v =
     2     3     2     8    11

>> v(8)=10
v =
     2     3     2     8    11     0     0    10
% If you set a positive index beyond the range of the vector,  Matlab
% expands the vector to accommodate.

>> v(10)
??? Attempted to access v(10); index out of bounds because numel(v)=5.
% But not if you try to access beyond the range.

>> v(-2)=3
??? Attempted to access v(-2); index must be a positive integer or logical.

% or if you try to set an invalid index.
```

2.5.2 Creating a Vector with Elements in Arithmetic Sequence

```
>> u=2:10
u =
     2     3     4     5     6     7     8     9    10
```

```
>> w= 3.16 : 0.5 : 6.0
w =
    3.1600      3.6600      4.1600      4.6600      5.1600      5.6600

>> ones(1,5)
ans =
    1      1      1      1      1

>> zeros(1,5)
ans =
    0      0      0      0      0
```

Boolean vectors.

```
>> q=[12,3,1,7,2]
q =
    12      3      1      7      2

>> p=[1,3,5,7,9]
p =
    1      3      5      7      9

>> p==q
ans =
    0      1      0      1      0
```

Subvectors.

```
>> w
w =
    3.1600      3.6600      4.1600      4.6600      5.1600      5.6600

>> w(2:4)
ans =
    3.6600      4.1600      4.6600

>> w(1:2:6)
ans =
    3.1600      4.1600      5.1600

>> q
q =
    12      3      1      7      2

>> q([3,3,1,5])
ans =
    1      1      12      2

>> q(2:4)=[4,1,6]
q =
    12      4      1      6      2
```

```
>> q<5
ans =
     0     1     1     0     1

>> q(q<5)
ans =
     4     1     2

>> % However, the following looks like it should be the same but gives an error
>> q([0,1,1,0,1])
??? Subscript indices must either be real positive integers or logicals.

>> % Why? Because [0,1,1,0,1], entered from the keyboard, is a vector of
>> % integers, not a vector of Booleans, and Matlab can't figure out that it
>> % needs to do a cast.  So you have to do an explicit cast.

>> q(logical([0,1,1,0,1]))
ans =
     4     1     2

>> % This is why a programming language should not have two basic constants
>> % (numeric 0 vs. logical 0) that print in the same way.
```

2.5.3   Basic Operations

```
>> u
u =
     2     3     4     5     6     7     8     9    10

>> 3*u
ans =
     6     9    12    15    18    21    24    27    30

>> 3+u
ans =
     5     6     7     8     9    10    11    12    13

>> u+(6:14)
ans =
     8    10    12    14    16    18    20    22    24

>> u+w
??? Error using ==> plus
Matrix dimensions must agree.
```

2.5.4   Element-by-Element Operations

In MATLAB, the notation u * v means matrix multiplication (discussed in Chapter 3). Therefore, if we want to multiply two vectors element-by-element, we use the special notation u .* v (period asterisk). Likewise with some other operations, such as exponentiation, that have a particular meaning for matrices, we use a leading dot to signify an element-by-element operation.

```
>> u
u =
     2     3     4     5     6     7     8     9    10

>> u.*(u+3)
ans =
    10    18    28    40    54    70    88   108   130

>> u./(u+3)
ans =
  Columns 1 through 7
    0.4000    0.5000    0.5714    0.6250    0.6667    0.7000
0.7273
  Columns 8 through 9
    0.7500    0.7692

>> u.^2
ans =
     4     9    16    25    36    49    64    81   100

>> % Numerical functions with no particular matrix significance can be applied
>> % without using the . notation
>> sin(u)
ans =
  Columns 1 through 7
    0.9093    0.1411   -0.7568   -0.9589   -0.2794    0.6570
0.9894
  Columns 8 through 9
    0.4121   -0.5440
```

2.5.5   Useful Vector Functions

```
>> u
u =
     2     3     4     5     6     7     8     9    10

>> mean(u)
ans =
     6

>> length(u) % Number of dimensions
ans =
     9

>> norm([3,4]) % Euclidean length
ans =
     5
```

```
>> norm(u)
ans =
   19.5959

>> ans^2
ans =
  384.0000

>> max(u)
ans =
   10

>> min(u)
ans =
    2

>> sum(u)
ans =
   54

>> sort([5,2,1,6,1,8])
ans =
    1     1     2     5     6     8

>> median([5,2,1,6,1,8])
ans =
   3.5000

>> dot([1,2,3],[3,2,1])
ans =
   10
```

2.5.6   Random Vectors

```
>> rand(1,5) % 1,5 because this is a random 1x5 matrix.
ans =
   0.8147    0.9058    0.1270    0.9134    0.6324

>> randperm(8) % random permutation of 1:8
ans =
    1     6     2     3     8     4     5     7
```

The following is an example calculation of the *correlation* of exam scores
and problem sets.

```
>> function cor = correlation(u,v)
     uprime = u - mean(u);
     vprime = v - mean(v);
     cor = dot(uprime/norm(uprime), vprime/norm(vprime));
   end
```

```
>> e=[85, 73, 68, 95, 77, 100]
e =
    85     73     68     95     77    100

>> p= [ 9.1, 6.2, 7.2, 9.9, 8.3, 8.6 ]
p =
    9.1000    6.2000    7.2000    9.9000    8.3000    8.6000

>> correlation(e,p)
ans =
    0.7536
```

2.5.7 Strings: Arrays of Characters

```
>> s='Call me Ishmael'
s =
Call me Ishmael

>> s(5:10)
ans =
 me Is
```

2.5.8 Sparse Vectors

To create a sparse vector with all 0s of length n, call sparse(1,n). (Again, we use the argument "1" because this is a $1 \times n$ matrix.) To turn an ordinary (full) vector v into a sparse vector s, call s=sparse(v); to go the other way, call v=full(s). Indexing, vector addition, scalar multiplication, and most vector functions work with sparse vectors exactly or nearly the same way as with full vectors.

```
>> v=sparse(1,10) % 1,10 because this is a sparse 1x10 matrix.
v =
   All zero sparse: 1-by-10

>> full(v)
ans =
     0     0     0     0     0     0     0     0     0     0

>> v(2)=3
v =
   (1,2)          3

>> u=sparse(1,10)
u =
   All zero sparse: 1-by-10

>> u(7)=4
u =
   (1,7)          4
```

```
>> x=u+v
x =
   (1,2)          3
   (1,7)          4
>> dot(x,u)
ans =
   (1,1)          16
>> % The dot product of two sparse vectors is a sparse vector of length 1.

>> norm(x)
ans =
      5

>> w=[1:10]
w =
      1      2      3      4      5      6      7      8      9     10

>> w+v
ans =
      1      5      3      4      5      6      7      8      9     10
% Note that sparse and full vectors can be combined; the result is a full
% vector
```

2.6 Plotting Vectors in MATLAB

One of the most useful and most complex aspects of MATLAB is its support for
two- and three-dimensional graphics. In this book, we discuss only the two-
dimensional plotting system, and only its most basic aspects.

The basic plotting function is plot. In its basic form, plot(x,y,c) takes as
arguments two k-dimensional vectors x and y and a control string c. It pro-
duces a plot of the points $\langle x[1], y[1] \rangle, \ldots, \langle x[k], y[k] \rangle$ in the display format indicated by control string c.

```
>> x=[1.23, 1.82, 2.45, 3.06, 3.91]
x =
    1.2300     1.8200     2.4500     3.0600     3.9100

>> y=[100, 99.4, 123.8, 110.6, 122]
y =
  100.0000    99.4000   123.8000   110.6000   122.0000

>> plot(x,y,'ro-')
```

The plot is output in a MATLAB window called "Figure 1," shown here in Figure 2.4. The figure can then be saved as a file in a variety of formats by using
the "Save" or "SaveAs" option in the GUI associated with the plot window.

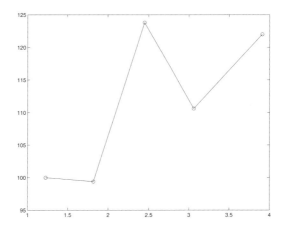

Figure 2.4. Simple plot.

The plotting routine chooses "nice" ranges and tick frequencies for the x and y axes. The "r" flag in the control string specifies the color red; the "o" flag specifies that the points are marked with circles, and the "-" flag specifies that the points are connected with a solid line. Table 2.1 shows other values for these specifiers.

The saveas function causes the current figure (denoted gcf) to be saved in a format that can be reloaded into MATLAB. The print operation causes the current figure to be exported into EPS format, suitable for use in LaTeX. MATLAB supports exporting into a variety of formats, including GIF, JPG, and so on.

	Color		Point Type		Line Type
b	blue	*	asterisk	–	solid
c	cyan	o	circle	--	dashed
g	green	x	cross	:	dotted
k	black	d	diamond	-.	dash-dot
m	magenta	.	point		
r	red	+	plus sign		
w	white	s	square		
y	yellow	<	left triangle		
		>	right triangle		
		^	up triangle		
		v	down triangle		
		p	5-point star		
		h	6-point star		

Table 2.1. Specifiers for plotting.

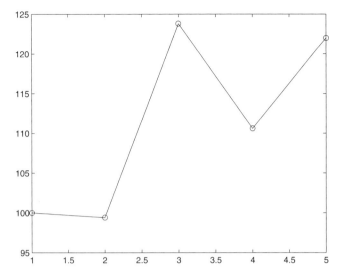

Figure 2.5. Using the default x-coordinates.

The control parameters can be typed in any order within the control string. If any of the control parameters is omitted, then a default value is used. The default color is blue, the default data point indicator is none, the default line type is no line if a data point indicator is included and a solid line if it is not. For example:

`plot(x,y)`	Points connected by blue line, no marker at points
`plot(x,y,'*g')`	Points marked by green asterisk, no line
`plot(x,y,'k:')`	Points connected by dotted black line, no marker at points
`plot(x,y,'--v')`	Blue points marked by downward triangle connected by blue dashed line.

If only one vector is specified, it is taken to be the vector of y-coordinates and is plotted against the vector 1:k. For example, `plot(y,'ro-')` gives the plot in Figure 2.5.

The next example illustrates the following three features:

- The plotted points need not be listed in increasing order of the x-coordinates. The lines connecting the points connect them in the order specified.

- The axes can be adjusted by using the `axis` function.

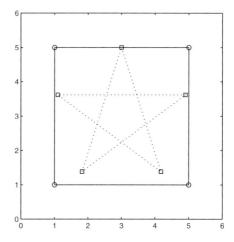

Figure 2.6. Plotting multiple datasets.

- Multiple sequences of data points can be plotted on a single graph by
  calling `plot` with additional arguments: The call `plot(x1, y1, c1, x2,
  y2, c2, ..., xn, yn, cn)` plots points x1,y1 with control c1, x2,y2 with
  control c2, and so on.

```
>> x=[1,5,5,1,1]
x =
     1     5     5     1     1

>> y=[1,1,5,5,1]
y =
     1     1     5     5     1

>>   a=[3,3+2*sin(4*pi/5),3+2*sin(8*pi/5),3+2*sin(12*pi/5), 3+2*sin(16*pi/5),3]
a =
    3.0000    4.1756    1.0979    4.9021    1.8244    3.0000

>> b=[5,3+2*cos(4*pi/5),3+2*cos(8*pi/5),3+2*cos(12*pi/5), 3+2*cos(16*pi/5),5]
b =
    5.0000    1.3820    3.6180    3.6180    1.3820    5.0000
>> plot(x,y,'ro-',a,b,'ks:')
>> axis equal
>> axis([0,6,0,6])
```

The final plot is shown in Figure 2.6. The `axis equal` command is useful
with geometric plots to ensure equal spacing in the x- and y-dimensions.

The MATLAB graphics package has many more features, including a GUI
interface; look them up as needed.

2.7   Vectors in Other Programming Languages

Almost every programming language supports one-dimensional and multidimensional arrays as primitive data structures.[7] We do, of course, have to be careful of the difference between 0-based indexing, used in such languages as C and Java, and 1-based indexing, used in this book and in MATLAB.

Programming language support for sparse vectors is much less common. In most languages, a sparse vector $\vec{v}$ must be implemented as a set of pairs $\{\langle a_1, c_1 \rangle, \ldots, \langle a_k, c_k \rangle\}$, where $c_1, \ldots, c_k$ are the nonzero components of $\vec{v}$ and $a_1, \ldots, a_k$ are the associated indices. For example, the vector $\vec{v} = [0, 0, 12, 0, -8, 12, 0, 0, 0, 0, 0]$ could be represented as the set $\{\langle 3, 12 \rangle, \langle 5, -8 \rangle, \langle 6, 12 \rangle\}$.

Sets themselves, of course, can be implemented in a number of ways in conventional programming languages; the choice of implementation here involves some trade-offs. Consider, for example, the simple operation of indexing into a sparse vector; that is, finding $\vec{v}[i]$, given the set of pairs. If the set is implemented as a list, then the solution is to go through the pairs until finding a pair $\langle a_j, c_j \rangle$ for which $a_j = i$, and return c_j. However, this requires time proportional to the number of nonzero elements, which can be large (the vector may still be sparse if the dimensionality is enormous). Alternatively, the set may be implemented as a hash table, which would support constant expected time indexing but would complicate other operations.

Exercises

Exercise 2.1. Compute each of the following expressions by hand. Do not use MATLAB.

(a) $3 \cdot \langle 1, -3, 2 \rangle$.

(b) $-2 \cdot \langle 4, 2, 0, 1 \rangle$.

(c) $\langle 1, -3, 2 \rangle + \langle 4, 2, 6 \rangle$.

(d) $\langle 0, 1, 5, 2 \rangle + \langle 1, -1, 1, -1 \rangle$.

(e) $\langle 1, -3 \rangle \bullet \langle 4, 2 \rangle$.

(f) $\langle 1, -3, 2 \rangle \bullet \langle 4, 2, 6 \rangle$.

Exercise 2.2. Compute $|\langle 1, 1, 2 \rangle|$ by hand. (You may leave this in the form $\sqrt{x}$.)

---

[7]Early versions of some programming languages, such as LISP and ML, did not include arrays, but these have mostly acquired arrays as they became increasingly used for production software. Prolog has been resistant to arrays; there are dialects that include arrays, but most do not.

Exercise 2.3. By hand, find the cosine of the angle between $\langle 1, -3, 0 \rangle$ and $\langle 2, 1, 3 \rangle$. (You may leave this in the form $x/(\sqrt{y}\sqrt{z})$.) Using MATLAB, find the actual angle.

Exercise 2.4. Let $\vec{u} = \langle 1, 4, -2 \rangle$, $\vec{v} = \langle 2, 3, 5 \rangle$, $\vec{w} = \langle -1, 4, 1 \rangle$.

(a) Verify by carrying out the calculation by hand that $\vec{u} \bullet (\vec{v} + \vec{w}) = (\vec{u} \bullet \vec{v}) + (\vec{u} \bullet \vec{w})$.

(b) Compute $(\vec{u} \bullet \vec{v}) \cdot \vec{w}$ and $(\vec{w} \bullet \vec{v}) \cdot \vec{u}$. Are they equal?

Exercise 2.5. By hand, find the correlation between:

(a) $\langle 1, 5, 3 \rangle$ and $\langle 2, 10, 6 \rangle$.

(b) $\langle 1, 5, 3 \rangle$ and $\langle 0, 1, 1 \rangle$.

(c) $\langle 1, 5, 3 \rangle$ and $\langle 5, 1, 3 \rangle$.

Problems

Problem 2.1. Find three two-dimensional vectors $\vec{u} \neq \vec{0}$, $\vec{v}$, $\vec{w}$ such that $\vec{u} \bullet \vec{v} = \vec{u} \bullet \vec{w}$, but $\vec{v} \neq \vec{w}$.

Problem 2.2. Prove the algebraic properties of the basic operations on vectors stated in Section 2.3.1.

Problem 2.3. Prove the algebraic properties of the dot product stated in Section 2.4.1.

Programming Assignments

Assignment 2.1 (Document vectors). Write a MATLAB function DocSimilarity(D,E) that computes the "similarity" of text documents D and E by using the vector model of documents. Specifically, the arguments D and E are each cell arrays of strings, each string being a word of the document, normalized to lowercase. (Cell arrays are heterogeneous arrays; see Section 3.10.5.) The function returns a number between 0 and 1, 0 meaning that the two documents have no significant words in common, 1 meaning that they have the identical significant words with the same frequency.

A word is considered "significant" if it has at least three letters and is not in the list of stop words provided in the file GetStopwords.m on the textbook website. A stop word is a very common word that should be ignored.

Your function should execute the following steps:

- Let LargeOdd be any reasonably large odd number that is not very close to a power of 256. The number 10,000,001 will do fine.

- Load in the cell array of stop words from GetStopwords.m

- Create three sparse vectors $\vec{S}, \vec{D}, \vec{E}$ of size LargeOdd as follows: for every word W, let i=hash(W,LargeOdd). You can find a hash function in the file hash.m on the textbook website. Then

    - $\vec{S}[i] = 1$ if W is on the list of stop words.
    - $\vec{D}[i]$ = the number of occurrences of W in D, if W is significant.
    - $\vec{E}[i]$ = the number of occurrences of W in E, if W is significant.

    (Create $\vec{S}$ first, then use it for a quick test for whether words in the documents are significant.) $\vec{D}$ and $\vec{E}$ are the document vectors (we omit the inverse document frequency).

- Return the quantity $\vec{D} \bullet \vec{E} / |\vec{D}| \cdot |\vec{E}|$

For instance,

```
>> D = { 'how', 'much', 'wood', 'could', 'a', 'woodchuck', 'chuck',
...
'if', 'a', 'woodchuck', 'could', 'chuck', 'wood' };
>> E = { 'all', 'the', 'wood', 'that', 'a', 'woodchuck', 'could',
...
'if', 'a', 'woodchuck', 'could', 'chuck', 'wood' };
>> DocSimilarity(D,E)
ans =
  0.9245
```

Note that the only significant words in these two texts are "chuck," "much," "wood," and "woodchuck."

You don't have to worry about hash collisions here because they are very infrequent, and the technique is completely imprecise in any case.

Assignment 2.2 (Plotting the harmonic series). The harmonic function $H(n)$ is defined as $H(n) = \sum_{i=1}^{n} 1/n$. For instance $H(4) = 1/1 + 1/2 + 1/3 + 1/4 = 25/12 = 2.0833$. For large n, $H(n)$ is approximately equal to $\ln(n) + \gamma$, where $\ln(n)$ is the natural logarithm and $\gamma = 0.5772$, known as Euler's constant.

Write a MATLAB function PlotHarmonic(N) that shows both $H(k)$ and $\ln(k) + \gamma$ for $k = 1, \ldots, N$.

Assignment 2.3 (Correlation to a noisy signal). Write a MATLAB function CorrelationNoisy(V,E) as follows. The input parameters are V, a vector, and E, a positive real. The function constructs a new vector U by adding a random number between $-E$ and E to each component of V (a different random choice for each component; use the function rand.) It then returns the correlation between V and U.

Experiment with $V = 1 : 100$ and various values of E. Run each value of E several times to get a range of correlations. What kind of values do you get for the correlation when $E = 10$? When $E = 100$? How large do you have to make E before you start to see examples with negative correlations?

Assignment 2.4 (Stars.). An n-pointed regular star can be constructed as follows:

- Choose $k > 1$ relatively prime to n.

- Place the ith vertex of the star at $x_i = \cos(2\pi ki/n)$, $y_i = \sin(2\pi ki/n)$ for $i = 0, n$. Note that $\langle x_n, y_n \rangle = \langle x_0, y_0 \rangle$, so the star closes.

Write a MATLAB function Star(N,K) which draws this N pointed star. Be sure to call axis equal so that the x- and y-axes are drawn at equal scales; otherwise, the star will be oddly squooshed.

Assignment 2.5 (Euler's Sieve). Euler's sieve is a souped-up version of the sieve of Eratosthenes, which finds the prime numbers. It works as follows:

```
L = the list of numbers from 2 to N;
P = 2; /* The first prime */
while (P^{2} < N) {
    L1 = the list of all X in L  such that P $\leq$ X $\leq$ N/P.
    L2 = P*L1;
    delete everything in L2 from L;
    P = the next value after P in L;
    }
return L;
```

For example, for $N = 27$, successive iterations proceed as follows:

```
Initialization
  L = [2 3 4 5 6 7 8 9 10 11 12 13 14 15 16 17 18 19 20 21 22 23 24 25 26 27]
  P = 2

First iteration
  L1 = [2 3 4 5 6 7 8 9 10 11 12 13]
  L2 = [4 6 8 10 12 14 16 18 20 22 24 26]
  L = [2 3 5 7 9 11 13 15 17 19 21 23 25 27]
  P = 3

Second iteration
  L1 = [3 5 7 9]
  L2 = [9 15 21 27]
```

```
L = [2 3 5 7 11 13 17 19 23 25]
P = 5

Third iteration
  L1 = [5]
  L2 = [25]
  L = [2 3 5 7 11 13 17 19 23]
```

(a) Write a MATLAB function `EulerSieve1(N)` that constructs the Euler sieve by implementing L, L1, L2 as arrays of integers, as above.

(b) Write a MATLAB function `EulerSieve2(N)` that constructs the Euler sieve by implementing L, L1, and L2 as Boolean arrays, where L[I] = 1 if I is currently in the set L. Thus, the final value returned in the example would now be the array

```
[0 1 1 0 1 0 1 0 0 0 1 0 1 0 0 0 1 0 1 0 0 0 1 0 0 0 0]
```

Chapter 3

Matrices

3.1 Definition of Matrices

An $m \times n$ *matrix* is a rectangular array of numbers with m rows, each of length n; equivalently, n columns, each of length m. For example,

$$A = \begin{bmatrix} 4 & 2 & -1 \\ 0 & 6 & 5 \end{bmatrix} \text{ is a } 2 \times 3 \text{ array;} \qquad B = \begin{bmatrix} 0 & -1.5 \\ 2.1 & 6 \\ -1.0 & -3.5 \\ 1.0 & 2.2 \end{bmatrix} \text{ is a } 4 \times 2 \text{ array.}$$

We use capital italic letters, such as M, for matrices. The element in the ith row, jth column in matrix M is denoted $M[i, j]$. In matrices A and B, $A[2, 1] = 0$; $A[1, 3] = -1$; $B[2, 1] = 2.1$; and $B[3, 2] = -3.5$.

There is no very standard notation for specifying a particular row or column of a matrix. Following MATLAB, we use the notation $M[i, :]$ for the ith row of M (a row vector) and $M[:, j]$ for the jth column. (MATLAB uses parentheses rather than square brackets, however.)

3.2 Applications of Matrices

The simplest types of $m \times n$ matrices are those whose m rows are just n-dimensional vectors or whose n columns are m-dimensional vectors of the types discussed in Chapter 2. In particular, an application of the third category discussed in Section 2.2.1, in which there is a numeric function of two entities or features, can be represented as a matrix in the obvious way: Given two sets of entities or features $\{O_1, \ldots, O_m\}$ and $\{P_1, \ldots, P_n\}$ and a numeric function $f(O, P)$, we construct the $m \times n$ matrix M such that $M[i, j] = f(O_i, P_j)$.

For example, in Application 2.4, the grocery shopping application, suppose we have a set of people { Amy, Bob, Carol } and a set of grocery items { gallon of milk, pound of butter, apple, carrot } then we can define a 3×4 array, S where

$S[i, j]$ is the number of items of type j on i's shopping list:

$$S = \begin{bmatrix} 0 & 2 & 1 & 0 \\ 1 & 1 & 0 & 0 \\ 1 & 0 & 0 & 6 \end{bmatrix}.$$

Thus, Amy is buying two pounds of butter and an apple; Bob is buying a gallon of milk and a pound of butter; and Carol is buying a gallon of milk and six carrots.

Note that each row of this matrix is the row vector corresponding to one person p; it shows p's shopping list indexed by item. Each column of this matrix is the column vector corresponding to one item i; it shows the quantity of that product being bought, indexed by person.

If the collection of grocery items indexed is large, then this will be a *sparse matrix*, that is, a matrix in which most of the entries are 0.

In Application 2.2 (time series), r different financial measures being tracked over m days could be recorded in an $m \times n$ matrix P, where $P[i, j]$ is the value of measure i on day j.

Particularly interesting are cases for which the two sets of entities, $\{O_1, \ldots, O_n\}$ and $\{P_1, \ldots, P_n\}$, are the same. In these cases, the resulting matrix is an $n \times n$ matrix, called a *square* matrix.

Application 3.1 (Distance graph). For instance, we could have a collection of n cities, and define an $n \times n$ matrix D, where $D[i, j]$ is the distance from i to j. In general, a directed graph (in the data structures sense) over n vertices can be represented as an $n \times n$ adjacency matrix, where $A[i, j] = 1$ if there is an arc from i to j and 0 if there is not. A graph with numeric labels on its arcs can be represented as an $n \times n$ matrix, where $A[i, j]$ is the label on the arc from i to j.

Application 3.2 (Functions over a rectangular region). A different category of matrix represents a numeric function over a planar region. This can be done by picking out a rectangular $m \times n$ grid of points, and constructing an $m \times n$ matrix M, where $M[i, j]$ is the value of the function at point i, j. In particular an *image*, either output for graphics or input for computer vision, is represented in terms of such a grid of pixels. A gray-scale image is represented by a single array I, where $I[i, j]$ is the intensity of light, often encoded as an 8-bit integer from 0 (black) to 255 (white), at the point with coordinates $\langle i, j \rangle$. A color image is represented by three matrices, representing the intensity of light at three different component frequencies.

From the standpoint of linear algebra, these kinds of arrays are actually atypical examples of matrices, since most of the important operations on images do not correspond very closely to the standard operations of linear algebra. Section 6.4.8 briefly discusses how geometric transformations are applied to pixel arrays.

3.3 Simple Operations on Matrices

The product of a matrix by a scalar and the sum of two matrices are computed component-by-component, as with vectors. That is,

$$(a \cdot M)[i, j] = a \cdot M[i, j];$$

for example,

$$3 \cdot \begin{bmatrix} 3 & 1 & 2 \\ 0 & -1 & 4 \end{bmatrix} = \begin{bmatrix} 3 \cdot 3 & 3 \cdot 1 & 3 \cdot 2 \\ 3 \cdot 0 & 3 \cdot -1 & 3 \cdot 4 \end{bmatrix} = \begin{bmatrix} 9 & 3 & 6 \\ 0 & -3 & 12 \end{bmatrix}.$$

And

$$(M + N)[i, j] = M[i, j] + N[i, j];$$

for example,

$$\begin{bmatrix} 3 & 1 & 2 \\ 0 & -1 & 4 \end{bmatrix} + \begin{bmatrix} 7 & -2 & 3 \\ 5 & 6 & 2 \end{bmatrix} = \begin{bmatrix} 3+7 & 1-2 & 2+3 \\ 0+5 & -1+6 & 4+2 \end{bmatrix} = \begin{bmatrix} 10 & -1 & 5 \\ 5 & 5 & 6 \end{bmatrix}.$$

The *transpose* of matrix M, denoted M^T, turns the rows of M into columns and vice versa. That is, if M is an $m \times n$ matrix, then M^T is an $n \times m$ matrix, and $M^T[i, j] = M[j, i]$. For instance,

$$\text{if } A = \begin{bmatrix} 4 & 2 & -1 \\ 0 & 6 & 5 \end{bmatrix}, \text{ then } A^T = \begin{bmatrix} 4 & 0 \\ 2 & 6 \\ -1 & 5 \end{bmatrix}.$$

The following properties are immediate from the definitions (in all of the following, M, N, and P are $m \times n$ matrices, and a and b are scalars):

$$M + N = N + M,$$
$$M + (N + P) = (M + N) + P,$$
$$a(M + N) = aM + aN,$$
$$(a + b)M = aM + bM,$$
$$(ab)M = a(bM),$$
$$(M^T)^T = M,$$
$$M^T + N^T = (M + N)^T,$$
$$aM^T = (aM)^T.$$

3.4 Multiplying a Matrix Times a Vector

Let M be an $m \times n$ matrix and let $\vec{v}$ be an n-dimensional vector. The product $M \cdot \vec{v}$ is an m-dimensional vector consisting of the dot products of each row of

M with $\vec{v}$. That is, $(M \cdot \vec{v})[i] = M[i,:] \bullet \vec{v} = \sum_{j=1}^{n} M[i,j] \cdot \vec{v}[j]$. For instance,

$$\text{if } M = \begin{bmatrix} 3 & 1 & 2 \\ 0 & -1 & 4 \end{bmatrix} \text{ and } \vec{v} = \langle 2, -1, 4 \rangle, \text{ then}$$

$$\begin{aligned} M\vec{v} &= \langle M[1,:] \bullet \vec{v}, M[2,:] \bullet \vec{v} \rangle \\ &= \langle (3 \cdot 2) + (1 \cdot -1) + (2 \cdot 4), (0 \cdot 2) + (-1 \cdot -1) + (4 \cdot 4) \rangle \\ &= \langle 13, 17 \rangle. \end{aligned}$$

Generally, for reasons to be discussed shortly, when we write out the product of a matrix times a vector, both vectors (the multiplicand and the product) are written as column vectors; thus,

$$\begin{bmatrix} 3 & 1 & 2 \\ 0 & -1 & 4 \end{bmatrix} \cdot \begin{bmatrix} 2 \\ -1 \\ 4 \end{bmatrix} = \begin{bmatrix} 13 \\ 17 \end{bmatrix}.$$

When written this way, we can use the "two-hands" method to compute the product: the left hand moves from left to right along each row of the matrix while the right hand moves from top to bottom along the vector. The order is critical here; $\vec{v} \cdot M$ means something quite different from $M \cdot \vec{v}$.

The following algebraic rules on multiplication of matrices times vectors follow immediately from the rules about dot products discussed in Section 2.4.1. In all the following, M and N are $m \times n$ matrices, $\vec{u}$ and $\vec{v}$ are n-dimensional vectors, and a is a scalar:

$$(M + N)\vec{v} = M\vec{v} + N\vec{v},$$
$$M(\vec{u} + \vec{v}) = M\vec{u} + M\vec{v},$$
$$(aM)\vec{v} = a(M\vec{v}) = M(a\vec{v}).$$

The product $M \cdot \vec{v}$ can also be described as the weighted sum of the columns of M, where the weights are the components of $\vec{v}$:

$$M \cdot \vec{v} = \vec{v}[1] \cdot M[:,1] + \ldots + \vec{v}[n] \cdot M[:,n];$$

for instance,

$$\begin{bmatrix} 3 & 1 & 2 \\ 0 & -1 & 4 \end{bmatrix} \cdot \begin{bmatrix} 2 \\ -1 \\ 4 \end{bmatrix} = 2 \cdot \begin{bmatrix} 3 \\ 0 \end{bmatrix} + -1 \cdot \begin{bmatrix} 1 \\ -1 \end{bmatrix} + 4 \cdot \begin{bmatrix} 2 \\ 4 \end{bmatrix} = \begin{bmatrix} 13 \\ 17 \end{bmatrix}.$$

The equivalence of these dual ways of describing $M \cdot v$—as the dot product of the rows of M with $\vec{v}$ or as the sum of the columns of M weighted by the components of $\vec{v}$—has deep consequences, as we see in Section 4.2.8.

3.4.1 Applications of Multiplying a Matrix Times a Vector

The geometric applications of multiplying a matrix by a vector are very important and interesting, but too complex for a short discussion here. These are discussed in Chapter 6.

The simplest applications of $M \cdot \vec{v}$ are those for which we are interested in the dot product of each of the rows with the vector separately. For instance, in Application 2.4 (grocery shopping), let B be a matrix of shopping baskets, where $B[i,j]$ is the number of item j that person i is buying, and let $\vec{p}$ be a vector of prices, where $\vec{p}[j]$ is the price of item j. Then $B \cdot \vec{p}$ is a vector of total cost for each person; that is, $(B \cdot \vec{p})[i]$ is the cost of the shopping basket for person i.

Similarly, let P be a matrix of prices of items at stores, where $P[i,j]$ is the price of item j at store i, and let $\vec{b}$ be the vector of a shopping basket, where $\vec{b}[j]$ is the number of item j to be bought. Then $P \cdot \vec{b}$ is the vector of the cost of the basket by store; that is $(P \cdot \vec{b})[i]$ is the cost of the shopping basket at store i.

More interesting, perhaps, are the applications in which M represents a transformation of the vector as a whole.

Application 3.3 (Transition matrix). Let $\vec{v}$ represent the populations of a collection of cities at a given time. Let M be an annual *transition* matrix for populations. That is, for any two cities i and j, if $i \neq j$, $M[i,j]$ is the fraction of the inhabitants of j who move to i. $M[i,i]$ is the fraction of the inhabitants of i who remain at i. Ignoring births and deaths, what is the population of city i after a year? First, there are the people who stayed in i; there are $M[i,i] \cdot \vec{v}[i]$ of these. Then there are the people who immigrated to i; from each city j, the number of people who have immigrated to i is $M[i,j]\vec{v}[j]$. Therefore, the total number of people in i is $\sum_j M[i,j] \cdot \vec{v}[j] = M[i,:] \bullet \vec{v}$, and $M\vec{v}$ is the vector of populations after a year.

For instance, suppose there are three cities A, B, and C, with the following transitions:

- Of the population of A, 70% remain in A; 20% move to B, and 10% move to C.

- Of the population of B, 25% move to A; 65% remain in B, and 10% move to C.

- Of the population of C, 5% move to A; 5% move to B, and 90% remain in C.

Thus, the transition matrix is

$$\begin{bmatrix} 0.7 & 0.25 & 0.05 \\ 0.2 & 0.65 & 0.05 \\ 0.1 & 0.1 & 0.9 \end{bmatrix}.$$

If initially there are 400,000 people in A, 200,000 in B, and 100,000 in C, then after a year, the population vector will be given by

$$\begin{bmatrix} 0.7 & 0.25 & 0.05 \\ 0.2 & 0.65 & 0.05 \\ 0.1 & 0.1 & 0.9 \end{bmatrix} \cdot \begin{bmatrix} 400,000 \\ 200,000 \\ 100,000 \end{bmatrix} = \begin{bmatrix} 335,000 \\ 215,000 \\ 150,000 \end{bmatrix}.$$

Note that each column of the matrix adds up to 1; this corresponds to the fact that total number of people remains constant. A matrix whose columns add to 1 is known as a *stochastic* matrix; we study these in greater depth in Chapter 10.

Application 3.4 (Smoothing a signal). Let $\vec{v}$ be a time sequence of a numeric quantity, as in Application 2.2. Suppose that we wish to eliminate noise from a signal. One standard way to do this is to estimate the true value of the signal at each point of time by using the weighted average of the signal at nearby points; this is known as *smoothing* the signal. The width of the range depends both on the frequencies of the noise and the signal and on the signal-to-noise ratio; ideally, we want to average over a range larger than the wavelength of the noise, smaller than the wavelength of the signal, and over a set of points large enough that the noise will average out to zero.

Let $\vec{d}[t]$ be the data at time t and let $\vec{s}[t]$ be the estimate of the signal. If the signal frequency is very high and the noise is fairly small, then we might estimate the signal at time t, $\vec{q}[t] = 1/3 \cdot (\vec{d}[t-1] + \vec{d}[t] + \vec{d}[t+1])$. Of course, at

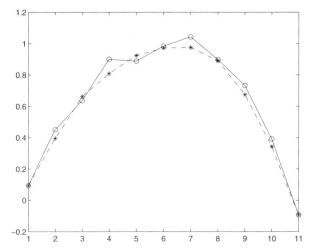

Figure 3.1. Smoothing. The noisy points are the circles on the solid line. The smoothed points are the asterisks on the dashed line.

the beginning of the time range, where $\vec{d}\,[t-1]$ is not recorded, and at the end, where $\vec{d}\,[t+1]$ is not recorded, we have to do something different; what we do is just set $\vec{q}\,[1] = \vec{d}\,[1]$ and $\vec{q}\,[k] = \vec{d}\,[k]$.

Each of these sums is a weighted average of the data; the entire transformation of signal to noise is thus the multiplication of the data vector by a smoothing matrix. For instance, consider the following artificial data. The true signal $\vec{s}$ is just the quadratic function $\vec{s}\,[i] = 1 - ((i-6)/5)^2$ for $i = 1,\ldots,11$; thus, $\vec{s} = \langle 0, 0.36, 0.64, 0.84, 0.96, 1, 0.96, 0.84, 0.64, 0.36, 0\rangle$. The noise $\vec{n}$ is generated by the MATLAB expression $(0.2 * \mathtt{rand(1,11)}) - 0.1$. The data $\vec{d} = \vec{s} + \vec{n} = \langle 0.0941,$ $0.4514, 0.6371, 0.9001, 0.8884, 0.9844, 1.0431, 0.8984, 0.7319, 0.3911, -0.0929\rangle$.

We smooth the data by carrying out the following multiplication (Figure 3.1):

$$
\begin{bmatrix}
1 & 0 & 0 & 0 & 0 & 0 & 0 & 0 & 0 & 0 & 0 \\
1/3 & 1/3 & 1/3 & 0 & 0 & 0 & 0 & 0 & 0 & 0 & 0 \\
0 & 1/3 & 1/3 & 1/3 & 0 & 0 & 0 & 0 & 0 & 0 & 0 \\
0 & 0 & 1/3 & 1/3 & 1/3 & 0 & 0 & 0 & 0 & 0 & 0 \\
0 & 0 & 0 & 1/3 & 1/3 & 1/3 & 0 & 0 & 0 & 0 & 0 \\
0 & 0 & 0 & 0 & 1/3 & 1/3 & 1/3 & 0 & 0 & 0 & 0 \\
0 & 0 & 0 & 0 & 0 & 1/3 & 1/3 & 1/3 & 0 & 0 & 0 \\
0 & 0 & 0 & 0 & 0 & 0 & 1/3 & 1/3 & 1/3 & 0 & 0 \\
0 & 0 & 0 & 0 & 0 & 0 & 0 & 1/3 & 1/3 & 1/3 & 0 \\
0 & 0 & 0 & 0 & 0 & 0 & 0 & 0 & 1/3 & 1/3 & 1/3 \\
0 & 0 & 0 & 0 & 0 & 0 & 0 & 0 & 0 & 0 & 1
\end{bmatrix}
\cdot
\begin{bmatrix}
0.0941 \\ 0.4514 \\ 0.6371 \\ 0.9001 \\ 0.8884 \\ 0.9844 \\ 1.0431 \\ 0.8984 \\ 0.7319 \\ 0.3911 \\ -0.0929
\end{bmatrix}
=
\begin{bmatrix}
0.0941 \\ 0.3942 \\ 0.6629 \\ 0.8085 \\ 0.9243 \\ 0.9720 \\ 0.9753 \\ 0.8912 \\ 0.6738 \\ 0.3434 \\ -0.0929
\end{bmatrix}.
$$

The correlation of the data with the ideal signal is 0.9841. The correlation of the smoothed data with the ideal signal is 0.9904.

Application 3.5 (Time-shifting a signal). Another operation on time sequences that can be modeled as matrix multiplication is time-shifting. Suppose you have a signal $\vec{s}$ of length n and wish to construct the same signal shifted q units later; that is, $\vec{z}\,[i] = \vec{z}\,[i-q]$. This can be viewed as multiplication by an $n \times n$ matrix M such that $M[i+q, i] = 1$ for $i = q+1,\ldots,n$ and $M[i, j] = 0$ for all $j \neq i + q$. For example, with $n = 6$, $q = 2$,

$$
\begin{bmatrix}
0 & 0 & 0 & 0 & 0 & 0 \\
0 & 0 & 0 & 0 & 0 & 0 \\
1 & 0 & 0 & 0 & 0 & 0 \\
0 & 1 & 0 & 0 & 0 & 0 \\
0 & 0 & 1 & 0 & 0 & 0 \\
0 & 0 & 0 & 1 & 0 & 0
\end{bmatrix}
\cdot
\begin{bmatrix}
2 \\ 8 \\ 5 \\ 7 \\ 1 \\ 4
\end{bmatrix}
=
\begin{bmatrix}
0 \\ 0 \\ 2 \\ 8 \\ 5 \\ 7
\end{bmatrix}.
$$

This may seem like quite an elaborate and verbose way of formulating a very simple operation, and if all you want to do is a time-shift, it certainly would be. But the point is that you can then combine it with other matrix operations and apply the results of matrix theory to these combinations.

Application 3.6 (Jacobians).   (*Note:* This application requires multivariable calculus. It is optional.)

Suppose $\mathbf{F}(\mathbf{p})$ is a differentiable function from the plane to itself. We write $\mathbf{F}_x$ and $\mathbf{F}_y$ for the x and y components of $\mathbf{F}$. For example, if $\mathbf{F}(\langle x, y \rangle) = \langle x + y, x^2 - y^2 \rangle$, then $\mathbf{F}_x(\langle x, y \rangle) = x + y$ and $\mathbf{F}_y(\langle x, y \rangle) = x^2 - y^2$.

We now take a particular point $\mathbf{a}$, draw tiny vectors with their tails at $\mathbf{a}$, and consider the mapping $\mathbf{F}$ does to those vectors. So let $\epsilon > 0$ be a very small distance, let $\vec{u}$ be a vector, and let $\mathbf{b} = \mathbf{a} + \epsilon \vec{u}$. Defining u_x and u_y as the coordinates of $\vec{u}$ in the x and y directions, respectively, we have $\vec{u} = u_x \hat{x} + u_y \hat{y}$, where $\hat{x}$ and $\hat{y}$ are the unit x and y vectors. Then $\mathbf{b} = \mathbf{a} + \epsilon u_x \hat{x} + \epsilon u_y \hat{y}$.

Now, what is $\mathbf{F}(\mathbf{b})$? By the definition of the partial derivative,

$$\mathbf{F}_x(\mathbf{a} + \epsilon u_x \hat{x}) - \mathbf{F}_x(\mathbf{a}) \approx \epsilon u_x \cdot \frac{\partial \mathbf{F}_x}{\partial x}$$

and

$$\mathbf{F}_x(\mathbf{a} + \epsilon u_x \hat{x} + \epsilon u_y \hat{y}) - \mathbf{F}_x(\mathbf{a} + \epsilon u_x \hat{x}) \approx \epsilon u_y \cdot \frac{\partial \mathbf{F}_x}{\partial y}.$$

Adding these together,

$$\mathbf{F}_x(\mathbf{b}) - \mathbf{F}_x(\mathbf{a}) = \mathbf{F}_x(\mathbf{a} + \epsilon u_x \hat{x} + \epsilon u_y \hat{y}) - \mathbf{F}_x(\mathbf{a}) \approx \epsilon \left(u_x \cdot \frac{\partial \mathbf{F}_x}{\partial x} + u_y \cdot \frac{\partial \mathbf{F}_x}{\partial y} \right).$$

Likewise,

$$\mathbf{F}_y(\mathbf{b}) - \mathbf{F}_y(\mathbf{a}) = \mathbf{F}_y(\mathbf{a} + \epsilon u_x \hat{x} + \epsilon u_y \hat{y}) - \mathbf{F}_y(\mathbf{a}) \approx \epsilon \left(u_x \cdot \frac{\partial \mathbf{F}_y}{\partial x} + u_y \cdot \frac{\partial \mathbf{F}_y}{\partial y} \right).$$

Let $\vec{w} = (\mathbf{F}(\mathbf{b}) - \mathbf{F}(\mathbf{a}))/\epsilon$ so $\mathbf{F}(\mathbf{b}) = \mathbf{F}(\mathbf{a}) + \epsilon \vec{w}$. Then we have

$$\vec{w} = \left(u_x \cdot \frac{\partial \mathbf{F}_x}{\partial x} + u_y \cdot \frac{\partial \mathbf{F}_x}{\partial y} \right) \hat{x} + \left(u_x \cdot \frac{\partial \mathbf{F}_y}{\partial x} + u_y \cdot \frac{\partial \mathbf{F}_y}{\partial y} \right) \hat{y} = \begin{bmatrix} \frac{\partial \mathbf{F}_x}{\partial x} & \frac{\partial \mathbf{F}_x}{\partial y} \\ \frac{\partial \mathbf{F}_y}{\partial x} & \frac{\partial \mathbf{F}_y}{\partial y} \end{bmatrix} \cdot \vec{u}.$$

The 2×2 matrix in this expression is known as the *Jacobian* of $\mathbf{F}$, notated $J(\mathbf{F})$.

For example, if $\mathbf{a} = \langle 2, 3 \rangle$, $\vec{u} = \langle 5, 2 \rangle$, and $\epsilon = 0.001$, then $\mathbf{b} = \langle 2.005, 3.002 \rangle$ and $\mathbf{F}(\mathbf{a}) = \langle 5, -5 \rangle$. The resulting Jacobian is

$$J(\mathbf{F}) = \begin{bmatrix} \frac{\partial \mathbf{F}_x}{\partial x} & \frac{\partial \mathbf{F}_x}{\partial y} \\ \frac{\partial \mathbf{F}_y}{\partial x} & \frac{\partial \mathbf{F}_y}{\partial y} \end{bmatrix} = \begin{bmatrix} 1 & 1 \\ x & -y \end{bmatrix},$$

so

$$\mathbf{F}(\mathbf{b}) \approx \mathbf{F}(\mathbf{a}) + J(\mathbf{F})|_{\mathbf{a}} \cdot \epsilon \vec{u} = \begin{bmatrix} 5 \\ -5 \end{bmatrix} + \begin{bmatrix} 1 & 1 \\ 2 & -3 \end{bmatrix} \begin{bmatrix} 0.005 \\ 0.002 \end{bmatrix} = \begin{bmatrix} 5.007 \\ -4.996 \end{bmatrix}.$$

We can generalize this to space of arbitrary dimension. Let $\mathbf{F}(\mathbf{p})$ be a differentiable function from $\mathbb{R}^n$ to $\mathbb{R}^m$. Let $\mathbf{a}$ be a point in $\mathbb{R}^n$, let $\epsilon > 0$ be small, let $\vec{u}$ be a vector in $\mathbb{R}^n$, and let $\mathbf{b} = \mathbf{a} + \epsilon \vec{u}$. Let $\hat{x}_1, \ldots, \hat{x}_n$ be the coordinate vectors of $\mathbb{R}^n$, and let $\mathbf{F}_1, \ldots, \mathbf{F}_m$ be the components of $\mathbf{F}$. Then

$$\mathbf{F}(\mathbf{b}) \approx \mathbf{F}(\mathbf{a}) + \begin{bmatrix} \frac{\partial \mathbf{F}_1}{\partial x_1} & \cdot & \cdot & \cdot & \frac{\partial \mathbf{F}_1}{\partial x_n} \\ & \cdot & \cdot & \cdot & \\ \cdot & & \cdot & & \cdot \\ \cdot & & & \cdot & \cdot \\ \frac{\partial \mathbf{F}_m}{\partial x_1} & \cdot & \cdot & \cdot & \frac{\partial \mathbf{F}_m}{\partial x_n} \end{bmatrix} \cdot \epsilon \vec{u}.$$

Again, this array is called the Jacobian.

3.5   Linear Transformation

Just as with the dot product, the fundamental significance of matrix multiplication is that it is a linear transformation.

Definition 3.1. Let $f(\vec{v})$ be a function from $\mathbb{R}^n$, the set of n-dimensional vectors, to $\mathbb{R}^m$, the set of m-dimensional vectors. Function f is a *linear transformation* if it satisfies the following two properties:

- For any vector $\vec{v}$ and scalar a, $f(a \cdot \vec{v}) = a \cdot f(\vec{v})$.

- For any vectors $\vec{v}$ and $\vec{u}$, $f(\vec{v} + \vec{u}) = f(\vec{v}) + f(\vec{u})$.

Theorem 3.2. *Let f be a linear transformation from $\mathbb{R}^n$ to $\mathbb{R}^m$. Then there exists a unique $m \times n$ matrix F such that, for all $\vec{v}$, $f(\vec{v}) = F \cdot \vec{v}$. We say that matrix F corresponds to transformation f, and vice versa.*

Note that Definition 3.1 is word-for-word identical to Definition 2.1 except that the range of the function f has been changed from the real numbers $\mathbb{R}$ to $\mathbb{R}^m$. Theorem 3.2 is the same as Theorem 2.2 except that the range has been changed, and "dot product with $\vec{f}$" has been changed to "multiplication by matrix F."

The proofs of the theorems are also essentially identical. Now we must imagine that we give a basket $\vec{b}$ over a space of n groceries, to a personal shopper, who reports back a vector $f(\vec{b})$ of the price of this basket at m different stores. The unit vector $\vec{e}^{\,i}$ corresponds to the basket containing one unit of the ith item. Then $f(\vec{e}^{\,i})$ is the vector showing the price of the ith item at all the stores. Construct an $m \times n$ matrix F so that $F[i, j] = f(\vec{e}^{\,j})[i]$. Now let $\vec{b} = \langle b_1, \ldots, b_n \rangle$ be any n-dimensional vector. We can write $\vec{b} = b_1 \vec{e}^{\,1} + \ldots + b_n \vec{e}^{\,n}$

so, by linearity, the ith component of $f(\vec{b})$ is

$$f(\vec{b})[i] = f(\sum_j b_j \cdot \vec{e}^{\,j})[i] = \sum_j b_j \cdot (f(\vec{e}^{\,j})[i]) = \sum_j b_j \cdot F[i,j] = F[i,:] \bullet \vec{b}.$$

Therefore, $f(\vec{b}) = F \cdot \vec{b}$.

The duality here between linear transformations and matrices is the key to linear algebra. Theorem 3.2 means that we can go back and forth between thinking about linear transformations over vector spaces and multiplying vectors by matrices, and transfer results from each view to the other.

The one critical operation for which the correspondence does not work very well is matrix transposition. Forming the transpose of a matrix is a simple operation on matrices and, as we shall see, an important one; however, it does not correspond to anything very simple in terms of the linear transformations involved.

Note that many simple and important functions on vectors are not linear transformations, such as the function $\max(\vec{v})$, which returns the largest component of vector $\vec{v}$; the function $\mathrm{sort}(\vec{v})$, which returns the components of $\vec{v}$ in increasing order; and the function $\mathrm{product}(\vec{v})$, which returns the product of the components of $\vec{v}$. The proof that none of these is a linear transformation is requested in Problem 3.4.

3.6   Systems of Linear Equations

A system of m linear equations in n unknowns is a set of equations of the form

$$a_{1,1}x_1 + a_{1,2}x_2 + \ldots + a_{1,n}x_n = c_1.$$
$$a_{2,1}x_1 + a_{2,2}x_2 + \ldots + a_{2,n}x_n = c_2.$$
$$\vdots$$
$$a_{m,1}x_1 + a_{m,2}x_2 + \ldots + a_{m,n}x_n = c_m.$$

Here all the $a_{i,j}$ and c_i are constants, and the x_i are variables. A *solution* to the system is a set of values x_i that satisfies all the equations. The values $a_{i,j}$ are called the *coefficients* of the system; the values c_j are called the *constant terms*.

For example, the following is a system of three equations in four unknowns:

$$1 \cdot w + 1 \cdot x + 1 \cdot y + 1 \cdot z = 2.$$
$$2 \cdot w + 0 \cdot x - 1 \cdot y + 3 \cdot z = -1.$$
$$-1 \cdot w + 2 \cdot x + 0 \cdot y + 1 \cdot z = 2.$$

One solution to this system is $w = 1, x = 2, y = 0, z = -1$, as can be seen by substition and checking the equations. Another solution is $w = 2, x = 3, y = -1, z = -2$.

A system of linear equations can be written as a problem in matrix multiplication. Let

$$A = \begin{bmatrix} a_{1,1} & \cdots & a_{1,n} \\ \vdots & & \vdots \\ a_{m,1} & \cdots & a_{m,n} \end{bmatrix} \qquad X = \begin{bmatrix} x_1 \\ \vdots \\ x_n \end{bmatrix} \qquad C = \begin{bmatrix} c_1 \\ \vdots \\ c_m \end{bmatrix}.$$

Then we can write the general system in the form $A \cdot X = C$. That is, if values $x_1, \ldots, x_n$ satisfy the system of equations, then the vector $\langle x_1, \ldots, x_n \rangle$ will satisfy the matrix equation. For example, the particular system above can be written as

$$\begin{bmatrix} 1 & 1 & 1 & 1 \\ 2 & 0 & -1 & 3 \\ -1 & 2 & 0 & 1 \end{bmatrix} \cdot \begin{bmatrix} w \\ x \\ y \\ z \end{bmatrix} = \begin{bmatrix} 2 \\ -1 \\ 2 \end{bmatrix}.$$

In Chapter 4, we develop a mathematical theory that allows us to categorize different types of systems of linear equations and their solutions. In particular, we prove that a system of linear equations has either one solution, infinitely many solutions, or no solutions. In Chapter 5, we present an algorithm for solving systems of linear equations.

3.6.1 Applications of Systems of Linear Equations

Applications of all kinds involve solving systems of linear equations. Obviously, we can take any of our applications of matrix multiplication and turn them around to get a system of linear equations. For example, in the shopping application (Application 2.4), we could ask, "Given a $m \times n$ price matrix P showing the prices of n items at m stores, and a vector $\vec{c}$ of the cost of an unknown basket at each store, find a basket $\vec{x}$ satisfying $P\vec{x} = \vec{c}$." In the population transfer (Application 3.3) we could ask, "Given a distribution of population at the end of the year $\vec{v}$ and the transition matrix for the year M, find the distribution of population $\vec{x}$ at the start of the year"; this is the solution to the problem $M\vec{x} = \vec{v}$. Further applications follow; more will arise in the course of the book.

Application 3.7 (Circuit analysis). A simple circuit of resistors and a power source, such as shown in Figure 3.2, gives rise to a system of linear equations. The variables are the voltages at each node and the currents through each branch. The equations are the equations for the voltage source and for ground, the equations corresponding to the resistor equation $V = I \cdot R$ (voltage difference equals current times resistance), and Kirchoff's circuit law, which states that the total current flowing into a node is 0. (Kirchoff's voltage law, stating that the sum of the voltage drops around a cycle is 0, is built into our decision

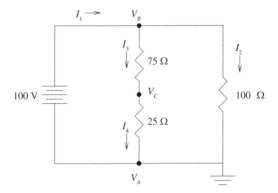

Figure 3.2. Circuit analysis.

to represent the voltage at each node as a variable rather than directly representing the voltage drop across each branch.) Applying these laws to the circuit in Figure 3.2, we get the system of eight equations in seven unknowns:

$$V_A = 0,$$
$$V_B - V_A = 100,$$
$$V_B - V_A = 100 \cdot I_2,$$
$$V_B - V_C = 75 \cdot I_3,$$
$$V_C - V_A = 25 \cdot I_4,$$
$$I_1 - I_2 - I_3 = 0,$$
$$I_3 - I_4 = 0,$$
$$I_4 + I_2 - I_1 = 0.$$

By moving all the variable terms to the right and converting to matrix form, we get

$$
\begin{bmatrix}
1 & 0 & 0 & 0 & 0 & 0 & 0 \\
-1 & 1 & 0 & 0 & 0 & 0 & 0 \\
-1 & 1 & 0 & 0 & -100 & 0 & 0 \\
0 & 1 & -1 & 0 & 0 & -75 & 0 \\
-1 & 0 & 1 & 0 & 0 & 0 & -25 \\
0 & 0 & 0 & 1 & -1 & -1 & 0 \\
0 & 0 & 0 & 0 & 0 & 1 & -1 \\
0 & 0 & 0 & -1 & 1 & 0 & 1
\end{bmatrix}
\cdot
\begin{bmatrix}
V_A \\
V_B \\
V_C \\
I_1 \\
I_2 \\
I_3 \\
I_4
\end{bmatrix}
=
\begin{bmatrix}
0 \\
100 \\
0 \\
0 \\
0 \\
0 \\
0 \\
0
\end{bmatrix}.
$$

The solution is $\langle 0, 100, 25, 2, 1, 1, 1 \rangle$.

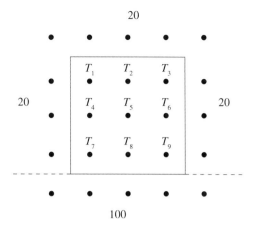

Figure 3.3. Heat distribution.

Application 3.8 (Temperature distribution). The distribution of temperature and other properties in a continuous material can be approximated by linear equations. For example, suppose that you have a square bar of metal that is heated on one side by a hot plate at $100°$C and the other three sides border the air at room temperature $20°$C. In a steady state, the temperature at each point is the average of the temperatures at the neighboring points. This is made precise in the partial differential equation $\nabla^2 T = 0$, but it can be approximated by choosing a uniform grid of points inside the metal and asserting the relation at every point.[1]

If we choose a grid of nine interior points, as shown in Figure 3.3, we get the following system of nine equations in nine unknowns:

$$T_1 = 1/4(20 + 20 + T_2 + T_4),$$
$$T_2 = 1/4(20 + T_1 + T_3 + T_5),$$
$$T_3 = 1/4(20 + 20 + T_2 + T_6),$$
$$T_4 = 1/4(20 + T_1 + T_5 + T_7),$$
$$T_5 = 1/4(T_2 + T_4 + T_6 + T_8),$$
$$T_6 = 1/4(20 + T_3 + T_5 + T_9),$$
$$T_7 = 1/4(20 + 100 + T_4 + T_8),$$
$$T_8 = 1/4(100 + T_5 + T_7 + T_9),$$
$$T_9 = 1/4(20 + 100 + T_6 + T_8).$$

[1]This example is from Philip Davis (1960).

These equations can be rewritten as the matrix equation

$$
\begin{bmatrix}
1 & -1/4 & 0 & -1/4 & 0 & 0 & 0 & 0 & 0 \\
-1/4 & 1 & -1/4 & 0 & -1/4 & 0 & 0 & 0 & 0 \\
0 & -1/4 & 1 & 0 & 0 & -1/4 & 0 & 0 & 0 \\
-1/4 & 0 & 0 & 1 & -1/4 & 0 & -1/4 & 0 & 0 \\
0 & -1/4 & 0 & -1/4 & 1 & -1/4 & 0 & -1/4 & 0 \\
0 & 0 & -1/4 & 0 & -1/4 & 1 & 0 & 0 & -1/4 \\
0 & 0 & 0 & -1/4 & 0 & 0 & 1 & -1/4 & 0 \\
0 & 0 & 0 & 0 & -1/4 & 0 & -1/4 & 1 & -1/4 \\
0 & 0 & 0 & 0 & 0 & -1/4 & 0 & -1/4 & 1
\end{bmatrix}
\cdot
\begin{bmatrix}
T_1 \\ T_2 \\ T_3 \\ T_4 \\ T_5 \\ T_6 \\ T_7 \\ T_8 \\ T_9
\end{bmatrix}
=
\begin{bmatrix}
10 \\ 5 \\ 10 \\ 5 \\ 0 \\ 5 \\ 30 \\ 25 \\ 30
\end{bmatrix}.
$$

The solution is $\langle 25.71, 27.85, 25.71, 35, 40, 35, 54.29, 62.14, 54.29 \rangle$.

Application 3.9 (Curve interpolation). Suppose we have a graph of q data points $\langle x_1, y_1 \rangle, \langle x_2, y_2 \rangle, \ldots, \langle x_q, y_q \rangle$ with independent variable x and dependent variable y, and we want to connect these with a smooth curve. This is known as *interpolating* a curve. One way to do this (not necessarily the best way) is to find a $(q-1)$ degree polynomial $y = t_{q-1}x^{q-1} + \ldots + t_1 x + t_0$ that fits the data. Finding the coefficients t_i can be viewed as solving a system of linear equations,[2] where the t_i are the variables, the coefficients are the powers of x_i, and the constants are the y_i.

For example, suppose we have the five data points, $\langle -3, 1 \rangle$, $\langle -1, 0 \rangle$, $\langle 0, 5 \rangle$, $\langle 2, 0 \rangle$, $\langle 4, 1 \rangle$. We then have the following system of equations:

$$
t_4(-3)^4 + t_3(-3)^3 + t_2(-3)^2 + t_1(3) + t_0 = 1,
$$
$$
t_4(-1)^4 + t_3(-1)^3 + t_2(-1)^2 + t_1(1) + t_0 = 0,
$$
$$
t_4(0)^4 + t_3(0)^3 + t_2(0)^2 + t_1(0) + t_0 = 5,
$$
$$
t_4(2)^4 + t_3(2)^3 + t_2(2)^2 + t_1(2) + t_0 = 0,
$$
$$
t_4(4)^4 + t_3(4)^3 + t_2(4)^2 + t_1(4) + t_0 = 1.
$$

In matrix form, we have

$$
\begin{bmatrix}
81 & -27 & 9 & -3 & 1 \\
1 & -1 & 1 & -1 & 1 \\
0 & 0 & 0 & 0 & 1 \\
16 & 8 & 4 & 2 & 1 \\
256 & 64 & 16 & 4 & 1
\end{bmatrix}
\cdot
\begin{bmatrix}
t_4 \\ t_3 \\ t_2 \\ t_1 \\ t_0
\end{bmatrix}
=
\begin{bmatrix}
1 \\ 0 \\ 5 \\ 0 \\ 1
\end{bmatrix}.
$$

The solution is $\langle 13/60, -26/60, -163/60, 176/60, 5 \rangle$.

---

[2]There is also a simple closed formula for finding the coefficients of an interpolating polynomial.

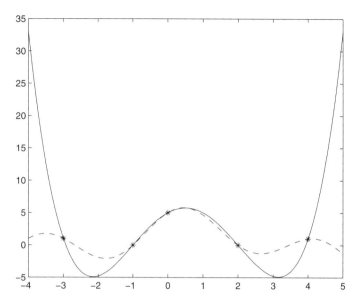

Figure 3.4. Polynomial interpolation (solid line) and sinusoidal interpolation (dashed line).

This technique is not restricted to polynomials. Let $f_1, \ldots, f_n$ be any set of n "basis" functions. (In the case of polynomials, these are the power functions $f_i(x) = x^i$.) If you wish to interpolate n data points by the linear sum $t_1 f_1(x) + \ldots + t_n f_n(x)$, then the weights t_i can be found by solving the associated system of linear equations $y_j = f_1(x_j) t_1 + \ldots + f_n(x_j) t_n$ for $j = 1, \ldots, n$. (Depending on the functions, the system may have no solution or the solution may not be unique.) In this system, the coefficients are the values $f_i(x_j)$, the variables are the t_i, and the constant terms are the y_j.

For instance, suppose that we use the five basis functions 1, $\sin(\pi x/2)$, $\sin(\pi x/4)$, $\cos(\pi x/2)$, $\cos(\pi x/6)$. Then the coefficients of the curve that interpolates the same set of data points satisfy the system

$$
\begin{bmatrix}
1 & 1 & -1/\sqrt{2} & 0 & 0 \\
1 & -1 & -1/\sqrt{2} & 0 & \sqrt{3}/2 \\
1 & 0 & 0 & 1 & 1 \\
1 & 0 & 1 & -1 & 0.5 \\
1 & 0 & 0 & 1 & -0.5
\end{bmatrix}
\cdot
\begin{bmatrix}
t_4 \\
t_3 \\
t_2 \\
t_1 \\
t_0
\end{bmatrix}
=
\begin{bmatrix}
1 \\
0 \\
5 \\
0 \\
1
\end{bmatrix}.
$$

The solution is $\langle 0.0217, 1.6547, 0.9566, 2.3116, 2.6667 \rangle$. These two curve interpolations are shown in Figure 3.4.

3.7   Matrix Multiplication

In view of Theorem 3.2, we can now look at basic operations on linear transformations and ask how these work out in terms of the corresponding matrices. In this section, we look at three operators: (1) multiplication of a transformation by a scalar and (2) addition of two transformations, which are easy, and (3) the *composition* of two transformations, which requires some work. The final important operator, the *inverse* of a transformation, is much more complicated and is discussed in Section 5.3.

First, we present two easy definitions. If $f(\vec{x})$ and $g(\vec{x})$ are linear transformations from $\mathbb{R}^n$ to $\mathbb{R}^m$ and a is a scalar, then we define the transformation $a \cdot f$ by the equation $(a \cdot f)(\vec{x}) = a \cdot f(\vec{x})$ and we define the transformation $f + g$ by the equation $(f + g)(\vec{x}) = f(\vec{x}) + g(\vec{x})$. It is easy to check that:

- $a \cdot f$ and $f + g$ are both linear transformations.

- If we let F be the matrix corresponding to f and G be the matrix corresponding to g, then $a \cdot F$ corresponds to $a \cdot f$ and $F + G$ corresponds to $f + g$.

We now turn to composition, which is not so simple. Suppose that $f(\vec{x})$ is a linear transformation from $\mathbb{R}^n$ to $\mathbb{R}^m$ and that $g(\vec{x})$ is a linear transformation from $\mathbb{R}^m$ to $\mathbb{R}^p$. Then the composition $g \circ f$ is the function from $\mathbb{R}^n$ to $\mathbb{R}^p$ defined as $(g \circ f)(\vec{x}) = g(f(\vec{x}))$. It is easy to show that $g \circ f$ is likewise a linear transformation, as follows:

$$
\begin{aligned}
(g \circ f)(a \cdot \vec{x}) &= g(f(a \cdot \vec{x})) && \text{(by definition)} \\
&= g(a \cdot f(\vec{x})) && \text{(since } f \text{ is linear)} \\
&= a \cdot g(f(\vec{x})) && \text{(since } g \text{ is linear)} \\
&= a \cdot (g \circ f)(\vec{x}) && \text{(by definition)}
\end{aligned}
$$

$$
\begin{aligned}
(g \circ f)(\vec{x} + \vec{y}) &= g(f(\vec{x} + \vec{y})) && \text{(by definition)} \\
&= g(f(\vec{x}) + f(\vec{y})) && \text{(since } f \text{ is linear)} \\
&= g(f(\vec{x})) + g(f(\vec{y})) && \text{(since } g \text{ is linear)} \\
&= (g \circ f)(\vec{x}) + (g \circ f)(\vec{y}) && \text{(by definition)}
\end{aligned}
$$

Therefore, by Theorem 3.2, there is a matrix M corresponding to the transformation $g \circ f$. The question is, how is M related to G and F, the matrices corresponding to functions f and g?

The answer is that M is the *matrix product* $G \cdot F$. We can determine how this product is computed by studying how it operates on a sample vector. To avoid too many subscripts and summation signs, we use an example here with specific numbers, where $n = 3$, $m = 2$, $p = 3$. Let G, F, and $\vec{v}$ be the following matrices:

$$G = \begin{bmatrix} 67 & 68 \\ 77 & 78 \\ 87 & 88 \end{bmatrix}, \qquad F = \begin{bmatrix} 31 & 32 & 33 \\ 41 & 42 & 43 \end{bmatrix}, \qquad \vec{v} = \begin{bmatrix} 10 \\ 11 \\ 12 \end{bmatrix}.$$

Let f and g be the transformations corresponding to F and G, and let H be the matrix corresponding to $g \circ f$. Then we have

$$H \cdot \vec{v} = (g \circ f)(\vec{v}) = g(f(\vec{v})) = G \cdot (F \cdot \vec{v})$$

$$= \begin{bmatrix} 67 & 68 \\ 77 & 78 \\ 87 & 88 \end{bmatrix} \cdot \left(\begin{bmatrix} 31 & 32 & 33 \\ 41 & 42 & 43 \end{bmatrix} \cdot \begin{bmatrix} 10 \\ 11 \\ 12 \end{bmatrix} \right)$$

$$= \begin{bmatrix} 67 & 68 \\ 77 & 78 \\ 87 & 88 \end{bmatrix} \cdot \begin{bmatrix} 31 \cdot 10 + 32 \cdot 11 + 33 \cdot 12 \\ 41 \cdot 10 + 42 \cdot 11 + 43 \cdot 12 \end{bmatrix}$$

$$= \begin{bmatrix} 67 \cdot (31 \cdot 10 + 32 \cdot 11 + 33 \cdot 12) + 68 \cdot (41 \cdot 10 + 42 \cdot 11 + 43 \cdot 12) \\ 77 \cdot (31 \cdot 10 + 32 \cdot 11 + 33 \cdot 12) + 78 \cdot (41 \cdot 10 + 42 \cdot 11 + 43 \cdot 12) \\ 87 \cdot (31 \cdot 10 + 32 \cdot 11 + 33 \cdot 12) + 88 \cdot (41 \cdot 10 + 42 \cdot 11 + 43 \cdot 12) \end{bmatrix}$$

$$= \begin{bmatrix} (67 \cdot 31 + 68 \cdot 41) \cdot 10 + (67 \cdot 32 + 68 \cdot 42) \cdot 11 + (67 \cdot 33 + 68 \cdot 43) \cdot 12 \\ (77 \cdot 31 + 78 \cdot 41) \cdot 10 + (77 \cdot 32 + 78 \cdot 42) \cdot 11 + (77 \cdot 33 + 78 \cdot 43) \cdot 12 \\ (87 \cdot 31 + 88 \cdot 41) \cdot 10 + (87 \cdot 32 + 88 \cdot 42) \cdot 11 + (87 \cdot 33 + 88 \cdot 43) \cdot 12 \end{bmatrix}$$

$$= \begin{bmatrix} 67 \cdot 31 + 68 \cdot 41 & 67 \cdot 32 + 68 \cdot 42 & 67 \cdot 33 + 68 \cdot 43 \\ 77 \cdot 31 + 78 \cdot 41 & 77 \cdot 32 + 78 \cdot 42 & 77 \cdot 33 + 78 \cdot 43 \\ 87 \cdot 31 + 88 \cdot 41 & 87 \cdot 32 + 88 \cdot 42 & 87 \cdot 33 + 88 \cdot 43 \end{bmatrix} \cdot \begin{bmatrix} 10 \\ 11 \\ 12 \end{bmatrix}.$$

Therefore,

$$G \cdot F = H = \begin{bmatrix} 67 \cdot 31 + 68 \cdot 41 & 67 \cdot 32 + 68 \cdot 42 & 67 \cdot 33 + 68 \cdot 43 \\ 77 \cdot 31 + 78 \cdot 41 & 77 \cdot 32 + 78 \cdot 42 & 77 \cdot 33 + 78 \cdot 43 \\ 87 \cdot 31 + 88 \cdot 41 & 87 \cdot 32 + 88 \cdot 42 & 87 \cdot 33 + 88 \cdot 43 \end{bmatrix}.$$

Each element of H, $H[i, j]$ corresponds to the dot product of the row $G[i,:]$ with the column $F[:, j]$. This example works in the same way for *any* matrices F and G.

Definition 3.3. Let F be an $m \times n$ matrix and let G be a $p \times m$ matrix. The product $G \cdot F$ is the $p \times n$ matrix defined by the rule

$$(G \cdot F)[i, j] = G[i,:] \bullet F[:, j] = \sum_{k=1}^{m} G[i, k] \cdot F[k, j].$$

Remember that the matrix product $G \cdot F$ is defined only if the number of columns in G is equal to the number of rows in F.

Theorem 3.4. *Let f be a linear transformation from $\mathbb{R}^m$ to $\mathbb{R}^k$, let g be a linear transformation from $\mathbb{R}^k$ to $\mathbb{R}^m$, and let F and G be the corresponding matrices. Then the product $G \cdot F$ is the matrix corresponding to the composition $g \circ f$.*

In manually computing the product of two matrices, $G \cdot F$, again we use the two hands method: The left hand moves left to right along the ith row G, the right hand moves top to bottom down the jth column of F, we multiply the two numbers and add them to a running total, and the final sum is the $[i, j]$th entry in $G \cdot F$.

A particular significant linear transformation is the *identity* on n-dimensional vectors; that is, the function i defined by $i(\vec{v}) = \vec{v}$ for all n-dimensional vectors $\vec{v}$. It is easy to show that the corresponding matrix, known as the identity matrix and denoted I_n, is the $n \times n$ matrix with 1s on the diagonal from the upper left to the lower right and 0s everywhere else. For $n = 6$, we then have

$$I_6 = \begin{bmatrix} 1 & 0 & 0 & 0 & 0 & 0 \\ 0 & 1 & 0 & 0 & 0 & 0 \\ 0 & 0 & 1 & 0 & 0 & 0 \\ 0 & 0 & 0 & 1 & 0 & 0 \\ 0 & 0 & 0 & 0 & 1 & 0 \\ 0 & 0 & 0 & 0 & 0 & 1 \end{bmatrix}.$$

Let's look at some examples of applications of matrix multiplication.

Example 3.5. A simple application of matrix multiplication involves computing all pairs of dot products of one collection of vectors by another collection of vectors. In the shopping application, for instance, suppose there are m stores, p items, and n shopping baskets. Let P be the $m \times p$ matrix with price vectors as rows; that is, $P[i, k]$ is the price of item k in shop i. Let B be the $p \times n$ matrix with baskets as columns; that is, $B[k, j]$ is the number of item k in basket j. Then the product $P \cdot B$ is the $m \times n$ matrix showing the price of each basket at each store; that is, $(P \cdot B)[i, j]$ is the price of basket j at store i.

Example 3.6. As an example of composition of transformations, suppose that a customer Joe is shopping for the ingredients of a number of recipes for a party. Suppose that there are n recipes, p types of groceries, and m stores. Let $\vec{d}$ be the "party vector," showing the number of servings of each dish that Joe is planning to make. Let R be the matrix of recipe ingredients; that is, $R[k, j]$ is the amount of ingredient k needed for one serving of recipe j. Then $R \cdot \vec{d}$ is the basket vector; that is, $(R \cdot \vec{d})[k]$ is the number of units of ingredient k that Joe needs to buy. Thus, multiplication by R corresponds to a linear transformation mapping a party vector to a shopping basket. Let P be the price matrix; that is, $P[i, k]$ is price of item k at store i. Then, for any basket $\vec{b}$, $P \cdot \vec{b}$ is the vector showing the cost of basket $\vec{b}$ at each store. Thus, P is a transformation from basket vectors to vectors of cost by store. Therefore, the product $P \cdot R$ is the transformation from party vectors to the vector showing the cost of the entire party by store. That is, if $\vec{p} = (P \cdot R) \cdot \vec{d} = P \cdot (R \cdot \vec{d})$, then $\vec{p}\,[i]$ is the price of buying all the ingredients for the party at store i.

Example 3.7. Combining Applications 3.4 and 3.5 (signal processing), if we want to carry out a smoothing operation and then a time shift, the combined operation corresponds to multiplication by the product $H \cdot M$, where M is the matrix for the smoothing and H is the matrix for the time-shift.

Example 3.8. In Application 3.3 (population transfer), suppose that A, B, and C are the matrices of population transfer in years 1, 2, and 3, respectively, and $\vec{v}$ is the population vector at the start of year 1. Then $A\vec{v}$ is the population vector at the end of year 1; $B \cdot (A\vec{v}) = (B \cdot A)\vec{v}$ is the population vector at the end of year 2; and $C \cdot ((B \cdot A) \cdot \vec{v}) = (C \cdot B \cdot A) \cdot \vec{v}$ is the population vector at the end of year 3. Thus, $B \cdot A$ is the transfer matrix for years 1 and 2 combined, and $C \cdot B \cdot A$ is the transfer matrix for years 1, 2, and 3 combined.

Example 3.9. In Application 3.6 (partial derivatives), it is easily shown that if $\mathbf{F}$ is a differentiable function from $\mathbb{R}^n$ to $\mathbb{R}^m$ and $\mathbf{G}$ is a differentiable function from $\mathbb{R}^m$ to $\mathbb{R}^p$, then the Jacobian of the composition $J(\mathbf{G} \circ \mathbf{F}) = J(\mathbf{G}) \cdot J(\mathbf{F})$.

3.8   Vectors as Matrices

The operation of multiplying $m \times n$ matrix M times n-dimensional vector $\vec{v}$ can be viewed as a special case of multiplying two matrices if $\vec{v}$ is associated with the $n \times 1$ matrix V, consisting of a single column. The operation of computing the dot product of two n-dimensional vectors $\vec{u} \bullet \vec{v}$ can be viewed as a special case of multiplying two matrices if $\vec{u}$ is associated with the $1 \times n$ matrix U, consisting of a single row, and $\vec{v}$ is associated with the $n \times 1$ matrix V, consisting of a single column.

Thus, sometimes it is convenient to associate vectors with rows, and sometimes it is convenient to associate them with columns; we may even have to do both in the same equation. In the mathematical literature, vectors are more often associated with $n \times 1$ matrices (i.e., column vectors), so we write the product of M by $\vec{v}$ as $M \cdot \vec{v}$, and the dot product of $\vec{u}$ and $\vec{v}$ as $\vec{u}^T \cdot \vec{v}$. (The notation $\vec{u}^T \cdot \vec{v}$ is sometimes used for the dot product even in contexts where there are no matrices involved; I prefer $\vec{u} \bullet \vec{v}$ in such cases.) MATLAB tries to be even-handed, but when forced to choose, it opts for row vectors. Thus, as discussed in Section 2.5, MATLAB functions with vector arguments such as dot(X,Y) or norm(X) generally work with either row or column vectors; built-in functions that return vectors such as size(A) generally return row rather than column vectors.

If U is a $1 \times m$ row vector and M is an $m \times n$ matrix, then the product $U \cdot M$ is a $1 \times n$ vector; this fact is sometimes useful. If U is a $1 \times n$ row vector and V is an $n \times 1$ column matrix, then the product $V \cdot U$ is an $n \times n$ matrix, consisting of all the products of an element of V times an element of U.

3.9   Algebraic Properties of Matrix Multiplication

The following basic algebraic properties of matrix multiplication are important. In the following discussion, A and B are $m \times n$ matrices; C and D are $n \times p$ matrices; E is a $p \times q$ matrix; and a is a scalar. I_n is the $n \times n$ identity matrix.

$$A \cdot (C \cdot E) = (A \cdot C) \cdot E, \qquad \text{(associative)} \qquad (3.1)$$
$$A \cdot (C + D) = (A \cdot C) + (A \cdot D), \qquad \text{(right distributive)} \qquad (3.2)$$
$$(A + B) \cdot C = (A \cdot C) + (B \cdot C), \qquad \text{(left distributive)} \qquad (3.3)$$
$$a \cdot (A \cdot B) = (a \cdot A) \cdot B = A \cdot (a \cdot B), \qquad (3.4)$$
$$A \cdot I_n = A, \qquad \text{(right identity)} \qquad (3.5)$$
$$I_n \cdot C = C, \qquad \text{(left identity)} \qquad (3.6)$$
$$(A \cdot B)^T = B^T \cdot A^T. \qquad (3.7)$$

All of these rules can be proven, without difficulty but somewhat drearily, from Definition 3.3 for matrix multiplication. For (3.1)–(3.6), however, it is even easier, and much more enlightening, to derive them by using the correspondence with linear transformation. (It is not possible to derive Rule (3.7) this way because, as already mentioned, the transpose does not correspond to anything simple in the world of linear transformations.)

For instance, to establish Rule (3.2), let c and d be linear transformations from $\mathbb{R}^p$ to $\mathbb{R}^n$; and let a be a linear transformation from $\mathbb{R}^n$ to $\mathbb{R}^m$. Then we have

$$\begin{aligned} (a \circ (c + d))(\vec{v}) &= a((c + d)(\vec{v})) & \text{(by definition of composition)} \\ &= a(c(\vec{v}) + d(\vec{v})) & \text{(by definition of } c + d) \\ &= a(c(\vec{v})) + a(d(\vec{v})) & \text{(by linearity of } a) \\ &= (a \circ c)(\vec{v}) + (a \circ d)(\vec{v}) & \text{(by definition of composition)} \\ &= ((a \circ c) + (a \circ d))(\vec{v}) & \text{(by definition of } (a \circ c) + (a \circ d)) \end{aligned}$$

Since $a \circ (c + d)$ corresponds to $A(C + D)$ and $(a \circ c) + (a \circ d)$ corresponds to $AC + AD$, we have $A(C + D) = AC + AD$.

Very important: matrix multiplication is not commutative.

That is, $A \cdot B$ is not in general equal to $B \cdot A$. To show this, we let A be an $m \times n$ matrix and B be a $p \times q$ matrix.

- AB is defined only if $n = p$ and BA is defined only if $q = m$.

- Suppose that $n = p$ and $q = m$, so that both products are defined. Then AB is an $m \times m$ matrix and BA is an $n \times n$ matrix, so the two products are the same shape only if $n = m$.

- Even if $m = n = p = q$, so that both products are defined and have the same shape, in most cases $AB \neq BA$. For example, for

$$A = \begin{bmatrix} 1 & 2 \\ 3 & 4 \end{bmatrix} \text{ and } B = \begin{bmatrix} 5 & 6 \\ 7 & 8 \end{bmatrix},$$

$$AB = \begin{bmatrix} 1 & 2 \\ 3 & 4 \end{bmatrix} \cdot \begin{bmatrix} 5 & 6 \\ 7 & 8 \end{bmatrix} = \begin{bmatrix} 19 & 22 \\ 43 & 50 \end{bmatrix}$$

but

$$BA = \begin{bmatrix} 5 & 6 \\ 7 & 8 \end{bmatrix} \cdot \begin{bmatrix} 1 & 2 \\ 3 & 4 \end{bmatrix} = \begin{bmatrix} 23 & 34 \\ 31 & 46 \end{bmatrix}.$$

3.9.1 Matrix Exponentiation

If A is a square $n \times n$ matrix, then the product $A \cdot A$ is defined and is likewise an $n \times n$ matrix. Therefore, we can continue multiplying by A repeatedly. The product $A \cdot A \cdot \ldots \cdot A$ (k times) is, naturally, denoted A^k.

As an example, suppose that in the population transition example (Application 3.3), the transition matrix M is constant year after year. Then after k years, the population vector will be $M^k \cdot \vec{v}$ where $\vec{v}$ is the population vector at the start.

For another example, consider the Fibonacci series, $1, 1, 2, 3, 5, 8, \ldots$ defined by the recurrence $F(1) = 1$, $F(2) = 1$, $F(i) = F(i-1) + F(i-2)$ for $n > 2$. This can be characterized in terms of matrix exponentiation as follows:[3] Consider the sequence of vectors $\vec{V}(1) = \langle 1, 1 \rangle$, $\vec{V}(2) = \langle 1, 2 \rangle$, $\vec{V}(3) = \langle 2, 3 \rangle$, and so on, where $\vec{V}(i)$ is the two-dimensional vector $\langle F(i), F(i+1) \rangle$. Then the recurrence condition can be expressed in the formula $\vec{V}(i+1) = M \cdot \vec{V}(i)$, where

$$M = \begin{bmatrix} 0 & 1 \\ 1 & 1 \end{bmatrix}.$$

Therefore, $\vec{V}(i) = M^{i-1} \vec{V}(1)$.

Application 3.10 (Paths in a directed graph). Let matrix A be the adjacency matrix for graph G; that is, $A[I, J] = 1$ if there is an edge from vertex I to vertex J in G and 0 otherwise. For example, the adjacency matrix for the graph

---

[3]This is not an effective way of computing Fibonacci numbers, just an interesting way.

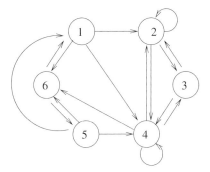

Figure 3.5. Directed graph.

in Figure 3.5 is

$$A - \begin{bmatrix} 0 & 1 & 0 & 1 & 0 & 1 \\ 0 & 1 & 1 & 1 & 0 & 0 \\ 0 & 1 & 0 & 1 & 0 & 0 \\ 0 & 1 & 1 & 1 & 0 & 1 \\ 1 & 0 & 0 & 1 & 0 & 1 \\ 1 & 0 & 0 & 0 & 1 & 0 \end{bmatrix}.$$

Thus we see that $A[1,4] = 1$ because there is an arc from vertex 1 to vertex 4.

Now, there is a path of length 2, $U \to V \to W$ in G just if $A[U,V]$ and $A[V,W]$ are both 1; that is, if $A[U,V] \cdot A[V,W] = 1$. Therefore, the number of paths of length 2 from U to W is equal to $\sum_{V=1}^{N} A[U,V] \cdot A[V,W] = A^2[U,W]$. That is, the matrix A^2 is the matrix such that for all U and W, $A^2[U,W]$ is the number of paths of length 2 from U to W.

In fact, in general, for any power k, $A^k[U,W]$ is the number of paths from U to W of length k. The proof is by induction: Suppose this is true for $k-1$. Now consider the paths of length k from U to W where the second vertex is V. If there is an arc from U to V then the number of such paths is equal to the number of paths of length $k-1$ from V to W, which by induction is $A^{k-1}[V,W]$. If there is no arc from U to V, then the number of such paths is 0. In either case, the number of such paths is $A[U,V] \cdot A^{k-1}[V,W]$. But the total number of paths of length k from U to W is the sum over all vertices V of the paths whose second vertex is V; that is, $\sum_V A[U,V] \cdot A^{k-1}[V,W] = (A \cdot A^{k-1})[U,W] = A^k[U,W]$.

3.10   Matrices in MATLAB

3.10.1   Inputting Matrices

When inputting matrices in MATLAB, items in a row are separated by commas or spaces, and rows are separated by semicolons or line breaks.

```
>> a=[11,12,13; 14,15,16]
a =
    11    12    13
    14    15    16

>> b=[
1 2 3
4 5 6]
b =
     1     2     3
     4     5     6

>> % Identity matrix (square)
>> eye(5)
ans =
     1     0     0     0     0
     0     1     0     0     0
     0     0     1     0     0
     0     0     0     1     0
     0     0     0     0     1

>> % Rectangular identity matrix
>> eye(5,3)
ans =
     1     0     0
     0     1     0
     0     0     1
     0     0     0
     0     0     0

>> % Square zero matrix
>> zeros(4)
ans =
     0     0     0     0
     0     0     0     0
     0     0     0     0
     0     0     0     0


>> % Rectangular zero matrix
>> zeros(2,4)
ans =
     0     0     0     0
     0     0     0     0

>> % Square matrix of 1's
>> ones(5)
ans =
     1     1     1     1     1
     1     1     1     1     1
     1     1     1     1     1
     1     1     1     1     1
     1     1     1     1     1
```

```
>> % Diagonal matrix
>> diag([1,4,9,16])
ans =
     1     0     0     0
     0     4     0     0
     0     0     9     0
     0     0     0    16
```

3.10.2   Extracting Submatrices

```
>> a
a =
    11    12    13
    14    15    16

>> a(1,2)
ans =
    12

>> a(2,3)
ans =
    16

>> a(2,3)=20
a =
    11    12    13
    14    15    20

>> a(3,5)=25
a =
    11    12    13     0     0
    14    15    20     0     0
     0     0     0     0    25

>> a(2,:)
ans =
    14    15    20     0     0

>> a(:,3)
ans =
    13
    20
     0

>> a(2:3,2:4)
ans =
    15    20     0
     0     0     0
```

3.10.3   Operations on Matrices

```
>> a=[1,1,2; 3,4,5]
a =
     1     1     2
     3     4     5

>> b=[1,0,-1; 2,2,1]
b =
     1     0    -1
     2     2     1

>> c=[0,1;2,2;-1,0]
c =
     0     1
     2     2
    -1     0

>> a+b
ans =
     2     1     1
     5     6     6

>> a*c
ans =
     0     3
     3    11

>> c*a
ans =
     3     4     5
     8    10    14
    -1    -1    -2

>> % Transpose
>> a'
ans =
     1     3
     1     4
     2     5

>> v=[1,2,3]
v =
     1     2     3

>> a*v'
ans =
     9
    26

>> size(a)
ans =
     2     3
```

```
>> size(v)
ans =
     1      3

>> % Element by element multiplication
>> a.*b
ans =
     1      0     -2
     6      8      5

>> % Scalar operations
>> 2*a
ans =
     2      2      4
     6      8     10

>> a+2
ans =
     3      3      4
     5      6      7

>> % Matrix exponentiation
>> p=a(:,1:2)
p =
     1      1
     3      4

>> p^2
ans =
     4      5
    15     19

>> p^5
ans =
         436            551
        1653           2089

>> f=[0,1;1,1]
f =
     0      1
     1      1

>> v=[1,1]'
v =
     1
     1

>> f*v
ans =
     1
     2
```

```
>> f^2*v
ans =
     2
     3

>> f^10*v
ans =
     89
    144

>> f^10
ans =
     34    55
     55    89
```

3.10.4   Sparse Matrices

Sparse matrices can be created with dimensions up to a large integer size.[4] The function "sparse" with a variety of types of arguments creates a sparse matrix. Many standard operations on full matrices can be applied to sparse matrices; some, such as computing rank, require specialized functions. The function `full(s)` converts a sparse matrix s to a full matrix.

```
% If A is a full matrix, then sparse(A) converts it to a sparse matrix
>> s=sparse(eye(6))
s =
   (1,1)        1
   (2,2)        1
   (3,3)        1
   (4,4)        1
   (5,5)        1
   (6,6)        1

% Note that a sparse matrix prints out by printing all the nonzero values
% with indices

% Indexing into a sparse matrix.
>> s(2,3) = 6
s =
   (1,1)        1
   (2,2)        1
   (2,3)        6
   (3,3)        1
   (4,4)        1
   (5,5)        1
   (6,6)        1
```

[4]According to the documentation, the dimension can be up to the maximum integer $2^{31}-1$, but, at least with the version of MATLAB that I have been running, I have found that I can get errors if the size is more than about 500,000,000. Moreover, the behavior is erratic around that value; the call `sparse(498182000)` sometimes succeeds and sometimes gives an error.

```
% Multiplying sparse matrices
>> s*s
ans =
   (1,1)          1
   (2,2)          1
   (2,3)         12
   (3,3)          1
   (4,4)          1
   (5,5)          1
   (6,6)          1

% Other operations
>> size(s)
ans =
     6     6

% Converting a sparse matrix to a full matrix
>> full(s)
ans =
     1     0     0     0     0     0
     0     1     6     0     0     0
     0     0     1     0     0     0
     0     0     0     1     0     0
     0     0     0     0     1     0
     0     0     0     0     0     1

% Other ways to create sparse matrices
% sparse(m,n) creates an m*n sparse 0 matrix

>> s=sparse(3,5)
s =
   All zero sparse: 3-by-5
>> s(1,3)=5
s =
   (1,3)          5

% sparse(i,j,v,m,n,maxnz): i, j, v are vectors of equal length
% creates an m*n sparse matrix S such that S(i(k), j(k)) = v(k)
% maxnz is an upper bound on the number of nonzero elements in this call

>> s=sparse([1,2,3], [1,5,4], [3.1, 2.6, 5.0], 7, 6, 10)
s =
   (1,1)        3.1000
   (3,4)        5.0000
   (2,5)        2.6000

% speye(n) creates the sparse n x n identity matrix

>> speye(6)
ans =
   (1,1)          1
   (2,2)          1
   (3,3)          1
```

```
     (4,4)           1
     (5,5)           1
     (6,6)           1
```

3.10.5   Cell Arrays

Cell arrays are heterogeneous arrays; that is, unlike regular arrays, the entities in a single cell array may vary in type and size. Cell arrays are created and indexed by using curly brackets. They are useful in creating "ragged arrays" such as collections of vectors, strings, or matrices of different kinds.

```
>> c={1, [1,2], 'Do-re-mi', [1,2,3;4,5,6]}
c =
     [1]     [1x2 double]     'Do-re-mi'     [2x3 double]

>> c{3}
ans =
Do-re-mi

>> text={'Four', 'score', 'and', 'seven', 'years', 'ago'}
text =
     'Four' 'score'     'and'     'seven'     'years'     'ago'

>> text{5}
ans =
years

>> m={eye(1),eye(2),eye(3),eye(4)}
m =
     [1]     [2x2 double]     [3x3 double]     [4x4 double]

>> m{2}
ans =
     1       0
     0       1
```

Exercises

Exercise 3.1.

$$\begin{bmatrix} 2 & -1 & 1 \\ 1 & 3 & 1 \end{bmatrix} + \begin{bmatrix} 0 & 3 & -2 \\ 1 & 2 & 3 \end{bmatrix}$$

Exercise 3.2.

$$\begin{bmatrix} 2 & -1 & 1 & 0 \\ 1 & 3 & 1 & -1 \\ -2 & 1 & 0 & 1 \end{bmatrix} \cdot \begin{bmatrix} 1 \\ -2 \\ 2 \\ 1 \end{bmatrix}$$

Exercise 3.3.

$$\begin{bmatrix} 1 & -2 & 1 \end{bmatrix} \cdot \begin{bmatrix} 2 & -1 & 1 & 0 \\ 1 & 3 & 1 & -1 \\ -2 & 1 & 0 & 1 \end{bmatrix}$$

Exercise 3.4.

$$\begin{bmatrix} 2 & -1 & 1 & 0 \\ 1 & 3 & 1 & -1 \\ -2 & 1 & 0 & 1 \end{bmatrix} \cdot \begin{bmatrix} 1 & 2 \\ -2 & 0 \\ 2 & -1 \\ 1 & 1 \end{bmatrix}$$

Exercise 3.5. Let

$$A = \begin{bmatrix} 1 \\ -2 \\ 1 \end{bmatrix}, \quad B = \begin{bmatrix} 3 & 1 & 2 \end{bmatrix}, \quad C = \begin{bmatrix} 2 \\ -1 \\ 2 \end{bmatrix}.$$

Compute the product of these three matrices, first as $(A \cdot B) \cdot C$ and second as $A \cdot (B \cdot C)$. Which method involves fewer integer multiplications?

Exercise 3.6. A *permutation matrix* is a square matrix such that each row and each column has one entry of 1 and the rest are 0s. For instance, the matrix

$$M = \begin{bmatrix} 0 & 0 & 1 & 0 & 0 & 0 \\ 0 & 0 & 0 & 0 & 0 & 1 \\ 1 & 0 & 0 & 0 & 0 & 0 \\ 0 & 1 & 0 & 0 & 0 & 0 \\ 0 & 0 & 0 & 0 & 1 & 0 \\ 0 & 0 & 0 & 1 & 0 & 0 \end{bmatrix}$$

is a 6×6 permutation matrix.

(a) Let V be the row vector $[1,2,3,4,5,6]$. What is $V \cdot M$? What is $M \cdot V^T$?

(b) Let U be the row vector $[4,5,6,3,2,1]$. Construct a permutation matrix such that $V \cdot M = U$. Construct a permutation matrix such that $U \cdot M = V$. What do you notice about these two matrices?

Exercise 3.7. Refer to the graph of Figure 3.5 (Section 3.9.1).

(a) Find the number of paths of length 6 from A to F.

(b) Draw a plot of the number of all paths in the graph of length K, as a function of K, for $K = 1, \ldots, 10$.

Problems

Problem 3.1. Refer to Rules (3.1)–(3.7) in Section 3.9.

(a) Prove each of Rules (3.1)–(3.7) directly from the numerical definition of matrix multiplication.

(b) Prove each of Rules (3.1)–(3.6), except Rule (3.2), using an argument based on linear transformations, analogous to the argument for Rule (3.2) given in the text.

Problem 3.2. What is the advantage of using the function `speye(N)` over writing the expression `sparse(eye(N))`?

Problem 3.3. Let M be an $n \times n$ permutation matrix as defined in exercise 3.6. Prove that $M \cdot M^T = I_n$. (*Hint:* What is the dot product of a row of M with a column of M^T?)

Problem 3.4. Show that none of the operations $\text{sort}(\vec{v})$, $\max(\vec{v})$ or $\text{product}(\vec{v})$ is a linear transformation.

Problem 3.5.

(a) Prove that the operation of taking the transpose of a matrix does not correspond to matrix multiplication. That is, prove that there does not exist an $n \times n$ matrix M such that, for every $n \times n$ matrix A, $M \cdot A = A^T$. (*Hint:* Use the fact that, for any two matrices U, V and any index i, $U \cdot (V[:, i]) = (U \cdot V)[:, i]$.)

Note that, if A is a pixel array, then A^T is the reflection of the image across the main diagonal; thus illustrating that geometric transformations on pixel arrays do not correspond to matrix multiplication. This topic is discussed further in Section 6.4.8.

(b) Transpose is a linear operator, however. That is, suppose we define a function $\Phi(A) = A^T$. Then Φ satisfies the equations

$$\Phi(A + B) = \Phi(A) + \Phi(B),$$
$$\Phi(c \cdot A) = c \cdot \Phi(A).$$

But we proved in Theorem 3.2 that every linear operator *does* correspond to matrix multiplication. Resolve the apparent contradiction with part (a) above. (This problem is tricky.)

Problem 3.6. A square matrix A is *nilpotent* if $A^p = 0$ for some power p.

(a) Prove that an adjacency matrix A for graph G is nilpotent if and only if G is a directed acyclic graph (DAG).

(b) Consider the smallest value of p for which $A^p = 0$. What is the significance of p in terms of the graph G?

Programming Assignments

Assignment 3.1 (Evaluating a linear classifier). Consider a classification problem of the following kind: given n numeric features of an entity, we want to predict whether the entity belongs to a specific category. As examples: Based on measurable geometric features, is a given image a picture of a camel? Based on test results, does a patient have diabetes? Based on financial data, is a prospective business a good investment?

A *linear classifier* consists of an n-dimensional weight vector $\vec{W}$ and a numeric threshold T. The classifier predicts that a given entity $\vec{X}$ belongs to the category just if $\vec{W} \bullet \vec{X} \geq T$.

Of all the different ways of evaluating how well a given classifier fits a given labeled dataset, the simplest is the *overall accuracy*, which is just the fraction of the instances in the dataset for which the classifier gets the right answer. Overall accuracy, however, is often a very unhelpful measure. Suppose we are trying to locate pictures of camels in a large collection of images collected across the Internet. Then, since images of camels constitute only a very small fraction of the collection, we can achieve a high degree of overall accuracy simply by rejecting all the images. Clearly, this is not a useful retrieval engine.

In this kind of case, the most commonly used measures are *precision* and *recall*. Suppose that we have a dataset consisting of collection D of m entities together with the associated correct labels $\vec{L}$. Specifically, D is an $m \times n$ matrix in which each row is the feature vector for one entity, and vector $\vec{L}$ is an m-dimensional column vector, where $\vec{L}[i] = 1$ if the ith entity is in the category and 0 otherwise.

Now let C be the set of entities that are actually in the category (i.e., labeled so by $\vec{L}$); let R be the set of entities that the classifier predicts are in the category; and let $Q = C \cap R$. Then precision is defined as $|Q|/|R|$ and recall is defined as $|Q|/|C|$.

In the camel example, the precision is the fraction of images that actually are camels out of all the images that the classifier identifies as camels, and the recall is the fraction of the images of camels in the collection that the classifier accepts as camels.

(a) Write a MATLAB function `evaluate(D,L,W,T)` that takes as arguments D, L, W, and T, as described above, and returns the overall accuracy, the precision, and the recall. For example, let $m = 6$ and $n = 4$; so,

$$D = \begin{bmatrix} 1 & 1 & 0 & 4 \\ 2 & 0 & 1 & 1 \\ 2 & 3 & 0 & 0 \\ 0 & 2 & 3 & 1 \\ 4 & 0 & 2 & 0 \\ 3 & 0 & 1 & 3 \end{bmatrix}, \quad L = \begin{bmatrix} 1 \\ 1 \\ 0 \\ 0 \\ 0 \\ 0 \end{bmatrix}, \quad W = \begin{bmatrix} 1 \\ 2 \\ 1 \\ 2 \end{bmatrix}, \quad T = 9.$$

Then the classifications returned by the classifier are $[1,0,0,1,0,1]$; the first, third, and fifth rows are correctly classified, and the rest are misclassified. Thus, the accuracy is $3/6 = 0.5$. The precision is $1/3$; of the three instances identified by the classifier, only one is correct. The recall is $1/2$; of the two actual instances of the category, one is identified by the classifier.

(b) Write a function `evaluate2(D,L,W,T)`. Here the input arguments D and L are as in part (a). W and T, however, represent a collection of q classifiers. W is an $n \times q$ matrix; T is a q-dimensional vector. For $j = 1,\ldots,q$, the column $W[;j]$ and the value $T[j]$ are the weight vector and threshold of a classifier. `E=evaluate2(D,L,W,T)` returns a $3 \times q$ matrix, where, for $j = 1,\ldots,q, E[1,j], E[2,j]$ and $E[3,j]$ are, respectively, the accuracy, precision, and recall of the jth classifier. For example, let D and L be as in part (a). Let $q = 2$ and let

$$W = \begin{bmatrix} 1 & 0 \\ 2 & 0 \\ 1 & 0 \\ 2 & 1 \end{bmatrix}, \qquad T = [9,2].$$

Then `evaluate2(D,L,W,T)` returns

$$\begin{bmatrix} 0.5 & 0.6667 \\ 0.3333 & 0.5 \\ 0.5 & 0.5 \end{bmatrix}.$$

Assignment 3.2. Write a function `TotalPaths(A)` that takes as argument an adjacency matrix `A` for graph `G` and returns the total number of different paths of any length in `G` if this is finite, or `Inf` if there are infinitely many paths. (*Note:* In the latter case, it should *return* `Inf`; it should *not* go into an infinite loop. *Hint:* It may be useful to do Problem 3.6 first.)

Assignment 3.3. The transition matrix model discussed in Application 3.3 can also be used to model diffusion of unevenly distributed material in space over time. As in the discussion of temperature distribution (Application 3.8), divide space up into a grid of points numbered $1,\ldots,N$. Let $\vec{v}_T[I]$ be the amount of material at point I at time T. Let $D[I,J]$ be a transition matrix that expresses the fraction of the material at J that flows to I in each unit of time. In particular, of course, $D[I,I]$ is the fraction of the material at I that remains at I. Then the amount of material at I at time $T+1$ is equal to $\sum_I D[I,J] \cdot \vec{v}_T[J]$, so $\vec{v}_{T+1} = M \cdot \vec{v}_T$.

Consider now a case similar to Application 3.8 in which we are considering a rectangular region, discretized into a grid of $M \times N$ interior points. Assume that the points are numbered left to right and top to bottom, as in Figure 3.3. The vector $\vec{v}$ is thus an $M \cdot N$ dimensional vector, with one dimension for each point, and D is then an $MN \times MN$ matrix. Let us say that the neighbors of a point are the four points to the left, to the right, above, and below. Assume the following values for $D[I, J]$. If $J \neq I$, then $D[I, J] = 0.1$ if J is a neighbor of I, and 0 otherwise. Matrix $D[I, I]$ is 0.6 if I is an interior point with four neighbors; 0.7 if I is a side point with three neighbors; and 0.8 if I is a corner point with two neighbors.

(a) Write a MATLAB function DiffusionMat(M,N) that returns the MN × MN diffusion matrix for an M × N grid of internal points.

(b) Write a MATLAB function Diffuse(Q,K) that takes as an input argument an M × N matrix representing the grid of mass distribution at time T = 0 and returns the M × N grid at time T = K. (*Hint:* Convert Q into a vector of length MN, calculate the answer in terms of another vector of length MN, and then convert back into an M × N matrix. The MATLAB function reshape will be helpful.)

(c) Write a function EquilibriumTime(V,EPSILON) that computes the time K that it takes for all the material to "smooth out," that is, to reach a state $\vec{V}_P$ such that the largest value in $\vec{V}_P$ minus the smallest value is less than EPSILON. Use a binary search. That is, in the first stage of the algorithm, compute in sequence the diffusion matrix D; then D^2, $D^4 = D^2 \cdot D^2$, $D^8 = D^4 \cdot D^4$ until you reach a value D^{2^k} such that $D^{2^k} \cdot \vec{v}_0$ satisfies the condition. Save all the powers of D you have computed in a three-dimensional array. The exact time P is thus greater than 2^{K-1} and less than or equal to 2^K.

Now, in the second stage of the algorithm, use a binary search to find the exact value of P. For instance, if $K = 5$, then P is somewhere between $16 = 2^4$ and $32 = 2^5$. Try halfway between, at $(16 + 32)/2 = 24$. If this satisfies the condition, search between 8 and 12; if it does not, search between 12 and 16. Then iterate. Note that at each stage of your search the upper bound U and the lower bound L differ by a power of 2, 2^I, and that you have already computed D^{2^I} in the first stage of the algorithm and saved it, so all you have to do now to compute the intermediate point is to multiply $D^L \cdot D^{2^I}$. Thus, you should be able to carry out the whole algorithm with at most $2\lceil \log_2 P \rceil$ matrix multiplications.

Chapter 4

_____

Vector Spaces

This chapter is about *vector spaces*, which are certain sets of n-dimensional vectors. The material in this chapter is theoretical—concepts, definitions, and theorems. We discuss algorithmic issues in Chapter 5, and then return at length to applications in Chapters 6 and 7.

There are two objectives in this chapter. The first is to discuss the structure of vector spaces; this framework will be useful in Chapters 5 and 6 and necessary in Chapter 7. The second is to give a characterization of the space of solutions to a system of linear equations; this follows easily from the analysis of vector spaces.

The chapter is divided into three parts, with increasing mathematical sophistication. Section 4.1 presents the material on vector spaces needed for the remainder of the book, as far as possible avoiding unnecessary abstraction and mathematical difficulty. Many students will no doubt find this section quite abstract and difficult enough, but it is important to master the material discussed here because as it will be needed later.

Section 4.2 goes over essentially the same ground; it discusses some further mathematical aspects and gives proofs of theorems. Students who like abstract math will, I think, find that this material is not only interesting and mathematically very elegant, but helpful in clarifying the basic concepts and their relations. Students who do not care for abstract math may prefer to skip this.

Finally, Section 4.3 (very optional) gives the most general definition of a vector space in the way that any self-respecting mathematician would give it. This section is included so that the self-respecting mathematicians will not burn this book in the town square.

4.1   Fundamentals of Vector Spaces

The material in this section will be needed for Chapters 5, 6, and 7.

4.1.1   Subspaces

A *subspace* of $\mathbb{R}^n$ is a particular type of set of n-dimensional vectors. Subspaces of $\mathbb{R}^n$ are a type of *vector space*; they are the only kind of vector space we consider in this book, except in Section 4.3.1 of this chapter.

Since we are dealing extensively with sets in this chapter, we will be using the standard set operators. $S \cup T$ is the union of S and T. $S \cap T$ is the intersection of S and T. $S \setminus T$ is the set difference S minus T; that is, the set of all elements in S but not in T. $\emptyset$ is the empty set, the set with no elements. To begin with, we need the idea of a linear sum:

Definition 4.1. Let V be a set of vectors in $\mathbb{R}^n$. A vector $\vec{u} \in \mathbb{R}^n$ is a *linear sum* over V if there exist $\vec{v}_1, \ldots, \vec{v}_m$ in V and scalars $a_1, \ldots, a_m$ such that $\vec{u} = a_1 \vec{v}_1 + \ldots + a_m \vec{v}_m$.

Definition 4.2. Let $\mathscr{S}$ be a set of vectors. The *span* of $\mathscr{S}$, denoted Span($\mathscr{S}$), is the set of linear sums over S.

Example 4.3. In $\mathbb{R}^3$ let $V = \{\vec{v}_1, \vec{v}_2\}$, where $\vec{v}_1 = \langle 0, 2, 0 \rangle$, and $\vec{v}_2 = \langle 1, 3, -1 \rangle$. Then Span($V$) is the set of all vectors $\vec{w}$ of the form $\vec{w} = a \cdot \langle 0, 2, 0 \rangle + b \cdot \langle 1, 3, -1 \rangle = \langle b, 2a + 3b, -b \rangle$. For instance, Span($V$) includes the vector $3\vec{v}_1 - 2\vec{v}_2 = \langle -2, 0, 2 \rangle$. It does not include the vector $\langle 4, 2, -2 \rangle$, however, because any vector $\vec{v}$ in Span(V) must satisfy $\vec{v}[3] = -\vec{v}[1]$.

Definition 4.4. A nonempty set of n-dimensional vectors, $\mathscr{S} \subset \mathbb{R}^n$ is a *subspace* of $\mathbb{R}^n$ if $\mathscr{S} = \text{Span}(V)$ for some set of vectors V.

Theorem 4.5 gives a useful technique for proving that a set $\mathscr{S}$ is a subspace.

Theorem 4.5. *A nonempty set of vectors $\mathscr{S} \subset \mathbb{R}^n$ is a subspace if it is closed under the vector operations. That is, $\mathscr{S}$ satisfies the following two properties:*

- *If $\vec{u}$ and $\vec{v}$ are in $\mathscr{S}$ then $\vec{u} + \vec{v}$ are in $\mathscr{S}$.*

- *If c is a scalar and $\vec{u}$ is in $\mathscr{S}$, then $c \cdot \vec{u}$ is in $\mathscr{S}$.*

The proof is the same as for Theorem 4.28 in Section 4.2.1.

Example 4.6. The entire space $\mathbb{R}^n$ is a (nonstrict) subspace of itself, called the "complete space."

Example 4.7. The set containing only the n-dimensional zero vector $\{\vec{0}\}$ is a subspace of $\mathbb{R}^n$ known as the "zero space."

Example 4.8. $\mathbb{R}^2$ has three kinds of subspaces:

1. the complete space,

2. the zero space,

3. for any line L through the origin, the set of vectors lying on L.

In the last subspace (3), the line $ax + by = 0$ is equal to $\text{Span}(\{\langle -b, a\rangle\})$. Note that a line that does not go through the origin is not a subspace. This is called an "affine space"; these are studied in Chapter 6.

Example 4.9. $\mathbb{R}^3$ has four kinds of subspaces: the three in Example 4.8 plus one more:

4. for any plane P through the origin, the vectors lying in P.

For instance, the line going equally in the x, y, z directions is equal to $\text{Span}(\{\langle 1, 1, 1\rangle\})$. The plane $x + y + z = 0$ is equal to $\text{Span}(\{\langle 1, 0, -1\rangle, \langle 0, 1, -1\rangle\})$.

Example 4.10. Consider a system of linear equations of m equations in n unknowns whose constant terms are all zero. The system can be expressed in the form $M\vec{x} = \vec{0}$; this is known as a *homogeneous* system of equations. Let $\mathscr{S}$ be the set of all solutions to this system. We can use Theorem 4.5 to show that $\mathscr{S}$ is a subspace of $\mathbb{R}^n$:

- If $\vec{u}$ and $\vec{v}$ are in $\mathscr{S}$, then $M\vec{u} = \vec{0}$ and $M\vec{v} = 0$. It follows that $M(\vec{u} + \vec{v}) = M\vec{u} + M\vec{v} = \vec{0} + \vec{0} = \vec{0}$, so $\vec{u} + \vec{v}$ is in $\mathscr{S}$.

- If $\vec{u}$ is in $\mathscr{S}$ and c is a scalar, then $M(c\vec{v}) = c \cdot M\vec{v} = c \cdot \vec{0} = \vec{0}$, so $c\vec{u}$ is in $\mathscr{S}$.

Example 4.11. Suppose that a retail store keeps a database with a record for each item. The record for item I is a three-dimensional vector, consisting of the price paid for I, the price at which I was sold, and the net profit or loss on I. Then a vector $\vec{v}$ is a legitimate entry if it just satisfies the equation $\vec{v}[3] = \vec{v}[2] - \vec{v}[1]$ (profit is selling price minus buying price). Therefore, the set of legitimate entries is a subspace of $\mathbb{R}^3$.

Of course, it is not quite accurate to say that any vector that satisfies the condition is actually a possible entry. Consider the vectors $\langle -2, -9, -7\rangle$, $\langle 2\pi, 9\pi, 7\pi\rangle$, and $\langle 2 \cdot 10^{30}, 9 \cdot 10^{30}, 7 \cdot 10^{30}\rangle$. Each of these vectors satisfies the condition, but is likely to raise suspicions in an auditor—it is quite unusual to buy or sell an item for a negative value, for an irrational value, or for more money than exists in the world. It would be more accurate to say that the set of legitimate values is a subset of the subspace of values satisfying $\vec{v}[3] = \vec{v}[2] - \vec{v}[1]$. However, we ignore that inconvenient reality here.

Example 4.12. Consider an alternative version of Example 4.11, where an international business records price, cost, and profit in a number of different currencies; and suppose, unrealistically, that the exchange rates are constant over time. Specifically, the database records the three amounts in dollars, euros, and yen. Suppose that 1 dollar = 0.8 euro = 90 yen. Then a valid entry is a 9-tuple satisfying the equations:

$$\vec{v}[3] = \vec{v}[2] - \vec{v}[1],$$
$$\vec{v}[4] = 0.8 \cdot \vec{v}[1],$$
$$\vec{v}[5] = 0.8 \cdot \vec{v}[2],$$
$$\vec{v}[6] = 0.8 \cdot \vec{v}[3],$$
$$\vec{v}[7] = 90 \cdot \vec{v}[1],$$
$$\vec{v}[8] = 90 \cdot \vec{v}[2],$$
$$\vec{v}[9] = 90 \cdot \vec{v}[3].$$

4.1.2 Coordinates, Bases, Linear Independence

Suppose we have a subspace V of $\mathbb{R}^n$ and we want to construct a *coordinate system* for V—that is, a notation that represents all and only vectors in V. That way, we can be sure that all of our records are valid vectors in V, since the notation can express *only* vectors in V.

The standard way to construct a coordinate system is by using a *basis* for V. A basis for V is a finite tuple $\mathcal{B} = \langle \vec{b}_1, \ldots, \vec{b}_m \rangle$, satisfying the two conditions delineated below. If $\vec{u}$ is a vector in V, then the *coordinates of $\vec{u}$ in basis $\mathcal{B}$*, denoted Coords($\vec{u}, \mathcal{B}$), is the sequence $\langle a_1, \ldots, a_m \rangle$ such that $\vec{u} = a_1 \vec{b}_1 + \ldots + a_m \vec{b}_m$. Now, for this to be a legitimate coordinate system, it must satisfy these two conditions:

- Any tuple of coordinates must represent a vector in V, and every vector in V must be representable as a tuple of coordinates over $\mathcal{B}$—that is, as a linear sum over $\mathcal{B}$. Therefore, we must choose $\mathcal{B}$ so that Span($\mathcal{B}$) = V.

- A single vector $\vec{u}$ has only one tuple of coordinates in $\mathcal{B}$; that is, any two different sets of coordinates give different vectors. This is guaranteed by requiring that $\mathcal{B}$ be *linearly independent*.

Definition 4.13. A set of vectors V is *linearly dependent* if one of the vectors in V is a linear sum of the rest. That is, for some $\vec{v} \in V$, $\vec{v} \in$ Span($V \setminus \{\vec{v}\}$). V is *linearly independent* if it is not linearly dependent.

We can now define a basis:

Definition 4.14. Let V be a subspace of $\mathbb{R}^n$. A finite subset $\mathcal{B} \subset V$ is a *basis* for V if

- Span($\mathscr{B}$) = $\mathcal{V}$, and

- $\mathscr{B}$ is linearly independent.

Theorem 4.15 states that a basis $\mathscr{B}$ for a subspace $\mathscr{S}$ defines a coordinate system with the desired properties.

Theorem 4.15. *Let $\mathscr{B} = \langle \vec{b}_1, \dots, \vec{b}_m \rangle$ be a basis for vector space $\mathscr{S}$ and let $\vec{v}$ be a vector in $\mathscr{S}$. Then*

(a) *For any vector $\vec{v} \in \mathscr{S}$, there exists a unique m-tuple of coordinates $\langle a_1, \dots, a_m \rangle$ = Coords$(\vec{v}, \mathscr{B})$ such that $\vec{v} = a_1 \vec{b}_1 + \dots + a_m \vec{b}_m$.*

(b) *For any m-tuple of coordinates $\langle a_1, \dots, a_m \rangle$ the linear sum $\vec{v} = a_1 \vec{b}_1 + \dots + a_m \vec{b}_m$ is in $\mathscr{S}$.*

The proof of (a) is the same as that given for Theorem 4.32 in Section 4.2.1; (b) follows directly from the definitions.

To be able to include the zero space in our theorems about subspaces, it is convenient to adopt the following conventions:

- The zero space is considered to be the span of the empty set: $\{\vec{0}\} = \mathrm{Span}(\varnothing)$. You get the zero vector for free, so to speak.

- The empty set is considered to be linearly independent.

- The set containing just the zero vector $\{\vec{0}\}$ is considered to be linearly dependent.

Examples of bases and coordinates. Any vector space, other than the zero space, has many different bases, including the following ones of interest:

- A basis for all of $\mathbb{R}^n$ is the set $\langle e^1, e^2, \dots, e^n \rangle$. (Recall that e^i is the vector in which the ith component is 1 and other components are 0.)

- A basis for the line $ax + by = 0$ is the singleton set $\{\langle b, -a \rangle\}$. In fact, if $\vec{v}$ is any nonzero vector in the line, then $\{\vec{v}\}$ is a basis for the line.

- A basis for the plane $x + y + z = 0$ is the pair of vectors $\{\langle 1, -1, 0 \rangle, \langle 1, 0, -1 \rangle\}$. Another basis is the pair $\{\langle 1, 1, -2 \rangle, \langle -2, 1, 1 \rangle\}$. In fact, any two noncolinear vectors in this plane form a basis.

- A basis for the space described in Example 4.11 is the pair $\{\langle 1, 0, -1 \rangle, \langle 0, 1, 1 \rangle\}$. Another basis is the pair $\{\langle 0, 1, 1 \rangle, \langle 1, 1, 0 \rangle\}$.

We leave it as an exercise to find a basis for the space described in Example 4.12.

Example 4.16. Considering the plane $x + y + z = 0$, we already identified two bases for this subspace:

- $\mathcal{B} = \{\vec{b}_1, \vec{b}_2\}$, where $\vec{b}_1 = \langle 1, 0, -1 \rangle$ and $\vec{b}_2 = \langle 0, 1, -1 \rangle$,

- $\mathcal{C} = \{\vec{c}_1, \vec{c}_2\}$, where $\vec{c}_1 = \langle 1, 1, -2 \rangle$ and $\vec{c}_2 = \langle -2, 1, 1 \rangle$.

We now use each of these as coordinate systems. Taking two particular vectors in the plane: $\vec{u} = \langle -5, 7, -2 \rangle$ and $\vec{v} = \langle 1, 8, -9 \rangle$, and measuring these in coordinate system $\mathcal{B}$, we find the following coordinates:

- $\text{Coords}(\vec{u}, \mathcal{B}) = \langle -5, 7 \rangle$ because $5 \cdot \vec{b}_1 + 7 \cdot \vec{b}_2 = \vec{u}$,

- $\text{Coords}(\vec{v}, \mathcal{B}) = \langle 1, 8 \rangle$ because $1 \cdot \vec{b}_1 + 8 \cdot \vec{b}_2 = \vec{v}$.

In contrast, measuring them in coordinate system $\mathcal{C}$, we find other coordinates:

- $\text{Coords}(\vec{u}, \mathcal{C}) = \langle 3, 4 \rangle$ because $3 \cdot \vec{c}_1 + 4 \cdot \vec{c}_2 = \vec{u}$.

- $\text{Coords}(\vec{v}, \mathcal{C}) = \langle 17/3, 7/3 \rangle$ because $17/3 \cdot \vec{c}_1 + 7/3 \cdot \vec{c}_2 = \vec{v}$.

It is easy to *check* that these are the correct coordinates by performing the indicated sums. We discuss in Chapter 7 how to *find* the coordinates of a vector relative to a given basis.

Properties of bases and coordinates. We have seen that any subspace has many different bases. But, as the examples suggest, the different bases of a subspace have a very important property in common: they all have the same number of elements.

Theorem 4.17. *Let V be a subspace of $\mathbb{R}^n$. Then any two bases for V have the same number of elements. This is known as the* dimension *of V, denoted* $\text{Dim}(V)$. *By convention, the dimension of the zero space is 0.*

The proof is as given for Corollary 4.34, in Section 4.2.2.

Now, suppose that V is an m-dimensional vector space, and $\mathcal{B}$ is a basis for V. Then for any vector $\vec{u}$ in V, the coordinates for $\vec{u}$ in terms of $\mathcal{B}$, $\text{Coords}(\vec{u}, \mathcal{B}) = \langle a_1, \ldots, a_m \rangle$, is itself an m-dimensional vector. The great thing about this coordinate notation is that you can carry out vector operations on the *actual* vectors by applying the operators to their *coordinates* in any basis.

Theorem 4.18. *Let V be a subspace of $\mathbb{R}^n$ and let $\mathcal{B}$ be a basis for V. Then*

- *For any $\vec{u}, \vec{w} \in V$, $\text{Coords}(\vec{u} + \vec{w}, \mathcal{B}) = \text{Coords}(\vec{u}, \mathcal{B}) + \text{Coords}(\vec{w}, \mathcal{B})$.*

- *For any $\vec{u} \in V$ and scalar a, $\text{Coords}(a \cdot \vec{u}, \mathcal{B}) = a \cdot \text{Coords}(\vec{u}, \mathcal{B})$.*

- *The coordinates of a linear sum is equal to the linear sum of the coordinates. That is,*

$$\text{Coords}(a_1 \cdot \vec{v}_1 + \ldots + a_n \cdot \vec{v}_n, \mathcal{B}) = a_1 \cdot \text{Coords}(\vec{v}_1, \mathcal{B}) + \ldots + a_n \cdot \text{Coords}(\vec{v}_n, \mathcal{B}).$$

The proof of Theorem 4.18 is left as an exercise (Problem 4.1).

For instance, let us go back to Example 4.16. Suppose that we have decided to use the coordinate system $\mathscr{C}$ and we have recorded that $\text{Coords}(\vec{u}, \mathscr{C}) = \langle 3, 4 \rangle$ and $\text{Coords}(\vec{v}, \mathscr{C}) = \langle 17/3, 7/3 \rangle$. We now want to compute $\text{Coords}(6\vec{v} - 2\vec{u}, \mathscr{C})$. One way would be to translate $\vec{u}$ and $\vec{v}$ back into their native form, do the computation, and then translate the result back into coordinates relative to $\mathscr{C}$. But Theorem 4.18 gives us a much simpler solution:

$$\text{Coords}(6\vec{v} - 2\vec{u}, \mathscr{C}) = 6\,\text{Coords}(\vec{v}, \mathscr{C}) - 2\,\text{Coords}(\vec{u}, \mathscr{C}) = \langle 28, 6 \rangle$$

4.1.3   Orthogonal and Orthonormal Basis

An *orthogonal basis* is one for which every pair of elements is orthogonal. For instance, the subspace $x + y + z = 0$ has the orthogonal basis $\mathscr{O} = \{\langle 1, -1, 0 \rangle, \langle 1, 1, -2 \rangle\}$.

An *orthonormal basis*[1] is an orthogonal basis in which every element has length 1. If $\mathscr{B}$ is an orthogonal basis for space $\mathscr{S}$, then you can get an orthonormal basis by dividing each element in $\mathscr{B}$ by its length. For instance, in the orthogonal basis $\mathscr{O}$ for the space $x + y + z = 0$, the first vector has length $\sqrt{2}$ and the second has length $\sqrt{6}$. Therefore, this turns into the orthonormal basis $\mathscr{N} = \langle 1/\sqrt{2}, -1/\sqrt{2}, 0 \rangle, \langle 1/\sqrt{6}, 1/\sqrt{6}, -2/\sqrt{6} \rangle$.

The obvious drawback of using orthonormal bases is that you get involved with irrational numbers. The advantage, however, is that if vectors in V are recorded in terms of their coordinates in an orthonormal basis, then lengths and dot products can also be computed without translating back to the standard coordinates.

Theorem 4.19. *Let V be a vector space and let $\mathscr{B}$ be an orthonormal basis for V. Then:*

- *For any $\vec{u}, \vec{w} \in V$, $\vec{u} \bullet \vec{w} = \text{Coords}(\vec{u}, \mathscr{B}) \bullet \text{Coords}(\vec{w}, \mathscr{B})$.*

- *For any $\vec{u} \in V$, $|\vec{u}| = |\text{Coords}(\vec{u}, \mathscr{B})|$.*

The proof is as given for Theorem 4.45 in Section 4.2.4.

We illustrate by using the subspace $x + y + z = 0$. Again, let $\vec{u} = \langle -5, 7, -2 \rangle$ and $\vec{v} = \langle 1, 8, -9 \rangle$. Then

$$\text{Coords}(\vec{u}, \mathscr{O}) = \langle -6, 1 \rangle \quad \text{and} \quad \text{Coords}(\vec{v}, \mathscr{O}) = \langle -7/2, 9/2 \rangle, \quad \text{so}$$

$$\text{Coords}(\vec{u}, \mathscr{N}) = \langle -6\sqrt{2}, \sqrt{6} \rangle = \langle -8.4852, 2.4495 \rangle \quad \text{and}$$

$$\text{Coords}(\vec{v}, \mathscr{N}) = \langle -7\sqrt{2}/2, 9\sqrt{6}/2 \rangle = \langle -4.9497, 11.0227 \rangle.$$

---

[1]Unhelpfully, in mathematical literature, the term "orthogonal basis" is often used to mean "orthonormal basis."

We leave it as an exercise (Exercise 4.3) to check that, as Theorem 4.19 claims,

$$|\vec{u}| = |\operatorname{Coords}(\vec{u}, \mathcal{N})| = \sqrt{78},$$

$$|\vec{v}| = |\operatorname{Coords}(\vec{v}, \mathcal{N})| = \sqrt{146}, \quad \text{and}$$

$$\vec{u} \bullet \vec{v} = \operatorname{Coords}(\vec{u}, \mathcal{N}) \bullet \operatorname{Coords}(\vec{v}, \mathcal{N}) = 69.$$

An orthonormal basis also has the important and elegant property stated in Theorem 4.20.

Theorem 4.20. *Let* $\mathcal{B} = \{\vec{b}_1, \ldots, \vec{b}_q\}$ *be an orthonormal basis for vector space* V. *If* $\vec{v}$ *is a vector in* V, *then the* ith *coordinate of* $\vec{v}$ *with respect to* $\mathcal{B}$, $\operatorname{Coords}(\vec{v}, \mathcal{B})[i]$ = $\vec{v} \bullet \vec{b}_i$.

The proof is as shown for Theorem 4.46 in Section 4.2.4.

A matrix M is orthonormal if its columns are orthonormal. Square orthonormal matrices have the properties given in Theorem 4.21, which are important in Chapter 7.

Theorem 4.21. *If* M *is an* $n \times n$ *orthonormal matrix, then* M^T *is also orthonormal and* $M \cdot M^T = M^T \cdot M = I_n$.

We leave the proof of Theorem 4.21 as an exercise (Problem 4.2).

4.1.4   Operations on Vector Spaces

We can combine two subspaces $\mathcal{U}$ and V of $\mathbb{R}^n$ in two important ways. First, we can take their intersection $\mathcal{U} \cap V$. Second, we can take their *direct sum*, denoted $\mathcal{U} \oplus V$ and defined as $\mathcal{U} \oplus V = \operatorname{Span}(\mathcal{U} \cup V)$. Both $\mathcal{U} \cap V$ and $\mathcal{U} \oplus V$ are themselves vector spaces.

In two dimensions, the direct sum of two distinct lines is the entire space, and their intersection is the zero space. In three-dimensional space, if $\mathcal{U}$ and V are different lines, then $\mathcal{U} \oplus V$ is the plane containing them. If $\mathcal{U}$ is a plane and V is a line outside $\mathcal{U}$, then $\mathcal{U} \oplus V$ is the entire space.

Theorem 4.22 states three important properties of these operations.

Theorem 4.22. *Let* $\mathcal{U}$ *and* V *be subspaces of* $\mathbb{R}^n$. *Then*

(a) *For any vector* $\vec{w}$ *in* $\mathcal{U} \oplus V$, *there exists* $\vec{u} \in \mathcal{U}$ *and* $\vec{v} \in V$ *such that* $\vec{w} = \vec{u} + \vec{v}$. *If* $\mathcal{U} \cap V = \{\vec{0}\}$, *then there is only one possible choice of* $\vec{u}$ *and* $\vec{v}$.

(b) *If* $\mathcal{U} \cap V = \{\vec{0}\}$, *then the union of a basis for* $\mathcal{U}$ *with a basis for* V *is a basis for* $\mathcal{U} \oplus V$.

(c) $\operatorname{Dim}(\mathcal{U} \oplus V) = \operatorname{Dim}(\mathcal{U}) + \operatorname{Dim}(V) - \operatorname{Dim}(\mathcal{U} \cap V)$.

The proof of part (c) is as given for Theorem 4.42 in Section 4.2.3. The proofs of parts (a) and (b) are left as exercises (Problem 4.3).

Two vector spaces $\mathcal{U}$ and $\mathcal{V}$ are *complements* if $\mathcal{U} \cap \mathcal{V} = \{\vec{0}\}$ and $\mathcal{U} \oplus \mathcal{V} = \mathbb{R}^n$. If $\mathcal{U}$ and $\mathcal{V}$ are complements, then any vector in $\mathbb{R}^n$ can be written in exactly one way as the sum of a vector in $\mathcal{U}$ plus a vector in $\mathcal{V}$. $\mathcal{U}$ and $\mathcal{V}$ are *orthogonal complements* if they are complements, and any vectors $\vec{u} \in \mathcal{U}$ and $\vec{v} \in \mathcal{V}$ are orthogonal. In fact, the orthogonal complement of $\mathcal{U}$ is just the set of all vectors $\vec{v}$ such that $\vec{c}$ is orthogonal to every vector in $\mathcal{U}$.

Theorem 4.23. *Any subspace of $\mathbb{R}^n$ has an orthogonal basis and an orthogonal complement.*

The proof is as shown for Theorem 4.49 in Section 4.2.4.

If $\mathcal{U}$ and $\mathcal{V}$ are complements, then $\mathcal{U} \oplus \mathcal{V} = \mathbb{R}^n$ so $\text{Dim}(\mathcal{U} \oplus \mathcal{V}) = n$, and $\mathcal{U} \cap \mathcal{V} = \{\vec{0}\}$ so $\text{Dim}(\mathcal{U} \cap \mathcal{V}) = 0$. By using the formulas in Theorem 4.22(c) and Theorem 4.23, we have $\text{Dim}(\mathcal{U}) + \text{Dim}(\mathcal{V}) = \text{Dim}(\mathcal{U} \oplus \mathcal{V}) + \text{Dim}(\mathcal{U} \cap \mathcal{V}) = n$. So, for example, in two dimensions, any two distinct lines through the origin are complements. In three dimensions a plane P through the origin and any line L through the origin not lying in P are complements.

4.1.5 Null Space, Image Space, and Rank

We can now use the theory of vector spaces that we developed so far to partially characterize what a linear transformation does.

Suppose that Γ is a linear transformation from $\mathbb{R}^n$ to $\mathbb{R}^m$. Then the *image* of Γ, $\text{Image}(\Gamma) = \Gamma(\mathbb{R}^n)$, is a subspace of $\mathbb{R}^m$. The *null space* (also called the *kernel*) of Γ is the set of vectors in $\mathbb{R}^n$ that Γ maps to $\vec{0}$:

$$\text{Null}(\Gamma) = \{\vec{v} \in \mathbb{R}^n \mid \Gamma(\vec{v}) = \vec{0}\}.$$

Now, let $\mathcal{V}$ be any subspace of $\mathbb{R}^n$ that is complementary to $\text{Null}(\Gamma)$. Since $\text{Null}(\Gamma)$ and $\mathcal{V}$ are complementary, any vector $\vec{p} \in \mathbb{R}^n$ is equal to $\vec{u} + \vec{v}$, where $\vec{u} \in \text{Null}(\Gamma)$ and $\vec{v} \in \mathcal{V}$. Then $\Gamma(\vec{p}) = \Gamma(\vec{u} + \vec{v}) = \Gamma(\vec{u}) + \Gamma(\vec{v}) = \Gamma(\vec{v})$. Thus, any vector in $\text{Image}(\Gamma)$ is in $\Gamma(\mathcal{V})$. In fact, we can show that Γ is a *bijection* between $\mathcal{V}$ and $\text{Image}(\Gamma)$; that is, for every vector $\vec{w} \in \text{Image}(\Gamma)$, there exists exactly one vector $\vec{v} \in V$ such that $\vec{w} = \Gamma(\vec{v})$.

So we can think about what Γ does in this way. We identify $\text{Null}(\Gamma)$, the set of vectors that Γ sends to $\vec{0}$. We choose some vector space $\mathcal{V}$ that is complementary to $\text{Null}(\Gamma)$. (Note that there are many ways to choose $\mathcal{V}$, although there is no choice about $\text{Null}(\Gamma)$. If we like, we may choose $\mathcal{V}$ to be the orthogonal complement to $\text{Null}(\Gamma)$; sometimes that is the best choice, but not always.) We can then divide any vector $\vec{p} \in \mathbb{R}^n$ into a component $\vec{u} \in \text{Null}(\Gamma)$ and a component

$\vec{v}$ in $\mathcal{V}$, such that $\vec{p} = \vec{v} + \vec{u}$. When Γ is applied to $\vec{p}$, it simply zeroes out the $\vec{u}$ component and sets $\Gamma(\vec{p}) = \Gamma(\vec{v})$.

How do we describe what Γ does to vectors in $\mathcal{V}$? The easiest way is to choose a basis for $\mathcal{V}$. If $\vec{b}_1, \ldots, \vec{b}_r$ is a basis for $\mathcal{V}$, then $\Gamma(\vec{b}_1), \ldots, \Gamma(\vec{b}_r)$ is a basis for Image(Γ). Moreover, if $\vec{v}$ is any vector in $\mathcal{V}$, then the coordinates of $\Gamma(\vec{v})$ with respect to the basis $\{\Gamma(\vec{b}_1), \ldots, \Gamma(\vec{b}_r)\}$ are the same as the coordinates of $\vec{v}$ with respect to the basis $\{\vec{b}_1, \ldots, \vec{b}_r\}$. The index r is the dimension both of $\mathcal{V}$ and of Image(Γ). This is known as the *rank* of Γ: Rank(Γ) = Dim(Image(Γ)).

Note that we now have *two* important uses for bases:

- to serve as the basis for a coordinate system.

- to use in characterizing linear transformations.

One further point deserves mention. Since $\mathcal{V}$ and Image(Γ) have bases of the same size, we have Dim($\mathcal{V}$) = Dim(Image(Γ)) = Rank(Γ). But since $\mathcal{V}$ and Null(Γ) are complementary subspaces of $\mathbb{R}^n$, we have Dim($\mathcal{V}$) + Dim(Null(Γ)) = n, so Dim(Null(Γ)) = n − Rank(Γ).

These functions apply to matrices in the same way; that is, if M is the matrix corresponding to transformation Γ, then Image(M) = Image(Γ), Null(M) = Null(Γ), and Rank(M) = Rank(Γ). Theorem 4.24 pertains to these properties of matrices.

Theorem 4.24. *For any matrix M:*

- Image(M) *is the space spanned by the columns of M. Therefore* Rank(M) = Dim(Image(M)) *is the dimension of the space spanned by the columns of M.*

- Rank(M) *is also the dimension of the space spanned by the* rows *of M. Therefore* Rank(M) = Rank(M^T).

The proof is as shown for Theorem 4.77 and Corollary 4.79 in Section 4.2.8.

Example 4.25. Let M be the matrix

$$M = \begin{bmatrix} 1 & 2 & -1 \\ 3 & 6 & -3 \end{bmatrix}.$$

Then Null(M) is the set of vectors $\langle x, y, z \rangle$ such that $x + 2y - z = 0$ and $3x + 6y - 3z = 0$. Since any vector that satisfies the first equation also satisfies the second, this is just the plane $x + 2y - z = 0$. We can choose for $\mathcal{V}$ any complement of Null(M). Since Null(M) is two-dimensional, $\mathcal{V}$ must be one-dimensional, so we may choose as a basis any vector $\vec{b}$ that does not lie in the plane Null(M). Let us choose $\vec{b} = \langle 1, 0, 0 \rangle$, the unit x-vector, and basis $\mathscr{B} = \{\vec{b}\}$, so $\mathcal{V} = \text{Span}(\mathscr{B}) = \{t \cdot \langle 1, 0, 0 \rangle\}$, the set of vectors on the x-axis.

Now $M \cdot \vec{b} = \langle 1, 3 \rangle$, and the image of M is given by

$$\text{Image}(M) = \{M \cdot \vec{v} \mid \vec{v} \in \mathcal{V}\} = \{M \cdot t \cdot \langle 1, 0, 0 \rangle\} = \text{Span}(M \cdot \vec{b}) = \{t \cdot \langle 1, 3 \rangle\}.$$

This is the line in the plane with slope 3.

Consider a vector in $\mathbb{R}^3$, say $\vec{w} = \langle 9, 2, 6 \rangle$. We can then find vectors $\vec{u} \in$ Null(M) and $\vec{v} \in \mathcal{V}$ such that $\vec{w} = \vec{u} + \vec{v}$. In this case, $\vec{u} = \langle 2, 2, 6 \rangle$ and $\vec{v} = \langle 7, 0, 0 \rangle$. Then $M \cdot \vec{w} = M \cdot \vec{v} = \langle 7, 21 \rangle$. Note that $\text{Coords}(\vec{v}, \mathcal{B}) = \text{Coords}(M \cdot \vec{v}, \{M \cdot \vec{b}\}) = \langle 7 \rangle$.

Example 4.26. Let M be the matrix

$$M = \begin{bmatrix} 1 & 2 & -1 \\ -1 & 0 & 1 \\ -1 & 2 & 1 \end{bmatrix}.$$

Then Null(M) is the one-dimensional line $\{t \cdot \vec{n}\}$, where $\vec{n}$ is the vector $\langle 1, 0, 1 \rangle$. The complementary space $\mathcal{V}$ is thus a two-dimensional plane. As a basis $\mathcal{B}$ for the complementary space, we may take any two noncolinear vectors that do not lie in Null(M), such as $\vec{b}_1 = \langle 0, 1, 1 \rangle$ and $\vec{b}_2 = \langle 1, 1, 0 \rangle$. Then

$$\mathcal{V} = \text{Span}(\mathcal{B}) = \{a_1 \cdot \vec{b}_1 + a_2 \cdot \vec{b}_2\} = \{\langle a_1, a1 + a_2, a_2 \rangle \mid a_1, a_2 \in \mathbb{R}\}.$$

We can check that these are indeed the null space and a complementary space by checking that

(a) $M \cdot \vec{n} = \vec{0}$,

(b) $M \cdot \vec{b}_1 \neq \vec{0}$, and $M \cdot \vec{b}_2 \neq \vec{0}$, so these are not in the null space,

(c) $\vec{b}_1$ and $\vec{b}_2$ are linearly independent.

It follows from (a) that Span$(\{\vec{n}\})$ is in the null space, and from (b) and (c) that Span$(\mathcal{B})$ is in a complementary space. But since we have now used up all three dimensions, it follows that Span$(\{\vec{n}\})$ is the complete null space and that Span$(\mathcal{B})$ is a complete complement.

Now $\{M \cdot \vec{b}_1, M \cdot \vec{b}_2\} = \{\langle 1, 1, 3 \rangle, \langle 3, -1, 1 \rangle\}$. So Image$(M) = \text{Span}(\{M \cdot \vec{b}_1, M \cdot \vec{b}_2\}) = \{a_1 \cdot \langle 1, 1, 3 \rangle + a_2 \cdot \langle 3, -1, 1 \rangle\}$, the plane spanned by $\langle 1, 1, 3 \rangle$ and $\langle 3, -1, 1 \rangle$.

Consider a vector in $\mathbb{R}^3$, say, $\vec{w} = \langle 5, 3, 4 \rangle$. We can then find vectors $\vec{u} \in$ Null(M) and $\vec{v} \in \mathcal{V}$ such that $\vec{w} = \vec{u} + \vec{v}$. In this case, $\vec{u} = \langle 3, 0, 3 \rangle$ and $\vec{v} = \langle 2, 3, 1 \rangle$. Then $M \cdot \vec{w} = M \cdot \vec{v} = \langle 7, -1, 5 \rangle$. Note that $\text{Coords}(\vec{v}, \mathcal{B}) = \text{Coords}(M\vec{v}, \{M \cdot \vec{b}_1, M \cdot \vec{b}_2\}) = \langle 1, 2 \rangle$.

4.1.6 Systems of Linear Equations

We now apply all this theory to systems of linear equations. Let M be an $m \times n$ matrix. First, we note that the null space of M is the set of vectors $\vec{x}$ such that

$M\vec{x} = \vec{0}$; that is, it is the set of solutions to the system $M\vec{x} = \vec{0}$ of m equations in n unknowns.

Second, let us think about the system of equations $M\vec{x} = \vec{c}$. This has a solution if $\vec{c}$ is in Image(M). Therefore, if Image(M) = $\mathbb{R}^m$—that is, Rank(M) = m—then every vector $\vec{c}$ in $\mathbb{R}^m$ is in Image(M), so the system of equations has a solution for every $\vec{c}$. If Rank(M) < m, then Image(M) is a proper subset of $\mathbb{R}^m$, so most vectors $\vec{c}$ that are in $\mathbb{R}^m$ are not in Image(M) and for those, the system $M\vec{x} = \vec{c}$ has no solution.

In contrast, now suppose $\vec{q}$ satisfies $M\vec{x} = \vec{c}$ and suppose that some nonzero vector $\vec{u}$ is in Null(M). Then $M(\vec{q} + \vec{u}) = M\vec{q} = \vec{c}$, so $\vec{q} + \vec{u}$ also satisfies the system $M\vec{x} = \vec{c}$. The converse is also true; if $\vec{q}$ and $\vec{p}$ are two solutions of the system $M\vec{x} = \vec{c}$, then

$$M(\vec{p} - \vec{q}) = M\vec{p} - M\vec{q} = \vec{c} - \vec{c} = \vec{0}.$$

So $\vec{p} - \vec{q}$ is in Null(M).

The set of all solutions to the system of equations $M\vec{x} = \vec{c}$ therefore has the form $\{\vec{b} + \vec{u} \mid \vec{u} \in \text{Null}(M)\}$. In general, a set of this form $\{\vec{p} + \vec{u} \mid \vec{u} \in \mathcal{U}\}$, where $\mathcal{U}$ is a subspace of $\mathbb{R}^n$, is called an *affine* space.

Therefore, two critical features of a matrix M determine the characteristics of the solutions to the system of equations $M\vec{x} = \vec{c}$. Each feature has two possibilities, giving four combinations.

Feature A. The image.

Case A.1. Image(M) is all of $\mathbb{R}^m$. This holds if Rank(M) = m. In this case, the system $M\vec{x} = \vec{c}$ has at least one solution for every $\vec{c}$.

Case A.2. Image(M) is a proper subset of $\mathbb{R}^m$. This holds if Rank(M) < m. In this case, the system $M\vec{x} = \vec{c}$ has no solution for any $\vec{c}$ that is in $\mathbb{R}^m$ but not in Image(M).

Feature B. The null space.

Case B.1. Null(M) is just the zero space. This holds if Rank(M) = n. In this case, the system $M\vec{x} = \vec{c}$ has at most one solution for any $\vec{c}$.

Case B.2. Null(M) contains nonzero vectors. This holds if Rank(M) < n. In this case, for any value $\vec{c}$, if the system $M\vec{x} = \vec{c}$ has any solutions, then it has infinitely many solutions.

Let us now consider systems of equations, $M\vec{x} = \vec{c}$, of m equations in n unknowns (that is, M is an $m \times n$ matrix). Keeping M fixed, but allowing $\vec{c}$ to vary, there are four categories, corresponding to the combinations of the cases above.

Category I. Image(M) = $\mathbb{R}^m$. Null(M) = $\{\vec{0}\}$. $n = m = $ Rank(M). For any $\vec{c}$ there exists exactly one solution.

Category II.   Image$(M) = \mathbb{R}^m$. Null$(M) \neq \{\vec{0}\}$. Dim(Null$(M)) = n - m > 0$. $m =$ Rank$(M) < n$. For any $\vec{c}$ there exist infinitely many solutions.

Category III.   Image$(M) \neq \mathbb{R}^m$. Null$(M) = \{\vec{0}\}$. $n =$ Rank$(M) < m$. Dim(Image$(M)) =$ Rank$(M) < m$, so Image(M) does not include all of $\mathbb{R}^m$. Null$(M) = \{\vec{0}\}$. For $\vec{c} \in$ Image(M), there exists a unique solution; for $\vec{c} \notin$ Image(M), there does not exist any solution.

Category IV.   Image$(M) \neq \mathbb{R}^m$. Null$(M) \neq \{\vec{0}\}$. Rank$(M) < m$, Rank$(M) < n$. Dim(Image$(M)) =$ Rank$(M) < m$, so Image(M) does not include all of $\mathbb{R}^m$. Dim(Null$(M)) = n -$ Rank$(M) > 0$. For $\vec{c} \in$ Image(M), there exists infinitely many solutions; for $\vec{c} \notin$ Image(M), there does not exist any solution.

4.1.7   Inverses

Let M be a Category I matrix; that is, an $n \times n$ matrix of rank n. Then the linear transformation $\Gamma(\vec{v}) = M \cdot \vec{v}$ is a bijection from $\mathbb{R}^n$ to itself; that is, for every $\vec{w} \in \mathbb{R}^n$ there exists exactly one $\vec{v} \in \mathbb{R}^n$, such that $\Gamma(\vec{v}) = \vec{w}$. Therefore, we can define an inverse Γ^{-1}, where $\Gamma^{-1}(\vec{w})$ is the unique $\vec{v}$ such that $\Gamma(\vec{v}) = \vec{w}$. In other words, $\Gamma^{-1}(\Gamma(\vec{v})) = \vec{v} = \Gamma(\Gamma^{-1}(\vec{v}))$.

It is easy to show (see Theorem 4.69 in Section 4.2.7) that in fact Γ^{-1} must be a linear transformation. Therefore, there exists an $n \times n$ matrix corresponding to Γ. This is called the *inverse* of M and is denoted M^{-1}. Since $\Gamma \circ \Gamma^{-1}$ and $\Gamma^{-1} \circ \Gamma$ are both the identity mapping from $\mathbb{R}^n$ to itself, it follows that $M \cdot M^{-1} = M^{-1} \cdot M = I^n$, the $n \times n$ identity matrix.

Now, suppose we have a system of linear equations $M \cdot \vec{x} = \vec{c}$. Multiplying both sides by M^{-1} gives you $M^{-1} M \vec{x} = \vec{x} = M^{-1} \vec{c}$, so we have solved the system! In practice, as we discuss in Chapter 5, this is not actually an effective way to solve one particular system of linear equations. However, this may be a good approach to solve systems of the form $M \vec{x} = \vec{c}$, with the same value of M but different values of $\vec{c}$: compute M^{-1} once and for all and then just do the matrix multiplication $M^{-1} \vec{c}$ for each different value of $\vec{c}$.

4.1.8   Null Space and Rank in MATLAB

The MATLAB function `rank(A)`, returns the rank of matrix A. The function `null(A)` returns a matrix whose columns are an orthonormal basis for the null space of matrix A. (These use a tolerance for closeness of the relevant quantities to zero, which can be set by using an optional parameter.) For example,

```
>> a=[1,2,3,4 ; 1,1,1,1 ; 2,3,4,5; 0,1,2,3]
a =
```

```
              1        2        3        4
              1        1        1        1
              2        3        4        5
              0        1        2        3

    >> rank(a)
    ans =
              2

    >> q=null(a)
    q   =
          0.4689      -0.2831
         -0.4142       0.7270
         -0.5783      -0.6046
          0.5236       0.1607

    >> a*q
    ans =
          1.0e-14  *
                 0       0.0888
           -0.0111       0.0444
                 0       0.1554
                 0       0.0389

    >> q'*q
    ans =
          1.0000      -0.0000
         -0.0000       1.0000
```

4.2   Proofs and Other Abstract Mathematics (Optional)

In this part, we give proofs for the theorems (except those whose proofs we leave as an exercise.) We also introduce some more abstract material.

4.2.1   Vector Spaces

Lemma 4.27. *If V is a subspace of $\mathbb{R}^n$, then $\mathrm{Span}(V) = V$.*

Proof: Since V is a subspace, there exists a set $\mathscr{U}$ of vectors in $\mathbb{R}^n$ such that $V = \mathrm{Span}(\mathscr{U})$. Let $\vec{w}$ be a vector in $\mathrm{Span}(V)$. Then $\vec{w}$ is a linear sum over V; that is, there exist vectors $\vec{v}_1, \ldots, \vec{v}_q \in V$ and scalars $a_1, \ldots, a_k$ such that $\vec{w} = a_1 \vec{v}_1 + \ldots + a_q \vec{v}_q$. Since $V = \mathrm{Span}(\mathscr{U})$, each of the $\vec{v}_i$ is a linear sum over $\mathscr{U}$; that is, $\vec{v}_i = b_{i,1} \vec{u}_{i,1} + \ldots + b_{i,p} \vec{u}_{i,p}$. Thus,

$$\vec{w} = a_1(b_{1,1} \vec{u}_{1,1} + \ldots + b_{1,p} \vec{u}_{1,p}) + \ldots + a_q(b_{q,1} \vec{u}_{q,1} + \ldots + b_{q,p} \vec{u}_{q,p}) = a_1 b_{1,1} \vec{u}_{1,1} + \ldots + a_q b_{q,p} \vec{u}_{q,p}.$$

Grouping together the repeated $\vec{u}$ vectors, we have an expression of $\vec{w}$ as a linear sum over $\mathscr{U}$. So $\vec{w}$ is in $\mathrm{Span}(\mathscr{U}) = V$.   □

Theorem 4.28. *A set of n-dimensional vectors $V \subset \mathbb{R}^n$ is a subspace of $\mathbb{R}^n$ if and only if it is closed under vector addition and scalar multiplication.*

Proof: Suppose V is a subspace of $\mathbb{R}^n$. By Lemma 4.27, $V = \text{Span}(V)$, so any linear sum of vectors in V is itself in V. In particular, if $\vec{v}$ and $\vec{u}$ are in V and a is a scalar, then $\vec{v} + \vec{u}$ and $a\vec{v}$ are in V.

Conversely, suppose that V is closed under addition and scalar multiplication. Then if $\vec{w}$ is a linear sum over V, $\vec{w} = a_1\vec{v}_1 + \ldots + a_q\vec{v}_q$, then each of the products $a_i\vec{v}_i$ is in V, and then, by induction, the partial sums

$$a_1\vec{v}_1 + a_2\vec{v}_2,$$
$$a_1\vec{v}_1 + a_2\vec{v}_2 + a_3\vec{v}_3,$$
$$\ldots,$$
$$a_1\vec{v}_1 + a_2\vec{v}_2 + \ldots + a_q\vec{v}_q$$

are all in V. So $\vec{w}$ is in V. Since V contains all its linear sums, $V = \text{Span}(V)$, so by Definition 4.4, V is a subspace of $\mathbb{R}^n$. □

Lemma 4.29. *Let V be a set of n-dimensional vectors and let $\vec{w}$ be a vector in $\text{Span}(V)$. Then $\text{Span}(V \cup \{\vec{w}\}) = \text{Span}(V)$.*

Proof: It is immediate that $\text{Span}(V \cup \{\vec{w}\}) \supset \text{Span}(V)$, since any linear sum over V is a linear sum over $V \cup \{\vec{w}\}$ with coefficient 0 on $\vec{w}$.

Since $\vec{w} \in \text{Span}(V)$, let $\vec{w} = a_1\vec{v}_1 + \ldots + a_m\vec{v}_m$. Let $\vec{z}$ be any vector in $\text{Span}(V \cup \{\vec{w}\})$. Then $\vec{z} = b_1\vec{v}_1 + \ldots + b_m\vec{v}_m + c\vec{w} = b_1\vec{v}_1 + \ldots + b_m\vec{v}_m + c(a_1\vec{v}_1 + \ldots + a_m\vec{v}_m) = (b_1 + ca_1)\vec{v}_1 + \ldots + (b_m + ca_m)\vec{v}_m$. So $\vec{z}$ is a linear sum of $\vec{v}_1, \ldots, \vec{v}_m$. □

4.2.2 Linear Independence and Bases

Theorem 4.30. *A set V is linearly dependent if and only if there exist distinct vectors $\vec{v}_1, \ldots, \vec{v}_m$ in V and scalars $a_1, \ldots, a_m$, which are not all equal to 0, such that $a_1\vec{v}_1 + \ldots + a_m\vec{v}_m = \vec{0}$.*

Proof: Suppose that V is linearly dependent. Then, by Definition 4.13, there exists some $\vec{u}$ in V that is a linear sum over the other vectors in V; that is, $\vec{u} = a_1\vec{v}_1 + \ldots + a_m\vec{v}_m$. Therefore, $\vec{0} = a_1\vec{v}_1 + \ldots + a_m\vec{v}_m - \vec{u}$. Note that the coefficient of $\vec{u}$ is -1, so the coefficients in this sum are not all zero, satisfying the condition of the theorem.

Conversely, suppose that V satisfies the conditions of the theorem; that is, $a_1\vec{v}_1 + \ldots + a_m\vec{v}_m = \vec{0}$, where the coefficients are not all 0. Let a_i be some nonzero coefficient. Then, we can solve for $\vec{v}_i$:

$$\vec{v}_i = -\frac{a_1}{a_i}\vec{v}_1 - \ldots - \frac{a_{i-1}}{a_i}\vec{v}_{i-1} - \frac{a_{i+1}}{a_i}\vec{v}_{i+1} - \ldots - \frac{a_m}{a_i}\vec{v}_m,$$

so $\vec{v}_i$ is a linear sum of the rest of the vectors. □

Corollary 4.31. *Let $\vec{v}_1, \vec{v}_2, \ldots, \vec{v}_k$ be a sequence of vectors such that for all i, $\vec{v}_i$ is not in $\mathrm{Span}(\{\vec{v}_1, \ldots, \vec{v}_{i-1}\})$. Then $\vec{v}_1, \vec{v}_2, \ldots, \vec{v}_k$ are linearly independent.*

Proof: of the contrapositive: Suppose that $\vec{v}_1, \vec{v}_2, \ldots, \vec{v}_k$ are linearly dependent. Then for some $a_1, \ldots, a_k$ not all zero, $a_1 \vec{v}_1 + \ldots + a_k \vec{v}_k = \vec{0}$. Let i be the largest index for which $a_i \neq 0$. Then $\vec{v}_i = (-1/a_i)(a_1 \vec{v}_1 + \ldots + a_{i-1} \vec{v}_{i-1})$, so $\vec{v}_p$ is in $\mathrm{Span}(\{\vec{v}_1, \ldots, \vec{v}_{i-1}\})$. □

Theorem 4.32. *If $\mathscr{B} = \langle \vec{b}_1, \ldots, \vec{b}_m \rangle$ is a basis for V, then any vector $\vec{v}$ in V has a unique tuple of coordinates relative to $\mathscr{B}$. That is, there is a unique sequence of coefficients $a_1, \ldots, a_m$ such that $\vec{v} = a_1 \vec{b}_1 + \ldots + a_m \vec{b}_m$.*

Proof: The fact that there exist such $a_1, \ldots, a_m$ is immediate from the fact that $\vec{v} \in \mathrm{Span}(\mathscr{B})$. To prove that the coordinate vector is unique, suppose that $\vec{v} = a_1 \vec{b}_1 + \ldots + a_m \vec{b}_m = c_1 \vec{b}_1 + \ldots + c_m \vec{b}_m$. Then $(a_1 - c_1)\vec{b}_1 + \ldots + (a_m - c_m)\vec{b}_m = \vec{0}$. Since the $\vec{b}_i$ are linearly independent, $a_i - c_i = 0$ for all i.

Lemma 4.33. *Let $\mathscr{B} = \{\vec{b}_1, \ldots, \vec{b}_m\}$ be a set of linearly independent n-dimensional vectors. Let $V = \vec{v}_1, \ldots, \vec{v}_p$ be a set of linearly independent vectors in $\mathrm{Span}(\mathscr{B})$. Then $p \leq m$.*

Proof: Since $\vec{v}_1 \in \mathrm{Span}(\mathscr{B})$, we can write $\vec{v}_1 = a_1 \vec{b}_1 + \ldots + a_m \vec{b}_m$, where one of the coefficients a_i is nonzero. Since the numbering on the b vectors does not matter, we can renumber them so that the coefficient a_1 is nonzero. (This renumbering is not significant, it just simplifies the wording of the proof.) Then we can solve for $\vec{b}_1$ and write

$$\vec{b}_1 = \frac{1}{a_1} \vec{v}_1 - \frac{a_2}{a_1} \vec{b}_2 - \ldots - \frac{a_m}{a_1} \vec{b}_m.$$

Thus, $\vec{b}_1$ is in $\mathrm{Span}(\vec{v}_1, \vec{b}_2, \ldots, \vec{b}_m)$ so by Lemma 4.29 $\mathrm{Span}(\vec{v}_1, \vec{b}_2, \ldots, \vec{b}_m) = \mathrm{Span}(\mathscr{B} \cup \{\vec{v}_1\}) = \mathrm{Span}(\mathscr{B})$.

Moving on, we have $\vec{v}_2 \in \mathrm{Span}(\mathscr{B})$; so $\vec{v}_2$ is in $\mathrm{Span}(\vec{v}_1, \vec{b}_2, \ldots, \vec{b}_m)$; so $\vec{v}_2 = c_1 \vec{v}_1 + c_2 \vec{b}_2 + \ldots + c_m \vec{b}_m$. Since $\vec{v}_2$ and $\vec{v}_1$ are linearly independent, it cannot be the case that c_1 is the only nonzero coefficient; at least one of the coefficients associated with $\vec{b}_2, \ldots, \vec{b}_m$ must be nonzero. By again renumbering, we can assume that the nonzero coefficient is c_2. Again, we can solve for $\vec{b}_2$ and show that $\vec{b}_2$ is a linear sum of $\vec{v}_1, \vec{v}_2, \vec{b}_3, \ldots, \vec{b}_m$. We now have that $\mathrm{Span}(\vec{v}_1, \vec{v}_2, \vec{b}_3, \ldots, \vec{b}_m) = \mathrm{Span}(\mathscr{B})$.

We can keep doing this, replacing a vector in $\mathscr{B}$ with vector $\vec{v}_i$ while maintaining the same span, until either we have moved all the $\vec{v}$ vectors into the set, if $p < m$, or we have replaced all the $\vec{b}$ vectors, if $p \geq m$. Now, suppose that $p > m$. At this point, we have $\mathrm{Span}(\vec{v}_1, \ldots, \vec{v}_m) = \mathrm{Span}(\mathscr{B})$, so for $i = m+1, \ldots, p$, $\vec{v}_i$ is in $\mathrm{Span}(\vec{v}_1, \ldots, \vec{v}_m)$. But that contradicts the assumption that the vectors $\vec{v}_1, \ldots, \vec{v}_p$ are linearly independent. Therefore, we must have $p \leq m$. □

Corollary 4.34. *Any two bases for the same vector space have the same number of vectors.*

Proof: Let $\mathcal{B}$ and $\mathcal{C}$ be bases for vector space V. Let m be the number of vectors in $\mathcal{B}$. By Lemma 4.33, since $\mathcal{C}$ is linearly independent, it cannot have more than m vectors. Since neither basis can have more vectors than the other, they must have equal numbers. □

Corollary 4.35. *A set of linearly independent vectors in $\mathbb{R}^n$ has at most n vectors.*

Proof: Immediate from Lemma 4.33 and the fact that $\{\vec{e}^{\,1}, \ldots, \vec{e}^{\,n}\}$ is a basis for $\mathbb{R}^n$. □

The number of vectors in a basis for vector space V is called the *dimension* of V, denoted "Dim(V)." By convention, the dimension of the zero space is zero.

Lemma 4.36. *Let $\mathcal{B}$ be a linearly independent set of n-dimensional vectors. If $\vec{v}$ is not a linear sum of $\mathcal{B}$, then $\mathcal{B} \cup \{\vec{v}\}$ is linearly independent.*

Proof: Proof of the contrapositive. Suppose that $\mathcal{B}$ is linearly independent and $\mathcal{B} \cup \{\vec{v}\}$ is not. Then there exist scalars $a_0, \ldots, a_m$, not all zero, such that $a_1 \vec{b}_1 + \ldots + a_m \vec{b}_m + a_0 \vec{v} = \vec{0}$. Since $\vec{b}_1, \ldots, \vec{b}_m$ are linearly independent, a_0 cannot be 0. Therefore, we can solve for $\vec{v}$,

$$\vec{v} = -\frac{a_1}{a_0} \vec{b}_1 - \ldots - \frac{a_m}{a_0} \vec{b}_m,$$

so $\vec{v}$ is a linear sum of $\vec{b}_1, \ldots, \vec{b}_m$. □

Theorem 4.37. *Let V be a subspace of $\mathbb{R}^n$ and let $\mathcal{U}$ be a set of linearly independent vectors in V. Then $\mathcal{U}$ can be extended to a basis for V; that is, there exists a basis $\mathcal{B}$ for V such that $\mathcal{U} \subset \mathcal{B}$.*

Proof: If Span($\mathcal{U}$) = V, then choose $\mathcal{B} = \mathcal{U}$. Otherwise, let $\vec{v}_1$ be a vector in V that is not in Span($\mathcal{U}$). Let $\mathcal{U}_1 = \mathcal{U} \cup \{\vec{v}_1\}$. By Lemma 4.36, $\mathcal{U}_1$ is linearly independent. Clearly, Span($\mathcal{U}_1$) $\subset V$.

If Span($\mathcal{U}_1$) = V, then we are done. If not, we choose a vector $\vec{v}_2$ in $V - (\mathcal{U} \cup \{\vec{v}_1\})$ and let $\mathcal{U}_2 = \mathcal{U}_1 \cup \{\vec{v}_2\}$. We can keep doing this until Span($\mathcal{U}_k$) = V for some k. Certainly it will have to stop before $\mathcal{U}_k$ gets more than n vectors, since there cannot be more than n linearly independent n-dimensional vectors. □

Corollary 4.38. *Any subspace of $\mathbb{R}^n$ has a basis.*

Proof: By Theorem 4.37, the empty set of vectors can be extended to a basis. □

Corollary 4.39. *Let V be a vector space of dimension m. Any set $\mathcal{B}$ of m linearly independent vectors in V spans V and thus is a basis for V.*

Proof: By Theorem 4.37, $\mathcal{B}$ can be extended to a basis for V. By Corollary 4.34, this extension must have m vectors; thus, it must be equal to $\mathcal{B}$. □

Theorem 4.40. *Let V be a vector space of dimension m, and let $\mathcal{B} = \{\vec{b}_1, \ldots, \vec{b}_p\}$ be a finite set of vectors that spans V. Then $p \geq m$, and there exists a subset of $\mathcal{B}$ that is a basis for V.*

Proof: If $\mathcal{B}$ is linearly independent, then it is a basis for V, so by Corollary 4.34 $p = m$. If $\mathcal{B}$ is not linearly independent, then let $\vec{v}$ be a vector in $\mathcal{B}$ that is the linear sum of other vectors in $\mathcal{B}$. It is immediate that we can delete $\vec{v}$; that is, $\mathcal{B} - \{\vec{v}\}$ still spans V. If the result is still not linearly independent, we can repeat this process. Eventually, we will reach a subset of $\mathcal{B}$ that is linearly independent and still spans V. This is a subset of $\mathcal{B}$ that is a basis for V, and so must contain m vectors. □

4.2.3   Sum of Vector Spaces

Throughout this section, let $\mathcal{U}$ and V be subspaces of $\mathbb{R}^n$.

Theorem 4.41. *The direct sum $\mathcal{U} \oplus V$ and the intersection $\mathcal{U} \cap V$ are vector spaces.*

Proof: Immediate from the definitions. □

Theorem 4.42. $\mathrm{Dim}(\mathcal{U}) \oplus \mathrm{Dim}(V) = \mathrm{Dim}(\mathcal{U}) + \mathrm{Dim}(V) - \mathrm{Dim}(\mathcal{U} \cap V)$

Proof: Let $p = \mathrm{Dim}(\mathcal{U})$, $q = \mathrm{Dim}(V)$, $r = \mathrm{Dim}(\mathcal{U} \cap V)$. Clearly, $r \leq p$ and $r \leq q$. Let $\mathcal{W} = \{\vec{w}_1, \ldots, \vec{w}_r\}$ be a basis for $\mathcal{U} \cap V$. Using Theorem 4.37, let $\mathcal{B} = \{\vec{b}_1, \ldots, \vec{b}_{p-r}\}$ be a set of vectors such that $\mathcal{W} \cup \mathcal{B}$ is a basis for $\mathcal{U}$, and let $\mathcal{C} = \{\vec{c}_1, \ldots, \vec{c}_{q-r}\}$ be a set of vectors such that $\mathcal{W} \cup \mathcal{C}$ is a basis for V. (If $p = r$, let $\mathcal{B} = \emptyset$; if $q = r$, let $\mathcal{C} = \emptyset$.)

Let $\mathcal{Q} = \mathcal{W} \cup \mathcal{B} \cup \mathcal{C}$. To show that $\mathcal{Q}$ is a basis for $\mathcal{U} \oplus V$, show that $\mathcal{Q}$ spans $\mathcal{U} \oplus V$ and that $\mathcal{Q}$ is linearly independent. Since every vector in $\mathcal{U} \oplus V$ is a linear sum over $\mathcal{U} \cup V$, every vector in $\mathcal{U}$ is a linear sum over $\mathcal{W} \cup \mathcal{B}$ and every vector in V is a linear sum over $\mathcal{W} \cup \mathcal{C}$, it follows that $\mathcal{U} \oplus V = \mathrm{Span}(\mathcal{Q})$.

To show that $\mathcal{Q}$ is linearly independent, suppose that

$$a_1 \vec{w}_1 + \ldots + a_r \vec{w}_r + d_1 \vec{b}_1 + \ldots + d_{p-r} \vec{b}_{p-r} + e_1 \vec{c}_1 + \ldots + e_{q-r} \vec{c}_{q-r} = \vec{0}.$$

Then

$$a_1 \vec{w}_1 + \ldots + a_r \vec{w}_r + d_1 \vec{b}_1 + \ldots + d_{p-r} \vec{b}_{p-r} = -e_1 \vec{c}_1 + \ldots + -e_{q-r} \vec{c}_{q-r}.$$

The left-hand side is a vector in $\mathcal{U}$ and the right-hand side is a vector in $\mathrm{Span}(\mathcal{C})$. Let us call this vector $\vec{z}$. Then $\vec{z} \in \mathrm{Span}(\mathcal{C}) \cap \mathcal{U}$, so $\vec{z} \in \mathcal{U} \cap V$. But the vectors in $\mathcal{C}$ are linearly independent of those in $\mathcal{U} \cap V$; hence, $\vec{z} = \vec{0}$. Therefore, all the coefficients a_i, d_i, and e_i are equal to 0. $\qquad\square$

Theorem 4.43. *Every subspace V of $\mathbb{R}^n$ has a complement.*

Proof: Using Corollary 4.38, let $\mathcal{B}$ be a basis for V. Using Theorem 4.37, let $\mathcal{Q}$ be an extension of $\mathcal{B}$ to a basis for $\mathbb{R}^n$. It is easily shown that $\mathrm{Span}(\mathcal{Q} \setminus \mathcal{B})$ is a complement for V. $\qquad\square$

4.2.4   Orthogonality

Recall that two vectors are *orthogonal* if their dot product is equal to 0.

Definition 4.44. Let $\mathcal{U}$ and V be subspaces of $\mathbb{R}^n$. $\mathcal{U}$ and V are *orthogonal* if, for every $\vec{u}$ in $\mathcal{U}$ and $\vec{v}$ in V, $\vec{u}$ is orthogonal to $\vec{v}$. $\mathcal{U}$ and V are *orthogonal complements* if they are orthogonal and they are complements. A finite set $\mathcal{B} = \vec{b}_1, \ldots, \vec{b}_q$ is *orthogonal* if, for $i \neq j$, $\vec{b}_i$ is orthogonal to $\vec{b}_j$. An *orthogonal basis* for subspace V is a basis for V that is orthogonal.

Theorem 4.45. *Let V be a vector space and let $\mathcal{B}$ be an orthonormal basis for V.*

- *For any $\vec{u}, \vec{w} \in V$, $\vec{u} \bullet \vec{w} = \mathrm{Coords}(\vec{u}, \mathcal{B}) \bullet \mathrm{Coords}(\vec{w}, \mathcal{B})$.*

- *For any $\vec{u} \in V$, $|\vec{u}| = |\mathrm{Coords}(\vec{u}, \mathcal{B})|$.*

Proof: Let $\mathcal{B} = \{\hat{b}_1, \ldots, \hat{b}_m\}$. Let $\mathrm{Coords}(\vec{u}, \mathcal{B}) = \langle u_1, \ldots, u_m \rangle$ and let $\mathrm{Coords}(\vec{v}, \mathcal{B}) = \langle v_1, \ldots, v_m \rangle$. Then

$$
\begin{aligned}
\vec{u} \bullet \vec{v} = {} & (u_1 \cdot \hat{b}_1 + \ldots + u_m \cdot \hat{b}_m) \bullet (v_1 \cdot \hat{b}_1 + \ldots + v_m \cdot \hat{b}_m) \\
= {} & u_1 \cdot v_1 \cdot (\hat{b}_1 \bullet \hat{b}_1) + u_1 \cdot v_2 \cdot (\hat{b}_1 \bullet \hat{b}_2) + \ldots + u_1 \cdot v_m \cdot (\hat{b}_1 \bullet \hat{b}_m) \\
& + u_2 \cdot v_1 \cdot (\hat{b}_2 \bullet \hat{b}_1) + \ldots + u_2 \cdot v_m \cdot (\hat{b}_2 \bullet \hat{b}_m) \\
& + \ldots \\
& + u_m \cdot v_1 \cdot (\hat{b}_m \bullet \hat{b}_1) + \ldots + u_m \cdot v_m \cdot (\hat{b}_m \bullet \hat{b}_m).
\end{aligned}
$$

But since $\mathcal{B}$ is orthonormal, the dot product $\hat{b}_i \bullet \hat{b}_j$ is equal to 0 if $i \neq j$ and equal to 1 if $i = j$. Therefore, the above sum reduces to

$$
u_1 \cdot v_1 + u_2 \cdot v_2 + \ldots + u_m \cdot v_m = \mathrm{Coords}(\vec{u}, \mathcal{B}) \bullet \mathrm{Coords}(\vec{v}, \mathcal{B}).
$$

For the second part of the theorem, note that

$$
|\vec{u}| = \sqrt{\vec{u} \bullet \vec{u}} = \sqrt{\mathrm{Coords}(\vec{u}, \mathcal{B}) \bullet \mathrm{Coords}(\vec{u}, \mathcal{B})} = |\mathrm{Coords}(\vec{u}, \mathcal{B})|. \qquad\square
$$

Theorem 4.46. *Let $\mathcal{B} = \{\vec{b}_1,\ldots,\vec{b}_q\}$ be an orthonormal basis for vector space V. If $\vec{v}$ is a vector in V, then for the ith coordinate of $\vec{v}$ with respect to $\mathcal{B}$,* $\mathrm{Coords}(\vec{v},\mathcal{B})[i] = \vec{v} \bullet \vec{b}_i$.

Proof: Let $c_i = \mathrm{Coords}(\vec{v},\mathcal{B})[i]$. Then $\vec{v} = c_1\vec{b}_1 + \ldots + c_q\vec{b}_q$, by definition of Coords. So

$$\vec{v} \bullet \vec{b}_i = (c_1\vec{b}_1 + \ldots + c_q\vec{b}_q) \bullet \vec{b}_i = c_1(\vec{b}_1 \bullet \vec{b}_i) + \ldots + c_q(\vec{b}_q \bullet \vec{b}_i) = c_i$$

because $\vec{b}_i \bullet \vec{b}_i = 1$ and $\vec{b}_i \bullet \vec{b}_j = 0$ for $j \neq i$. $\qquad\square$

Definition 4.47. Let $\mathcal{B} = \vec{b}_1,\ldots,\vec{b}_q$ be an orthogonal set of n-dimensional vectors, and let $\vec{v}$ be an n-dimensional vector.

Define $\mathrm{Project}(\vec{v},\mathcal{B})$ as

$$\mathrm{Project}(\vec{v},\mathcal{B}) = \frac{\vec{v} \bullet \vec{b}_1}{\vec{b}_1 \bullet \vec{b}_1} \cdot \vec{b}_1 + \ldots + \frac{\vec{v} \bullet \vec{b}_q}{\vec{b}_q \bullet \vec{b}_q} \cdot \vec{b}_q.$$

Note that $\mathrm{Project}(\vec{v},\mathcal{B})$ is in $\mathrm{Span}(\mathcal{B})$. In fact, $\mathrm{Project}(\vec{v},\mathcal{B})$ is the point in $\mathrm{Span}(\mathcal{B})$ that is closest to $\vec{v}$.

Lemma 4.48. *Let $\mathcal{B} = \vec{b}_1,\ldots,\vec{b}_q$ be an orthogonal set of n-dimensional vectors, and let $\vec{v}$ be an n-dimensional vector that is not in $\mathrm{Span}(\mathcal{B})$. Let $\vec{u} = \vec{v} - \mathrm{Project}(\vec{v},\mathcal{B})$, and note that $\vec{u}$ is in $\mathrm{Span}(\mathcal{B} \cup \{\vec{v}\})$. Then $\vec{u}$ is orthogonal to $\vec{b}_1,\ldots,\vec{b}_q$.*

Proof: For any i,

$$\vec{u} \bullet \vec{b}_i = (\vec{v} - \frac{\vec{v} \bullet \vec{b}_1}{\vec{b}_1 \bullet \vec{b}_1} \cdot \vec{b}_1 - \ldots - \frac{\vec{v} \bullet \vec{b}_q}{\vec{b}_q \bullet \vec{b}_q} \cdot \vec{b}_q) \bullet \vec{b}_i$$

$$= \vec{v} \bullet \vec{b}_i - \frac{\vec{v} \bullet \vec{b}_1}{\vec{b}_1 \bullet \vec{b}_1} \cdot \vec{b}_1 \bullet \vec{b}_i - \ldots - \frac{\vec{v} \bullet \vec{b}_q}{\vec{b}_q \bullet \vec{b}_q} \cdot \vec{b}_q \bullet \vec{b}_i.$$

Since the $\vec{b}$s are orthogonal, $\vec{b}_j \bullet \vec{b}_i = 0$ for all $j \neq i$, so all these terms disappear except

$$\vec{v} \bullet \vec{b}_i - \frac{\vec{v} \bullet \vec{b}_i}{\vec{b}_i \bullet \vec{b}_i} \cdot \vec{b}_i \bullet \vec{b}_i = 0. \qquad\square$$

Theorem 4.49. *Any vector space V has an orthogonal basis and an orthogonal complement.*

Proof: Carry out the following algorithm, known as *Gram-Schmidt orthogonalization*:

{ let $\mathscr{C} = \{\vec{c}_1, \ldots, \vec{c}_q\}$ be any basis for $\mathcal{V}$;
  extend $\mathscr{C}$ to a basis $\{\vec{c}_1, \ldots, \vec{c}_n\}$ for $\mathbb{R}^n$;
  $\mathscr{B} = \emptyset$;
  $\mathscr{D} = \emptyset$;
  for $(i = 1, \ldots, q)$ {
   let $\vec{u}_i = \vec{c}_i - \text{Project}(\vec{c}_i, \mathscr{B})$;
   add $\vec{u}_i$ to $\mathscr{B}$; }
  for $(i = q+1, \ldots, n)$ {
   let $\vec{u}_i = \vec{c}_i - \text{Project}(\vec{c}_i, \mathscr{B} \cup \mathscr{D})$;
   add $\vec{u}_i$ to $\mathscr{D}$; }
}

It then follows immediately from Lemma 4.48 that $\mathscr{B}$ is an orthogonal basis for $\mathcal{V}$ and that $\text{Span}(\mathscr{D})$ is the orthogonal complement for $\mathcal{V}$. $\qquad\square$

Example 4.50. In $\mathbb{R}^4$, let $\mathscr{C}$ be the set of two vectors $\langle 1,1,1,1 \rangle$ and $\langle 0,0,1,1 \rangle$. Extend $\mathscr{B}$ to a basis for $\mathbb{R}^4$ by adding the two vectors $\vec{e}^{\,1}, \vec{e}^{\,3}$. Then

$$\vec{u}_1 = \vec{c}_1 = \langle 1,1,1,1 \rangle,$$

$$\vec{u}_2 = \vec{c}_2 - \frac{\vec{u}_1 \bullet \vec{c}_2}{\vec{u}_1 \bullet \vec{u}_1} \cdot \vec{u}_1 = \langle 0,0,1,1 \rangle - \frac{2}{4} \cdot \langle 1,1,1,1 \rangle = \langle -1/2, -1/2, 1/2, 1/2 \rangle,$$

$$\mathscr{B} = \{\vec{u}_1, \vec{u}_2\},$$

$$\vec{u}_3 = \vec{c}_3 - \frac{\vec{u}_1 \bullet \vec{c}_3}{\vec{u}_1 \bullet \vec{u}_1} \cdot \vec{u}_1 - \frac{\vec{u}_2 \bullet \vec{c}_3}{\vec{u}_2 \bullet \vec{u}_2} \cdot \vec{u}_2$$

$$= \langle 1,0,0,0 \rangle - \frac{1}{4}\langle 1,1,1,1 \rangle - \frac{-1/2}{1}\langle -1/2,-1/2,1/2,1/2 \rangle$$

$$= \langle 1/2, -1/2, 0, 0 \rangle,$$

$$\vec{u}_4 = \vec{c}_4 - \frac{\vec{u}_1 \bullet \vec{c}_4}{\vec{u}_1 \bullet \vec{u}_1} \cdot \vec{u}_1 - \frac{\vec{u}_2 \bullet \vec{c}_4}{\vec{u}_2 \bullet \vec{u}_2} \cdot \vec{u}_2 - \frac{\vec{u}_3 \bullet \vec{c}_4}{\vec{u}_3 \bullet \vec{u}_3} \cdot \vec{u}_3$$

$$= \langle 0,0,1,0 \rangle - \frac{1}{4}\langle 1,1,1,1 \rangle - \frac{1/2}{1}\langle -1/2,-1/2,1/2,1/2 \rangle - \frac{0}{1/2}\langle 1/2,-1/2,0,0 \rangle$$

$$= \langle 0,0,1/2,-1/2 \rangle,$$

$$\mathscr{D} = \{\vec{u}_3, \vec{u}_4\}.$$

Corollary 4.51. *The set of all vectors orthogonal to $\mathcal{V}$ is complementary to $\mathcal{V}$ and is therefore the orthogonal complement of $\mathcal{V}$.*

Proof: Clearly, any vector orthogonal to $\mathcal{V}$ is in the orthogonal complement of $\mathcal{V}$, and vice versa. What is nontrivial here is that there exists an orthogonal complement, which is guaranteed by Theorem 4.49.

4.2.5   Functions

We now turn to linear transformations, but we begin with some simple observations about functions in general, from any domain to any range.

Definition 4.52. Let f be a function from domain D to range E. Let S be a subset of D. Then $f(S) = \{f(s) \mid s \in S\}$, the set of values obtained by applying f to elements of S. We define Image$(f) = f(D)$, the set of values obtained by applying f to any element in the domain D.

Definition 4.53. Let f be a function from domain D to range E. Let S be a subset of D and let T be a subset of E. We say that f is a *surjection* from S onto T (or f maps S *onto* T) if T is a subset of $f(S)$. We say that f is an *injection* over S (or f is one-to-one on S) if f maps each element of S to a different element of E; that is, if s_1, s_2 are in S and $s_1 \neq s_2$, then $f(s_1) \neq f(s_2)$. We say that f is a *bijection* from S to T, if $f(S) = T$ and f is an injection over S.

Definition 4.54. Let f be a function from D to E, and let g be a function from E to D. Suppose that, for every element d of D, $g(f(d)) = d$. Then we say that g is a *left inverse* of f and that f is a *right inverse* of g.

Example 4.55. Let $D = \{a, b, c\}$ and let $E = \{1, 2, 3, 4\}$. The function $f : D \rightarrow E$ defined by $f(a) = 4, f(b) = 2, f(c) = 3$ is an injection. The function $g : D \rightarrow E$ defined by $g(a) = 4, g(b) = 2, g(c) = 4$ is not an injection because $g(a) = g(c)$. There obviously cannot be a surjection from D to E because there are more elements in E than D. The function $j : E \rightarrow D$ defined by $j(1) = b, j(2) = b, j(3) = c, j(4) = a$ is a surjection. The function $h : E \rightarrow D$ defined by $h(1) = a, h(2) = b, h(3) = b, h(4) = a$ is not a surjection because there is no value x such that $j(x) = c$. There cannot be an injection from E to D because there are more elements in E than in D; this is known as the *pigeon-hole principle*.

$$\text{Image}(f) = \{2, 3, 4\}, \text{Image}(g) = \{2, 4\}, \text{Image}(j) = \{a, b, c\}, \text{Image}(h) = \{a, b\}.$$

Function j is a left inverse of f and f is a right inverse of j because $j(f(a)) = j(4) = a$; $j(f(b)) = j(2) = b$, and $j(f(c)) = j(3) = c$.

Example 4.56. Consider functions from $\mathbb{Z}$, the set of integers, to itself. The function $f(x) = 2x$ is an injection—if $2x = 2y$, then $x = y$—but not a surjection—if z is odd, then there is no x such that $f(x) = z$. The function $g(x) = \lfloor x/2 \rfloor$ is a surjection—for any z, $g(2z) = z$ and $g(2z + 1) = z$—but not an injection—for any z, $g(2z) = g(2z + 1)$. Image(f) is the set of all even integers. Image(g) is the set of all integers. Function g is a left inverse of f, and f is a right inverse of g.

Example 4.57. Of course, the case that actually interests us here is linear transformations. Let $D = \mathbb{R}^3$ and let $E = \mathbb{R}^2$. Let $f : D \rightarrow E$ be the function

$$f(\vec{v}) = \begin{bmatrix} 1 & 0 & 0 \\ 0 & 1 & 0 \end{bmatrix} \cdot \vec{v}.$$

Function f is a surjection because for any vector $\langle x, y \rangle \in E$, $f(\langle x, y, 0 \rangle) = \langle x, y \rangle$. It is not an injection because $f(\langle x, y, 1 \rangle) = f(\langle x, y, 0 \rangle) = \langle x, y \rangle$.

Let $g : D \rightarrow E$ be the function

$$g(\vec{v}) = \begin{bmatrix} 1 & 0 & 0 \\ 1 & 0 & 0 \end{bmatrix} \cdot \vec{v}.$$

Function g is not a surjection because for any vector $\langle x, y, z \rangle \in D$, $g(\langle x, y, z \rangle) = \langle x, x \rangle$, so for any $y \neq x$, the vector $\langle x, y \rangle$ is not in Image(g). Nor is g an injection.

Let $h : E \rightarrow D$ be the function

$$h(\vec{v}) = \begin{bmatrix} 1 & 0 \\ 0 & 1 \\ 1 & 1 \end{bmatrix} \cdot \vec{v}.$$

Function h is a injection because, if $h(\langle w, x \rangle) = h(\langle y, z \rangle)$, then we have $h(\langle w, x \rangle) = \langle w, x, w + x \rangle$ and $h(\langle y, z \rangle) = \langle y, z, y + z \rangle$, so $w = y$ and $x = z$. It is not an surjection because for any vector $\langle x, y, z \rangle \in D$, if $z \neq x + y$, then $\langle x, y, z \rangle$ is not in Image(f).

Let $j : E \rightarrow D$ be the function

$$j(\vec{v}) = \begin{bmatrix} 1 & 0 \\ 1 & 0 \\ 1 & 0 \end{bmatrix} \cdot \vec{v}.$$

Function j is not an injection because, $h(\langle 4, 2 \rangle) = h(\langle 4, 5 \rangle) = \langle 4, 4, 4 \rangle$. It is certainly not a surjection.

Function h is a right inverse for f, and f is a left inverse for h because for any vector $\langle x, y \rangle$ in E,

$$f(h(\langle x, y \rangle)) = f(\langle x, y, x + y \rangle) = \langle x, y \rangle.$$

Not coincidentally, the corresponding matrix product is the identity

$$\begin{bmatrix} 1 & 0 & 0 \\ 0 & 1 & 0 \end{bmatrix} \cdot \begin{bmatrix} 1 & 0 \\ 0 & 1 \\ 1 & 1 \end{bmatrix} = \begin{bmatrix} 1 & 0 \\ 0 & 1 \end{bmatrix}.$$

Theorem 4.58. *If g is a left inverse for f, then f is an injection over D and g is a surjection onto D.*

Proof: Assume g is a left inverse for f; then $g(f(d)) = d$ for all d in D. Then for every d in D there exists r in E for which $g(r) = d$; namely $r = f(d)$. Thus g is a surjection onto D.

If f is not an injection, then there exist d and e in D such that $d \neq e$ but $f(d) = f(e)$. Since $g(f(d)) = g(f(e))$ we cannot have both $g(f(d)) = d$ and $g(f(e)) = e$. □

Definition 4.59. If g is both a left inverse and a right inverse for f, then g is said to be the *full inverse*, or simply the *inverse*, of f. We write $g = f^{-1}$. (Note that there can exist only one inverse.)

Theorem 4.60. *Function f has an inverse if and only if f is a bijection.*

Proof: By Theorem 4.58, if f has an inverse, then f is a bijection. Conversely, if f is a bijection, then, for each r in E, there is a unique d in D for which $f(d) = r$; we define the inverse g such that $g(r) = d$. □

Theorem 4.61. *If g is a left inverse of f and h is a right inverse of f, then $g = h$ and $g = f^{-1}$.*

Proof: Let i_D be the identity function over D and let i_E be the identity function over E. Then $g = g \circ i_E = g \circ (f \circ h) = (g \circ f) \circ h = i_D \circ h = h$. □

If that seems so clever as to be suspicious, write it out longhand as follows. Let e be any element of E and consider $g(f(h(e)))$. On the one hand, since h is the right inverse of f, $f(h(e)) = e$, so $g(f(h(e))) = g(e)$. On the other hand, since g is the left inverse of f, $g(f(d)) = d$ for all d in D. In particular, this holds for $d = h(e)$. Thus, $g(f(h(e))) = h(e)$. Since g and h have the same value for all arguments, they are the same function.

4.2.6   Linear Transformations

We now return to linear transformations. In this section, $f(\vec{v})$ is a linear transformation from $\mathbb{R}^n$ to $\mathbb{R}^m$, and F is the corresponding matrix.

Theorem 4.62. *For any linear mapping f from $\mathbb{R}^n$ to $\mathbb{R}^m$, Image(f) and Null(f) are vector spaces. Thus Image(f) is a subspace of $\mathbb{R}^m$ and Null(f) is a subspace of $\mathbb{R}^n$.*

Proof: We leave this as an exercise (Problem 4.4). □

Theorem 4.63. *For any finite set of n-dimensional vectors $\mathcal{B}$, Span$(f(\mathcal{B})) = f(\text{Span}(\mathcal{B}))$.*

Proof: Immediate from the fact that $f(a_1 \vec{b}_1 + \ldots + a_q \vec{b}_q) = a_1 f(\vec{b}_1) + \ldots + a_q f(\vec{b}_q)$. □

Corollary 4.64. *For any subspace V of $\mathbb{R}^n$, $\mathrm{Dim}(f(V)) \leq \min(m, \mathrm{Dim}(V))$. In particular, $\mathrm{Rank}(f) \leq \min(m, n)$.*

Proof: Since $f(V)$ is a subset of $\mathbb{R}^m$, it follows that $\mathrm{Dim}(f(V)) \leq m$.

Let $\vec{b}_1, \ldots, \vec{b}_q$ be a basis for V, where $q = \mathrm{Dim}(V)$. Then, by Theorem 4.63, $f(\vec{b}_1), \ldots, f(\vec{b}_q)$ spans $f(V)$ (they may not be linearly independent). Hence, by Theorem 4.40, $\mathrm{Dim}(f(V)) \leq q$.

For the second statement, we have $\mathrm{Rank}(f) = \mathrm{Dim}(f(V))$, where $V = \mathbb{R}^n$, so $\mathrm{Rank}(f) = \mathrm{Dim}(f(\mathbb{R}^n)) \leq \mathrm{Dim}(\mathbb{R}^n) = n$. □

Theorem 4.65. *Let $\mathcal{B}$ be a linearly independent set, and let $V = \mathrm{Span}(\mathcal{B})$. If f is an injection over V, then $f(\mathcal{B})$ is a linearly independent set.*

Proof: Proof of the contrapositive. Suppose that f is an injection and $f(\mathcal{B})$ is linearly dependent. Let $\mathcal{B} = \{\vec{b}_1, \ldots, \vec{b}_q\}$. Then for some $a_1, \ldots, a_q$ not all equal to 0, $\vec{0} = a_1 f(\vec{b}_1) + \ldots + a_q f(\vec{b}_q) = f(a_1 \vec{b}_1 + \ldots + a_q \vec{b}_q)$. Since $\vec{0} = f(\vec{0})$ and f is an injection, we must have $a_1 \vec{b}_1 + \ldots + a_q \vec{b}_q = \vec{0}$, so $\vec{b}_1, \ldots, \vec{b}_q$ are linearly dependent. □

Corollary 4.66. *If f has a left inverse, then $m \geq n$.*

Proof: By Theorem 4.58, if f has a left inverse, then f is an injection. Let $\mathcal{B}$ be a basis for $\mathbb{R}^n$. By Theorem 4.65, $f(\mathcal{B})$, which is a set of vectors in $\mathbb{R}^m$, is linearly independent, so by Corollary 4.35, $n \leq m$. □

Theorem 4.67. *Let V be a complement to $\mathrm{Null}(f)$. Then f is a bijection between V and $\mathrm{Image}(f)$.*

Proof: Let $\vec{u}$ be any vector in $\mathrm{Image}(f)$; thus, $\vec{u} = f(\vec{w})$ for some $\vec{w}$ in $\mathbb{R}^n$. Since V is a complement of $\mathrm{Null}(f)$, we have $\vec{w} = \vec{v} + \vec{n}$ for some $\vec{v}$ in V and some $\vec{n}$ in $\mathrm{Null}(f)$. Then $\vec{u} = f(\vec{w}) = f(\vec{v} + \vec{n}) = f(\vec{v}) + f(\vec{n}) = f(\vec{v}) + \vec{0} = f(\vec{v})$ so f is a surjection from V onto $\mathrm{Image}(f)$.

Conversely, suppose that $f(\vec{v}_1) = f(\vec{v}_2)$ for some $\vec{v}_1, \vec{v}_2$ in V. Then $f(\vec{v}_1 - \vec{v}_2) = \vec{0}$, so $\vec{v}_1 - \vec{v}_2$ is in $\mathrm{Null}(f)$. But $\vec{v}_1 - \vec{v}_2$ is also in V, which is a complement of $\mathrm{Null}(f)$; so $\vec{v}_1 - \vec{v}_2 = \vec{0}$, or $\vec{v}_1 = \vec{v}_2$. Thus, f is an injection over V. □

Corollary 4.68. *$\mathrm{Dim}(\mathrm{Image}(f)) + \mathrm{Dim}(\mathrm{Null}(f)) = n$.*

Proof: Let V be the complement of $\mathrm{Null}(f)$. By Theorems 4.67 and 4.65, $\mathrm{Dim}(\mathrm{Image}(f)) = \mathrm{Dim}(V)$, and by Theorem 4.42, $\mathrm{Dim}(\mathrm{Null}(f)) + \mathrm{Dim}(V) = n$. □

4.2.7 Inverses

Theorem 4.69. *If f is invertible, then f^{-1} is a linear transformation and $m = n$.*

Proof: Since f is linear we have $f(a \cdot f^{-1}(\vec{v})) = a \cdot f(f^{-1}(\vec{v})) = a\vec{v}$ and $f(f^{-1}(\vec{v}) + f^{-1}(\vec{u})) = f(f^{-1}(\vec{v})) + f(f^{-1}(\vec{u})) = \vec{v} + \vec{u}$. Applying f^{-1} to both sides of both equations, we have $f^{-1}(a\vec{v}) = af^{-1}(\vec{v})$ and $f^{-1}(\vec{v} + \vec{u}) = f^{-1}(\vec{v}) + f^{-1}(\vec{u})$, so f^{-1} is a linear transformation.

By Corollary 4.66, since f has a left inverse, we have $m \geq n$, but since f^{-1} has a left inverse, we also have $n \geq m$, so $m = n$. □

A matrix that has no full inverse is said to be *singular*. A matrix that has an inverse is said to be *nonsingular*.

Theorem 4.70. *If f is a linear transformation and an injection from $\mathbb{R}^n$ to $\mathbb{R}^m$, then there exists a left inverse g that is a linear transformation.*

Proof: Let V be a complement for Image(f) in $\mathbb{R}^m$. Define the function $g(\vec{w})$ from $\mathbb{R}^m$ to $\mathbb{R}^n$ by the following steps:

1. Let $\vec{w} = \vec{u} + \vec{v}$, where $\vec{u}$ is in Image(f) and $\vec{v}$ is in V (this is unique, since these spaces are complements).

2. Let $\vec{x}$ be the vector in $\mathbb{R}^n$ such that $f(\vec{x}) = \vec{u}$ (this exists, since $\vec{u}$ is in Image(f), and is unique, since f is an injection).

3. Let $g(\vec{w}) = \vec{x}$.

It is then easily shown that g is a left inverse for f and that g is a linear transformation. □

Corollary 4.71. *If matrix F is an injection, then F has an left inverse G such that $G \cdot F = I_n$.*

Theorem 4.72. *If f is a linear transformation and a surjection from $\mathbb{R}^n$ to $\mathbb{R}^m$, then there exists a right inverse h that is a linear transformation.*

Proof: Let V be a complement for Null(f) in $\mathbb{R}^n$. By Theorem 4.67, f is a bijection between V and $\mathbb{R}^m$, so for any $\vec{x}$ in $\mathbb{R}^m$, there exists a unique $\vec{v}$ in V such that $f(\vec{v}) = \vec{x}$. Define $h(\vec{x}) = \vec{v}$. It is easily shown that h is a right inverse and a linear transformation. □

Corollary 4.73. *If matrix F is an surjection, then F has an right inverse H such that $F \cdot H = I_m$.*

4.2.8 Systems of Linear Equations

We are now able to characterize the space of solutions to systems of linear equations with a given matrix F of coefficients.

Theorem 4.74. *The system of linear equations $F \cdot \vec{x} = \vec{b}$ has a solution only if $\vec{b}$ is in* Image(f).

Proof: Immediate from the meaning of Image(f) and the correspondence between f and F. $\qquad\square$

Theorem 4.75. *The system of linear equations $F \cdot \vec{x} = \vec{b}$ has a solution for every $\vec{b}$ if and only if* Rank$(F) = m$.

Proof: $m = $ Rank$(f) = $ Dim(Image(f)) if and only if Image$(f) = \mathbb{R}^m$. $\qquad\square$

Definition 4.76. Let F be an $m \times n$ matrix. The *row space* of F is the span of its rows, Span$(F[1,:], F[2,:], \ldots, F[m,:])$; this is a subspace of $\mathbb{R}^n$. The *column space* of F is the span of its columns, Span$(F[:,1], F[:,2], \ldots, F[:,n])$; this is a subspace of $\mathbb{R}^m$.

Theorem 4.77. *The image of F is the span of its columns:* Image$(F) = $ Span$(F[:, 1], \ldots, F[:,n])$.

Proof: Immediate from the fact that $F \cdot \vec{v} = \vec{v}[1] \cdot F[:,1] + \ldots + \vec{v}[n] \cdot F[:,n]$. $\qquad\square$

Theorem 4.78. Null(F) *is the orthogonal complement of the row space of F.*

Proof: Vector $\vec{v}$ is in Null(F) if and only if $F \cdot \vec{v} = \vec{0}$. But $F \cdot \vec{v} = \langle F[1,:] \bullet \vec{v}, \ldots, F[m,:] \bullet \vec{v} \rangle$, the vector consisting of the dot product of each row with $\vec{v}$, so this product is $\vec{0}$ only if $\vec{v}$ is orthogonal to each of the rows in F—that is, if $\vec{v}$ is in the orthogonal complement of the row space of F. $\qquad\square$

Note that the previous two theorems come from the two different views of matrix multiplication mentioned in Section 3.7.

Corollary 4.79. *Let R be the row space of matrix F. Then* Rank$(F) = $ Dim(Image(F)) $= $ Dim$(R) = n - $ Dim(Null(F)).

Proof: By Theorems 4.78 and 4.42, we have Dim$(R) + $ Dim(Null(F)) $= n$. By Corollary 4.68, we have Dim(Image(F)) $+ $ Dim(Null(F)) $= n$. $\qquad\square$

Putting all this together, we can amplify the description of the categories of matrices given in Section 4.1.6.

Category I. Rank(F) = m = n. F is a nonsingular square matrix. The rows of F and the columns of F are both bases for $\mathbb{R}^n$. F is a bijection from $\mathbb{R}^n$ to itself. Image(F) = $\mathbb{R}^n$. Null(F) is the zero space. F has a full inverse. The system $F \cdot \vec{x} = \vec{b}$ has a unique solution for all $\vec{b}$.

Category II. Rank(F) = $m < n$. The rows of F are linearly independent but do not span $\mathbb{R}^n$. The columns of F span $\mathbb{R}^m$ but are linearly dependent. F is a surjection from $\mathbb{R}^n$ to $\mathbb{R}^m$ but not an injection. Image(F) = $\mathbb{R}^m$. Null(F) is not the zero space. Dim(Null(F)) = $n - m$. F has a right inverse. The system $F \cdot \vec{x} = \vec{b}$ has infinitely many solutions for every $\vec{b}$. The system of equations is *underconstrained*.

Category III. Rank(F) = $n < m$. The columns of F are linearly independent but do not span $\mathbb{R}^m$. The rows of F span $\mathbb{R}^n$ but are linearly dependent. F is an injection from $\mathbb{R}^n$ to $\mathbb{R}^m$ but not an surjection. Image(F) $\neq \mathbb{R}^m$. Dim(Image(F)) = n. Null(F) is the zero space. F has a left inverse. The system $F \cdot \vec{x} = \vec{b}$ does not have solutions for every $\vec{b}$; for any value of $\vec{b}$, it has at most one solution. The system of equations is *overconstrained*.

Category IV. Rank(F) < m, and Rank(F) < n. The rows and the columns are each linearly dependent; the rows do not span $\mathbb{R}^n$ and the columns do not span $\mathbb{R}^m$. F is neither an injection nor a surjection. Image(F) $\neq \mathbb{R}^m$. Dim(Image(F)) < min(m, n). Null(F) is not the zero space. Dim(Null(F)) > max($n - m, 0$). F has neither a left inverse nor a right inverse. The system $F \cdot \vec{x} = \vec{b}$ does not have solutions for every $\vec{b}$. If $F \cdot \vec{x} = \vec{b}$ has a solution for a particular value of $\vec{b}$, then it has infinitely many solutions for that value.

In principle, if we can compute a right inverse or a full inverse H for F, then we can solve the system of linear equations $F\vec{x} = \vec{b}$ as $H\vec{b}$ since $FH\vec{b} = \vec{b}$. If we can compute a left image G for F, and if the system $F\vec{x} = \vec{b}$ has a solution, then by applying G to both sides of the equation, we see that $\vec{x} = GF\vec{x} = G\vec{b}$. Note, however, that $G\vec{b}$ is not necessarily a solution to the equation, since $FG\vec{b}$ is not necessarily equal to $\vec{b}$; all we can be sure of is that *if* a solution exists, then it must be equal to $G\vec{b}$. However, as we see in Chapter 5, these are not in practice the best way to solve systems of linear equations, even aside from the fact that they do not apply to systems in Category IV.

4.3   Vector Spaces in General (Very Optional)

4.3.1   The General Definition of Vector Spaces

In this section we give a much more general definition of a vector space, which is standard in the mathematical literature, and we illustrate it with a few examples. This section is quite optional, and nothing in this section will be needed in the remainder of this book.

First, we need to define a field. A *field* is a set on which we can define addition and multiplication operators that have some of the properties that we are used to with real numbers.

Definition 4.80. A *field* consists of a set of elements, one of which is named 0 and another of which is named 1, and two binary functions, named + and *, satisfying the following axioms.

1. $0 \neq 1$.

2. For all a and b, $a + b = b + a$.

3. For all a, b, and c, $(a + b) + c = a + (b + c)$.

4. For all a, $a + 0 = a$.

5. For all a, there exists an element b such that $a + b = 0$.

6. For all a and b, $a \cdot b = b \cdot a$.

7. For all a, b, and c, $(a \cdot b) \cdot c = a \cdot (b \cdot c)$.

8. For all a, $a \cdot 1 = a$.

9. For all $a \neq 0$, there exists an element b such that $a \cdot b = 1$.

10. For all a, b, and c, $(a + b) \cdot c = (a \cdot c) + (b \cdot c)$.

The set of real numbers is a field. Other examples of fields include

- the rational numbers;

- the complex numbers;

- for some fixed integer r, all numbers of the form $p + q\sqrt{r}$, where p and q are rational;

- for any prime p, the numbers $0, \dots, p - 1$, where + and *, respectively, are interpreted as addition and multiplication modulo p (all the axioms except (9), the existence of a multiplicative inverse, are immediate; proving (9) takes a little more work);

- the set of all rational polynomials; that is, all fractions $f(x)/g(x)$, where $f(x)$ and $g(x)$ are polynomials and $g(x) \neq 0$.

We can now give the general definition of a vector space.

Definition 4.81. Let $\mathscr{F}$ be a field; the elements of $\mathscr{F}$ are called *scalars*. A *vector space over $\mathscr{F}$* consists of a collection of elements, called *vectors*, one of which is called $\vec{0}$, and two operations. The first operation is named + and takes two vectors as arguments. The second operation is named · and takes a scalar and a vector as arguments. The operations satisfy the following axioms:

- For any vectors $\vec{u}$ and $\vec{v}$, $\vec{u} + \vec{v} = \vec{v} + \vec{u}$.

- For any vectors $\vec{u}$, $\vec{v}$, and $\vec{w}$, $(\vec{u} + \vec{v}) + \vec{w} = \vec{u} + (\vec{v} + \vec{w})$.

- For any vector $\vec{u}$, $\vec{u} + \vec{0} = \vec{u}$.

- For any scalars a, b, and vector $\vec{v}$, $a \cdot (b \cdot \vec{v}) = (a \cdot b) \cdot \vec{v}$.

- For any vector $\vec{u}$, $0 \cdot \vec{u} = \vec{0}$ and $1 \cdot \vec{u} = \vec{u}$.

- For any scalars a, b, and vector $\vec{v}$, $(a + b) \cdot \vec{v} = (a \cdot \vec{v}) + (b \cdot \vec{v})$.

- For any scalar a and vectors $\vec{u}, \vec{v}$, $a \cdot (\vec{u} + \vec{v}) = (a \cdot \vec{u}) + (a \cdot \vec{v})$.

Here are a few examples of vector spaces (in each of these, the definition of vector sum and of scalar multiplication are obvious):

- For any field $\mathscr{F}$ and fixed n, define a vector to be an n-tuple of elements of $\mathscr{F}$.

- For any field $\mathscr{F}$ and any set S, define a vector to be a function from S to $\mathscr{F}$.

- Define a vector f to be a differentiable function from $\mathbb{R}$ to $\mathbb{R}$.

- Define a vector f to be a function from $\mathbb{R}$ to $\mathbb{R}$ such that the integral $\int_{-\infty}^{\infty} f^2(x)\,dx$ exists and is finite.

Particularly note that all discussions of vectors and vector spaces in this book, except for this section, assume that the vectors are in $\mathbb{R}^n$. Therefore, it is not safe to assume that the theorems in this book apply, or can be modified to apply, to any other kind of vectors; nor that the definitions are correct or even meaningful. For example, for vectors over the complex numbers, the definition of the dot product has to be slightly changed. For continuous functions viewed as vectors, the definition of the dot product has to be substantially changed. For vectors over finite fields, the concept of a dot product is not useful.

Exercises

Exercise 4.1. Which of the following sets of vectors are vector spaces? Justify your answer in each case.

(a) $\mathbb{R}^2 \cup \mathbb{R}^3$.

(b) $\{\langle x, 2x \rangle \mid x \in \mathbb{R}\}$.

(c) $\{\langle k, 2k \rangle \mid k \text{ is an integer }\}$.

(d) $\{\langle x, 2x + 1 \rangle \mid x \in \mathbb{R}\}$.

(e) $\{\langle x, 0 \rangle \mid x \in \mathbb{R}\} \cup \{\langle 0, y \rangle \mid y \in \mathbb{R}\}$.

(f) $\text{Span}(\{\langle 1, 3, 2 \rangle, \langle -1, 0, 3 \rangle\})$.

Exercise 4.2. For each of the following sets of three-dimensional vectors, state whether or not it is linearly independent. Justify your answers.

(a) $\{\langle 1, 2, 3 \rangle\}$.

(b) $\{\langle 1, 1, 0 \rangle, \langle 0, 1, 1 \rangle\}$.

(c) $\{\langle 1, 1, 0 \rangle, \langle 2, 2, 1 \rangle\}$.

(d) $\{\langle 1, 1, 0 \rangle, \langle 2, 2, 0 \rangle\}$.

(e) $\{\langle 2.3, 4.5, -1.2 \rangle, \langle 3.7, 1.2, 4.3 \rangle, \langle 0, 1.4, 2.5 \rangle, \langle -2.9, -3.1, 1.8 \rangle\}$. (*Hint:* It should not be necessary to do any calculations involving the components.)

Exercise 4.3. In the example illustrating Theorem 4.19 (Section 4.1.3), verify that, as claimed,

$$|\vec{u}| = |\text{Coords}(\vec{u}, \mathcal{N})| = \sqrt{78},$$
$$|\vec{v}| = |\text{Coords}(\vec{v}, \mathcal{N})| = \sqrt{146}, \quad \text{and}$$
$$\vec{u} \bullet \vec{v} = \text{Coords}(\vec{u}, \mathcal{N}) \bullet \text{Coords}(\vec{v}, \mathcal{N}) = 69.$$

Exercise 4.4. Let $\mathcal{U} = \text{Span}(\{\langle 1,0,0,0 \rangle, \langle 0,1,0,0 \rangle\})$ and $\mathcal{V} = \text{Span}(\{\langle 1,1,0,0 \rangle, \langle 0,0,1,1 \rangle\})$ be subspaces of $\mathbb{R}^4$.

(a) Give a basis for $\mathcal{U} \oplus \mathcal{V}$.

(b) Give a basis for $\mathcal{U} \cap \mathcal{V}$.

Exercise 4.5. (Do this exercise by hand, in terms of square roots, if you can; if not, use MATLAB.) Let $\mathcal{U} = \text{Span}(\{\langle 1,0,1,0 \rangle, \langle 1,1,1,1 \rangle\})$ be a subspace of $\mathbb{R}^4$.

(a) Find an orthonormal basis for $\mathcal{U}$.

(b) Extend your answer in (a) to an orthonormal basis for $\mathbb{R}^4$.

Exercise 4.6. What is the dimension of the space spanned by the following four vectors? (Do this both by hand and by using MATLAB.)

$$\langle 1,2,2,-1\rangle, \langle 0,1,-1,0\rangle, \langle 1,4,0,-1\rangle, \langle 2,1,1,-1\rangle.$$

Exercise 4.7. Let $\mathcal{V} = \text{Span}(\{\langle 1,1,0,3\rangle; \langle 2,4,-3,1\rangle; \langle 7,9,-6,5\rangle\})$. Using MATLAB, determine whether the vector $\langle 1,3,-4,0\rangle$ is in $\mathcal{V}$.

Problems

Problem 4.1. Prove Theorem 4.18. The statement of this theorem is in Section 4.1.2, but you should master the corresponding material in Section 4.2.2 before attempting the proof.

Problem 4.2. Prove Theorem 4.21. The statement of this theorem is in Section 4.1.3, but you should master the corresponding material in Section 4.2.4 before attempting the proof.

Problem 4.3. Prove Theorem 4.22, parts a and b. The statement of this theorem is in Section 4.1.4, but you should master the corresponding material in Sections 4.2.1 and 4.2.3 before attempting the proof.

Problem 4.4. Prove Theorem 4.62. (*Hint:* Consider Example 4.10.)

Problem 4.5. (Use MATLAB for this problem. Since the answer is not at all unique, you should submit the code you used to compute it, not just the numeric answer.) Let

$$\mathcal{U} = \text{Span}(\{\langle 1,2,3,4,3,2,1\rangle, \langle 4,0,2,0,-2,0,-4\rangle, \langle 3,1,2,0,1,-1,0\rangle\})$$

and

$$\mathcal{V} = \text{Span}(\{\langle 2,1,3,4,0,3,-3\rangle, \langle 4,3,5,4,4,1,1\rangle, \langle 1,1,1,1,1,1,2\rangle\})$$

be subspaces of $\mathbb{R}^7$.

(a) Find an orthogonal basis for $\mathcal{U} \cap \mathcal{V}$.

(b) Extend your answer in (a) to an orthogonal basis for $\mathcal{U} \oplus \mathcal{V}$.

(c) Check your answer.

Problem 4.6. As discussed in Problem 3.6, an $n \times n$ matrix A is nilpotent if $A^p = 0$ for some power p. Prove that if $A^p = 0$ and A has rank r, then $p \cdot (n-r) \geq n$.

Programming Assignments

Assignment 4.1 (Extracting a basis). Write a MATLAB function FindBasis(A) that does the following. The input A is an $m \times n$ matrix, which we will view as m n-dimensional vectors. The value returned is a $p \times n$ matrix containing a subset (or all) of the rows of A that form a basis for the subspace spanned by the rows of A. For instance, if

$$A = \begin{bmatrix} 1 & 0 & 1 & 2 \\ 2 & 0 & 2 & 4 \\ 1 & -1 & 1 & 1 \end{bmatrix},$$

then FindBasis(A) should return either

$$\begin{bmatrix} 1 & 0 & 1 & 2 \\ 1 & -1 & 1 & 1 \end{bmatrix} \quad \text{or} \quad \begin{bmatrix} 2 & 0 & 2 & 4 \\ 1 & -1 & 1 & 1 \end{bmatrix}.$$

If

$$A = \begin{bmatrix} 1 & 0 & 1 & 2 \\ 0 & 1 & 1 & 1 \\ 1 & -1 & 0 & 1 \end{bmatrix},$$

then FindBasis(A) should return a 2×4 matrix with any two of the rows of A. If

$$A = \begin{bmatrix} 1 & 0 & 1 & 2 \\ 0 & 1 & 1 & 1 \\ 1 & -1 & 1 & 1 \end{bmatrix},$$

then FindBasis(A) should return A.

Chapter 5

Algorithms

The theory developed in Chapter 4 is very elegant. Unfortunately, it presents almost no algorithms to compute actual answers to the questions raised. (The only algorithm in that chapter is Gram-Schmidt orthogonalization.)

This chapter presents two algorithms. The first is *Gaussian elimination*, which reduces a matrix to *row-echelon form*, and allows us to solve a system of linear equations, to compute the rank of a matrix, and to compute the coordinates of a given vector relative to a given basis. The second is the computation of the inverse of a matrix.

5.1   Gaussian Elimination: Examples

We begin with two examples of solving systems of simultaneous equations by using Gaussian elimination.

Example 5.1. Consider the following system of equations:

$$2x - 4y + 2z = -8, \tag{5.1}$$
$$2x + 2y + z = 5, \tag{5.2}$$
$$x + y - 2z = 5. \tag{5.3}$$

Step 1. Subtract Equation (5.1) from Equation (5.2), giving $6y - z = 13$.

Step 2. Subtract one half of Equation (5.1) from Equation (5.3), giving $3y - 3z = 9$. The system is now

$$2x - 4y + 2z = -8, \tag{5.4}$$
$$6y - z = 13, \tag{5.5}$$
$$3y - 3z = 9. \tag{5.6}$$

Step 3. Subtract one half of Equation (5.5) from Equation (5.6), giving $(-5/2)z = 5/2$. The system is now

$$2x - 4y + 2z = -8, \tag{5.7}$$
$$6y - z = 13, \tag{5.8}$$
$$(-5/2)z = 5/2, \tag{5.9}$$

which is easily solved. By Equation (5.9), we have $z = -1$. By substituting this into Equation (5.8), we get $y = 2$. Then, by substituting these two values into Equation (5.7), we get $x = 1$.

Example 5.2. More complex issues arise in the following system of four equations in four unknowns:

$$w + x + y + z = 5, \tag{5.10}$$
$$w + x + 2y + 2z = 7, \tag{5.11}$$
$$x - y + z = -2, \tag{5.12}$$
$$w + 2y = 7. \tag{5.13}$$

Step 1. Subtract Equation (5.10) from Equation (5.11), giving $y + z = 2$.

Step 2. Subtract Equation (5.10) from Equation (5.13), giving $-x + y - z = 2$. Then multiply this by -1, giving $x - y + z = -2$. The system is now

$$w + x + y + z = 5, \tag{5.14}$$
$$y + z = 2, \tag{5.15}$$
$$x - y + z = -2, \tag{5.16}$$
$$x - y + z = -2. \tag{5.17}$$

Step 3. Switch Equation (5.15) and Equation (5.16) (just a reordering). The system is now

$$w + x + y + z = 5, \tag{5.18}$$
$$x - y + z = -2, \tag{5.19}$$
$$y + z = 2, \tag{5.20}$$
$$x - y + z = -2. \tag{5.21}$$

Step 4. Subtract Equation (5.19) from Equation (5.21). The system is now

$$w + x + y + z = 5, \tag{5.22}$$
$$x - y + z = -2, \tag{5.23}$$
$$y + z = 2, \tag{5.24}$$
$$0 = 0.$$

This system is now easily solved. Starting with Equation (5.24), we can assign z to be an arbitrary value, and y to be $2 - z$; we will choose $z = 0$, $y = 2$. By Equation (5.23), $x = 0$, and by Equation (5.22), $w = 3$.

5.2 Gaussian Elimination: Discussion

The procedure used to solve a system S of linear equations in Section 5.1 is divided into two parts:

- Use the *Gaussian elimination algorithm* to convert the input system into an *equivalent* system R in *row echelon form*.

- Use a simple procedure to find the solutions to R.

Definition 5.3. Two systems of linear equations $S1$ and $S2$ are *equivalent* if they have the same solution space.

Definition 5.4. An n-dimensional vector $\vec{v}$ has k *leading zeros* if either

- $\vec{v} = \vec{0}$ and $k = n$; or

- $\vec{v}[i] = 0$ for $i = 1, \ldots, k$ and $\vec{v}[k+1] \neq 0$.

Definition 5.5. Let A be an $m \times n$ matrix. A is in *row echelon form* if, for each row $A[i,:]$, either $A[i,:] = \vec{0}$ or $A[i,:]$ has more leading zeros than $A[i-1,:]$. A system of linear equations $A\vec{x} = \vec{c}$ is in row echelon form if A is in row echelon form.

If the system of equations $A\vec{x} = \vec{c}$ is in row echelon form, then a solution can easily be found by using Algorithm 5.1, SolveRowEchelon:

It is easy to see that the value $\vec{x}$ returned by procedure SolveRowEchelon satisfies the system of equations. The procedure fills in values of $\vec{x}$ from bottom to top. At each iteration of the loop, the procedure deals with the ith equation. Because the system is in row-echelon form, the variable corresponding to the first nonzero coefficient is not yet set, and can therefore be set in a way that satisfies the equation.

The Gaussian elimination procedure to convert a system of equations to row-echelon form is given in Algorithm 5.2.

We can easily see that the following loop invariant holds after the ith iteration of the main loop of procedure GaussianEliminate. Let j be the index of the first nonzero component of $A[i,:]$ (that is, $j = L(A, i)$). Then,

- $A[i, j] = 1$,

- for all $p > i$ and all $q \leq j$, $A[p, q] = 0$.

---

function L(A, i)
    return the number of leading zeros in $A[i, :]$.

    /* SolveRowEchelon($A, \vec{c}$) takes as arguments an $m \times n$ matrix A in row
    echelon form and an m-dimensional vector $\vec{c}$. If the system of equations
    $A\vec{x} = \vec{c}$ has a solution, then the algorithm returns one such solution. If
    the system has no such solution, then the algorithm returns **false** */

procedure SolveRowEchelon($A, \vec{c}$)
    if (there exists i for which $A[i, :] = \vec{0}$ and $\vec{c}[i] \neq 0$)
    then return false; /* The ith equation is $0 \cdot x[1] + \ldots + 0 \cdot x[n] = c \neq 0$ */
    else if $A = 0$ then return any n-dimensional vector; /* the system is $0 \cdot \vec{x} = \vec{0}$ */
    else let t be the largest value for which $A[t, :] \neq \vec{0}$
    let $\vec{x}$ be an n-dimensional vector;
    for ($i \leftarrow t$ **downto** 1)
        $p \leftarrow$ L(A, i)+1;
        if ($i = m$) **then** $q \leftarrow n$ **else** $q \leftarrow$ L($A, i + 1$);
        if ($p < q$)
            then for (q **downto** $p + 1$)
                    $\vec{x}[j] \leftarrow$ an arbitrary value;
                endfor
        endif
        $\vec{x}[p] \leftarrow (\vec{c}[i] - \sum_{j=p+1}^{n} A[i, j] \cdot \vec{x}[j])/A[i, p]$;
    endfor
    return $\vec{x}$;
endif
end SolveRowEchelon

Algorithm 5.1. Solving a system of equations in row-echelon form.

Thus, when the loop exits (either because $i = m$ or all the remaining rows are now $\vec{0}$), the matrix A will be in row-echelon form.

To show that the input and output sets of equations are equivalent, we note that the execution of the algorithm consists of a sequence of three different kinds of operations on the system of equations:

1. Change the order of the equations.

2. Multiply an equation by a nonzero scalar.

3. For some p, i, replace the pair of equations

$$A[i, :] \cdot \vec{x} = \vec{c}[i], \tag{5.25}$$
$$A[p, :] \cdot \vec{x} = \vec{c}[p] \tag{5.26}$$

```
    /* GaussianEliminate(A, c⃗) takes an m × n matrix A and an m-dimensional
    vector c⃗. It converts the system of equations Ax⃗ = c⃗ to an equivalent sys-
    tem in row-echelon form. */

procedure GaussianEliminate(A, c⃗)
for (i ← 1 to m) /* i is the row we are working on */
    if (A[p, :] = 0⃗ for all p ≥ i) then exitloop;
    j ← the smallest value such that A[p, j] ≠ 0 for some p ≥ i; /* j is the column we are working on. */
    q ← argmax_{p≥i}|A[p, j]|;
        /* q is the index of the row p ≥ i with the largest value (in absolute value) in the jth column. */
    if (q ≠ i) then { swap(A[i, :], A[q, :]);
                      swap(c⃗[i], c⃗[q]); }
    c⃗[i] ← c⃗[i]/A[i, j]; /* Divide equation for i by A[i, j]. */
    A[i, :] ← A[i, :]/A[i, j];
    /* Subtract the appropriate multiple of the ith equation from each of the lower equations
       to zero out the jth column. */
    for (p ← i + 1 to m)
        c⃗[p] ← c⃗[p] − A[p, j] · c⃗[i];
        A[p, :] ← A[p, :] − A[p, j] · A[i, :];
    endfor
endfor
return A, c⃗
end GaussianEliminate
```

Algorithm 5.2. Gaussian elimination.

by the pair of equations

$$A[i, :] \cdot \vec{x} = \vec{c}[i], \tag{5.27}$$

$$(A[p, :] + \alpha \cdot A[i, :]) \cdot \vec{x} = \vec{c}[p] + \alpha \cdot \vec{c}[i]. \tag{5.28}$$

(Note that Equation (5.27) is just a repeat of Equation (5.25).)

It is trivial that operations (1) and (2) do not change the solution space. To show that operation (3) does not change the solution space, note that, if Equation (5.25) and (5.26) are true for a particular $\vec{x}$, then Equation (5.28) can be derived by multiplying Equation (5.25) by α and adding it to Equation (5.26); and if Equation (5.27) and Equation (5.28) are true, then Equation (5.26) can be derived by multiplying Equation (5.27) by $-\alpha$ and adding it to Equation (5.28). Thus, the two pairs of equations are equivalent.

We can now put the two procedures together into a method for solving systems of linear equations (Algorithm 5.3).

Based on the discussion above, we can state several theorems.

> /* SolveLinearEqns($A, \vec{c}$) takes as arguments an $m \times n$ matrix A and an m-dimensional vector $\vec{c}$. If the system has a solution, then it returns one solution; otherwise, it returns **false** */
>
> **procedure** SolveLinearEqns($A, \vec{c}$)
> **return** SolveRowEchelon(GaussianEliminate($A, \vec{c}$))

Algorithm 5.3. Solving a system of linear equations.

Theorem 5.6. *Let A be an $m \times n$ matrix A and let $\vec{c}$ be an m-dimensional vector. Let $A', \vec{c}\,'$ be the values returned by* GaussianEliminate$(A, \vec{c})$. *Then the system of equations $A'\vec{x} = \vec{c}\,'$ is in row-echelon form and is equivalent to the system $A\vec{x} = \vec{c}$.*

Theorem 5.7. *Let A be an $m \times n$ matrix in row-echelon form and let $\vec{c}$ be an m-dimensional vector. If the system of equations $A\vec{x} = \vec{c}$ has a solution, then* SolveRowEchelon$(A, \vec{c})$ *returns some solution to the system. If the system has no solution, then* SolveRowEchelon$(A, \vec{c})$ *returns* false.

Theorem 5.8. *Let $A\vec{x} = \vec{c}$ be a system of linear equations. If this system has a solution, then* SolveLinearEqns$(A, \vec{c})$ *returns a solution. If it does not, then* SolveLinearEqns$(A, \vec{c})$ *returns* false.

The procedure SolveLinearEqns can also be used to determine whether a vector $\vec{c}$ is in the vector space spanned by a given collection of vectors $\mathcal{B}$ and to find the coordinates of $\vec{c}$ in $\mathcal{B}$.

Theorem 5.9. *Let $\mathcal{B} = \{\vec{b}_1, \ldots, \vec{b}_k\}$ be a set of vectors and let $\vec{c}$ be a vector. Let B be the matrix whose columns are $\vec{b}_1, \ldots, \vec{b}_k$. Then $\vec{c}$ is in* Span$(\mathcal{B})$ *if and only if the system $B\vec{x} = \vec{c}$ has a solution; if it does, then any solution $\vec{x}$ is a coordinate vector for $\vec{c}$ in $\mathcal{B}$.*

Proof: Immediate from the fact that $B\vec{x}$ is the weighted sum of the columns of B: $B\vec{x} = \vec{x}[1]B[:, 1] + \ldots + \vec{x}[n]B[:, n]$. □

5.2.1   Gaussian Elimination on Matrices

The Gaussian elimination procedure can also be used to reduce a matrix A (rather than a system of equations) to row-echelon form; one can either delete the references to the constant vector $\vec{c}$ in Algorithm 5.2 or, more simply, pass an arbitrary value for $\vec{c}$ and ignore the value returned. For this purpose, we treat the procedure GaussianEliminate(A) as a function of one argument returning one value. The following theorems then follow directly.

Theorem 5.10. *Let A be an m × n matrix, and let A′ = GaussianEliminate(A).
Then A′ is in row-echelon form and the row span of A′ is equal to the row span
of A.*

Proof: The proof is exactly analogous to the proof of Theorem 5.6. In particular,
the proof that A' has the same row span as A works in exactly the same way as
the proof that the system $A'\vec{x} = \vec{c}\,'$ is equivalent to the system $A\vec{x} = \vec{c}$.   □

Theorem 5.11. *If matrix A is in row-echelon form, then the nonzero rows of A
are linearly independent.*

Proof: Let $A[i,:]$ be a nonzero row, and let q =L(A, i)+1 be the position of the
first nonzero value in $A[i,:]$. Since $A[i, q] = 1$ and $A[p, q] = 0$ for all $q > i$, clearly
$A[i,:]$ is not a linear sum of the vectors $A[i + 1,:], \ldots, A[m,:]$. Applying Corollary 4.31 to the nonzero rows in backward order, it follows that the rows are
linearly independent.   □

Corollary 5.12. *Let A be any matrix. Then the nonzero rows of* GaussianElim-
inate(A) *are a basis for the row span of A, and* Rank(A) *is equal to the number of
nonzero rows of* GaussianEliminate(A).

Proof: Immediate from Theorems 5.10 and 5.11, together with the definitions
of basis and rank.   □

5.2.2   Maximum Element Row Interchange

There is one aspect of the Gaussian elimination algorithm that we have not yet
explained. The second step of the main **for** loop finds the row q that maximizes
$|A[q, i]|$, and the third step swaps that with the ith row. Now, it is clear that to
achieve row echelon form, if $A[i, i] = 0$, then we have to swap it with some row
q, where $A[q, i] \neq 0$. But why particularly the row with the largest absolute
value?

In fact, if computers could do exact real-valued arithmetic—which, by the
way, is the idealization used in all the above theorems—then it wouldn't matter
which row we chose; any row q where $A[q, i] \neq 0$ would work just as well. But
computers do not (normally) do exact real computation; they do floating-point
arithmetic, which is subject to roundoff error. The point of the "maximum element row interchange" operation is that it reduces the error due to roundoff.
We discuss the general issue of roundoff a little further in Section 5.6; for a detailed discussion of the maximum element row interchange rule, see (Trefethen
and Bau, 1997).

n	2	3	4	5	6	10	20
Success rate	59%	15%	7.6%	5.2%	3.7%	1.9%	0.48%

Table 5.1. Frequency with which row-echelon reduction of a random singular matrix produces a final row equal to zero.

5.2.3 Testing on Zero

In a number of places in the above algorithms, there is a test whether a value is equal to zero. For instance, SolveRowEchelon includes the instruction "**if** (there exists i for which $A[i,:] = \vec{0}$ and $\vec{c}[i] \neq 0$)"; GaussianEliminate contains "**if** ($A[p,:] = \vec{0}$ for all $p \geq i$) **then exitloop**;" and subroutine L for SolveRowEchelon states, "return the number of leading zeroes." Some numerical analysts may object very strongly to these tests, saying that in actual computation you never test on zero, since in actual computation you never get zeroes because of roundoff errors and noise. (Some programming languages do not allow testing for equality between floating-point numbers, for that very reason.) The thinking is that Theorems 5.6–5.12 are therefore meaningless; they refer to conditions that are never met.

Thus, we should simply carry out the computations without testing on zero; if we divide by zero at some stage, we will get Inf or NaN, but that is OK in MATLAB or any other system that meets the IEEE floating-point standard (Overton, 2001). When we have computed the final answer, then we evaluate it; if it contains either a very large value or Inf or NaN, then it may well not be meaningful.

The following experiment illustrates the difficulty in obtaining true zero with floating-point arithmetic. We generate a random $n \times (n-1)$ matrix A and a random $n-1$ dimensional column vector $\vec{v}$. We set the nth column of A to be $A \cdot \vec{v}$. Now we have a random matrix[1] of rank $n-1$. We use the built-in MATLAB function lu to reduce A to row-echelon form, and test whether the last row is actually equal to the zero vector. Table 5.1 shows the success rate of this test for each value of n. By contrast, the MATLAB function rank(A), which uses a nonzero threshhold, always got the correct answer of $n-1$ in the experiments I ran, including 50×50 and 100×100 matrices.

Nonetheless, these tests are included here for two reasons:

1. They don't do any harm. If they do not succeed, then we haven't lost anything by carrying them out. The only case for which they arguably give a wrong result is when the true value is very small but due to roundoff is computed as 0. In this case, including the test causes the algorithm to report that the matrix is singular, whereas if the test is eliminated, the algorithm returns an answer with some values being Inf or NaN. But this case is *extremely* rare, and the added value given by this answer is quite small.

---

[1] Other ways of generating random singular matrices will give different distributions. There is no unique, obviously proper, notion of a "random singular matrix."

2. As we discuss in Section 5.6, viewing these algorithms in terms of what can actually be accomplished with floating-point operations is only one of several viewpoints one might take. If we consider the algorithm from the standpoint of pure mathematics or exact computation, then these tests and Theorems 5.6–5.12 are valid.

5.3   Computing a Matrix Inverse

We now turn to the problem of inverting an $n \times n$ nonsingular matrix. (We omit the problem of finding a left or right inverse of a nonsquare matrix; the techniques are analogous.)

The algorithm for computing the inverse modifies the Gaussian elimination algorithm in two ways. First, we continue carrying out the row transformation operations to reduce the matrix row-echelon form to the identity. Second, at the same time that we are carrying out operations to reduce the input matrix A to the identity, we carry out the same operations on a matrix C that starts as the identity. When we have carried out enough operations to reduce A to the identity, then the final state of C is A^{-1}.

The explanation of the modified version of the Gaussian elimination algorithm involves a few steps. First, we observe that, if R is in row-echelon form, we can reduce it to the identity by repeatedly subtracting $R[i,j] \cdot R[j,:]$ from $R[i,:]$ for all $j > i$. Since $R[j,j] = 1$ for all j, this operation sets $R[i,j]$ to be 0. When we are done with all these, we have

- for $i > j$, $R[i,j]$ was set to 0 in the initial translation of A to the row-echelon form R,

- for $i = j$, $R[i,j]$ was set to 1 in the initial translation of A to the row-echelon form R,

- for $i < j$, $R[i,j]$ has been set to 0 on this second pass of operations.

So R is now the identity (see Example 5.13).

Second, we observe that each of the operations that procedure Gaussian-Eliminate(A) carries out corresponds to multiplication on the left by a specific matrix, as seen in the three cases presented here.

1. Switching rows p and q of matrix A corresponds to forming the product $S^{p,q} A$, where $S^{p,q}$ is the matrix defined by

$$S^{p,q}[i,j] = \begin{cases} 1 & \text{if } i = j, i \neq p, i \neq q, \\ 1 & \text{if } i = p, j = q, \\ 1 & \text{if } i = q, j = p, \\ 0 & \text{otherwise.} \end{cases}$$

For example, with $n = 5$, $p = 1$, $q = 4$, we have

$$S^{1,4} = \begin{bmatrix} 0 & 0 & 0 & 1 & 0 \\ 0 & 1 & 0 & 0 & 0 \\ 0 & 0 & 1 & 0 & 0 \\ 1 & 0 & 0 & 0 & 0 \\ 0 & 0 & 0 & 0 & 1 \end{bmatrix}.$$

2. Multiplying row p by α corresponds to forming the product $T^{p,\alpha} A$, where $T^{p,\alpha}$ is the matrix defined by

$$T^{p,\alpha}[i, j] = \begin{cases} 1 & \text{if } i = j, i \neq p, \\ \alpha & \text{if } i = j = p, \\ 0 & \text{otherwise.} \end{cases}$$

For example, with $p = 2$, $\alpha = 3.7$, we have

$$T^{2,3.7} = \begin{bmatrix} 1 & 0 & 0 & 0 & 0 \\ 0 & 3.7 & 0 & 0 & 0 \\ 0 & 0 & 1 & 0 & 0 \\ 0 & 0 & 0 & 1 & 0 \\ 0 & 0 & 0 & 0 & 1 \end{bmatrix}.$$

3. Adding α times row p to row q corresponds to forming the product $P^{p,q,\alpha} A$, where $P^{p,q,\alpha}$ is the matrix defined by

$$P^{p,q,\alpha}[i, j] = \begin{cases} 1 & \text{if } i = j, \\ \alpha & \text{if } i = q, j = p, \\ 0 & \text{otherwise.} \end{cases}$$

For example, with $p = 2$, $q = 5$, $\alpha = -4.6$, we have

$$P^{2,5,4.6} = \begin{bmatrix} 1 & 0 & 0 & 0 & 0 \\ 0 & 1 & 0 & 0 & 0 \\ 0 & 0 & 1 & 0 & 0 \\ 0 & 0 & 0 & 1 & 0 \\ 0 & -4.6 & 0 & 0 & 1 \end{bmatrix}.$$

Therefore, the sequence of operations used to transform matrix A into its row-echelon form R and to transform R into I corresponds to a sequence of matrix multiplications $R = M_t \cdot M_{t-1} \cdot \ldots \cdot M_2 \cdot M_1 A$, where M_1 is the matrix corresponding to the first operation performed by the Gaussian elimination algorithm, M_2 is the second operation, and so on. Therefore if we form the product $C = M_t \cdot M_{t-1} \ldots M_2 \cdot M_1$ then that is a matrix such that $CA = I$.

But (third) we don't have to actually generate these transformation matrices at all. Since $C = C \cdot I = M_t \cdot M_{t-1} \cdot \ldots \cdot M_2 \cdot M_1 \cdot I$, where I is the identity, we can get the effect of all these matrix multiplication just by initializing matrix C to be the identity, and then performing the same sequence of operations on C that we are performing on A. Thus, we are implicitly computing the above matrix product. The resultant modified algorithm is shown in Algorithm 5.4.

(This argument may seem strange; we first replace the operations by the matrices S, T, and P, and then go back and replace the matrices by the same operations we started with. What is the point of that? The point is we need to show that what the sequence of operations in the algorithm does *overall* in converting matrix A to row-echelon form R consists of multiplying A on the left by a matrix C. To do this, we show that each operation corresponds to multiplication by a matrix, and therefore the sequence of operations, whether applied to A or applied to I^n, correspond to multiplication by the product of these. What the matrices corresponding to the individual operations actually *are* ends up not mattering.)

Algorithm 5.4 is used for computing the inverse. Note that if in the course of the initial conversion to row-echelon form, the leading zeroes in one row increase from the previous row by more than 1, then the algorithm can immediately return with **fail**, signaling that the matrix is singular.

Example 5.13. We illustrate Algorithm 5.4 by computing the inverse of the matrix

$$A = \begin{bmatrix} 2 & 0 & 1 \\ -1 & 1 & 1 \\ 0 & 1 & 2 \end{bmatrix}.$$

We first convert to row-echelon form.

$$B = \begin{bmatrix} 2 & 0 & 1 \\ -1 & 1 & 1 \\ 0 & 1 & 2 \end{bmatrix}, \qquad C = \begin{bmatrix} 1 & 0 & 0 \\ 0 & 1 & 0 \\ 0 & 0 & 1 \end{bmatrix};$$

$$B[1,:] = (1/2) \cdot B[1,:], \qquad C[1,:] = (1/2) \cdot C[1,:];$$

$$B = \begin{bmatrix} 1 & 0 & 1/2 \\ -1 & 1 & 1 \\ 0 & 1 & 2 \end{bmatrix}, \qquad C = \begin{bmatrix} 1/2 & 0 & 0 \\ 0 & 1 & 0 \\ 0 & 0 & 1 \end{bmatrix};$$

$$B[2,:] = B[2,:] + B[1,:], \qquad C[2,:] = C[2,:] + C[1,:];$$

$$B = \begin{bmatrix} 1 & 0 & 1/2 \\ 0 & 1 & 3/2 \\ 0 & 1 & 2 \end{bmatrix}, \qquad C = \begin{bmatrix} 1/2 & 0 & 0 \\ 1/2 & 1 & 0 \\ 0 & 0 & 1 \end{bmatrix};$$

/* Inverse(A) takes an $n \times n$ matrix A and returns A^{-1}. */

procedure Inverse(A)
$B = A$; $C = I^n$; /* Initialize C to be the identity.
                    At each stage of the algorithm, $CB = A$. */
for ($i \leftarrow 1$ to n)
   $q \leftarrow \text{argmax}_{p \geq i} |B[q, i]|$;
     /* q is the index of the row p from i with the largest value (in absolute value) in the ith column. */
   if $B[q, i] = 0$ **then return fail**; /* A is singular */
   if ($q \neq i$) **then** { swap($B[i, :], B[q, :]$);
                       swap($C[i, :], C[q, :]$); }
   $C[i, :] \leftarrow C[i, :]/B[i, i]$; /* Divide row i in both matrices by $B[i, i]$. */
   $B[i, :] \leftarrow B[i, :]/B[i, i]$;
   /* In both matrices, subtract the appropriate multiple of the ith row
   from each of the lower rows to zero out the ith column of B. */
   for ($j \leftarrow i + 1$ to n)
      $C[j, :] \leftarrow C[j, :] - B[j, i] \cdot C[i, :]$;
      $B[j, :] \leftarrow B[j, :] - B[j, i] \cdot B[i, :]$;
   endfor
endfor
   /* In both matrices, subtract the appropriate multiple of the ith row
   from each of the upper rows to zero out the ith column of B. */
for ($i \leftarrow 2$ to n)
   for ($j \leftarrow 1$ to $i - 1$)
      $C[j, :] \leftarrow C[j, :] - B[j, i] \cdot C[i, :]$;
      $B[j, :] \leftarrow B[j, :] - B[j, i] \cdot B[i, :]$;
   endfor
endfor
return C;
end Inverse.

Algorithm 5.4. Computing the matrix inverse.

$$B[3, :] = B[3, :] - B[2, :], \qquad C[3, :] = C[3, :] - C[2, :];$$

$$B = \begin{bmatrix} 1 & 0 & 1/2 \\ 0 & 1 & 3/2 \\ 0 & 0 & 1/2 \end{bmatrix}, \qquad C = \begin{bmatrix} 1/2 & 0 & 0 \\ 1/2 & 1 & 0 \\ -1/2 & -1 & 1 \end{bmatrix};$$

$$B[3, :] = B[3, :] \cdot 2, \qquad C[3, :] = C[3, :] \cdot 2;$$

$$B = \begin{bmatrix} 1 & 0 & 1/2 \\ 0 & 1 & 3/2 \\ 0 & 0 & 1 \end{bmatrix}, \qquad C = \begin{bmatrix} 1/2 & 0 & 0 \\ 1/2 & 1 & 0 \\ -1 & -2 & 2 \end{bmatrix}.$$

We now continue on, converting B to the identity. We omit the steps where the factor is 0.

$$B[1,:] = B[1,:] - (1/2)B[3,:], \qquad\qquad C[1,:] = C[1,:] - (1/2)C[3,:];$$

$$B = \begin{bmatrix} 1 & 0 & 0 \\ 0 & 1 & 3/2 \\ 0 & 0 & 1 \end{bmatrix}, \qquad\qquad C = \begin{bmatrix} 1 & 1 & -1 \\ 1/2 & 1 & 0 \\ -1 & -2 & 2 \end{bmatrix};$$

$$B[2,:] = B[2,:] - (3/2)B[3,:], \qquad\qquad C[2,:] = C[2,:] - (3/2)C[3,:];$$

$$B = \begin{bmatrix} 1 & 0 & 0 \\ 0 & 1 & 0 \\ 0 & 0 & 1 \end{bmatrix}, \qquad\qquad C = \begin{bmatrix} 1 & 1 & -1 \\ 2 & 4 & -3 \\ -1 & -2 & 2 \end{bmatrix}.$$

We can now return

$$A^{-1} = C = \begin{bmatrix} 1 & 1 & -1 \\ 2 & 4 & -3 \\ -1 & -2 & 2 \end{bmatrix}.$$

5.4 Inverse and Systems of Equations in MATLAB

The MATLAB utility for computing the inverse of matrix M could not be much simpler; just write `inv(M)` or `M^-1`. The MATLAB utility for computing the solution to the equations $M\vec{x} = \vec{c}$ could not be any simpler; just write `M\C`. (Note that $\vec{c}$ here must be a *column* vector). Of course, any half-decent programming environment provides a library function for solving systems of linear equations;[2] but it is not every language in which it is a single character.

The value returned by `M\C` depends on the characteristics of the system of equations. Let M be an $m \times n$ matrix.

Case 1. If $m = n = \text{Rank}(M)$ (i.e., M is a nonsingular square matrix), then `M\C` returns the unique solution to the system of equations $M\vec{X} = \vec{C}$.

Case 2. If $m < n$ and $\text{Rank}(M) = m$, then `M\C` returns a solution to the equation $M\vec{X} = \vec{C}$ with the maximal number of zeroes.

Case 3. If $m > n$ and $\text{Rank}(M) = n$, and the system $M\vec{X} = \vec{C}$ has a solution, then `M\C` returns that solution.

---

[2]Not all libraries have a very *good* function.

Case 4. If $m > n$ and $\text{Rank}(M) = n$, and the system $M\vec{X} = \vec{C}$ does not have a solution, then M\C returns the *least squares* best fit to the equations; that is, it returns the value of $\vec{X}$ such that $|M \cdot \vec{X} - \vec{C}|$ is minimal. This is discussed further in Section 14.5.

For example, if

$$M = \begin{bmatrix} 1 & 1 \\ 2 & 1 \\ 1 & 2 \end{bmatrix} \text{ and } C = \begin{bmatrix} 1 \\ 1 \\ 1 \end{bmatrix},$$

then in MATLAB, X=M\C returns

$$X = \begin{bmatrix} 0.3636 \\ 0.3636 \end{bmatrix}.$$

Then norm(M*X-C) has the value 0.3015; this is smaller for this value of X than for any other value.

Case 5. If $\text{Rank}(M) < \min(m, n)$ and the system has no solution, then MATLAB will issue a warning and return an answer full of NaNs and Infs.

Case 6. If $\text{Rank}(M) < \min(m, n)$ and the system has a solution, then the result of M\C is difficult to predict. It may return an answer with the warning "Rank deficient matrix," or it may return an answer full of NaNs and Infs with the warning "Matrix is singular."

But why are underdetermined systems of equations treated differently in Case 2 and Case 6? As the following examples of each of the cases demonstrate, if we give MATLAB the equation "$x + y = 2$," it happily returns the answer $x = 2, y = 0$; but if we give it the system of equations

$$x + y = 2,$$
$$x + y = 2,$$

it huffily tells us that the matrix is singular, and gives the answer $x = $ NaN, $y = $ NaN. The reason for this difference is that the single equation is *well-conditioned*—a small change in the constant or coefficients gives a correspondingly small change in the answer—whereas the system of two equations is *ill-conditioned*—a small change in the coefficients or constant may make the solution drastically different or may even make the system inconsistent. This is discussed further in Section 5.5.

The MATLAB function lu(M) returns a pair of matrices $[L, U]$. U is the row-echelon reduction of M. L is the matrix corresponding to the product of the operations used in reducing M to U. Thus, $L \cdot U = M$. The pair $[L, U]$ is known as the *LU factorization* of M. We use matrix A from Example 5.4 as our matrix, for illustration.

```
>> m=[2,0,1; -1,1,1; 0,1,2]

>> m
m =
     2     0     1
    -1     1     1
     0     1     2

>> inv(m)
ans =
     1     1    -1
     2     4    -3
    -1    -2     2

>> m^-1
ans =
     1     1    -1
     2     4    -3
    -1    -2     2

>> c=[2;5;7]
c =
     2
     5
     7

>> m\c
ans =
     0
     3
     2

% Cases of m\c:

% Case 1: Square, nonsingular. m=n=Rank(M)

>> [1,2;2,1]\[4;5]
ans =
     2
     1

% Case 2: m=Rank(M) < n.

>> [1,1,1]\[3]
ans =
     3
     0
     0
>> [1,1,1;1,2,3]\[3;6]
ans =
    1.5000
         0
    1.5000
```

```
% Case 3: Rank(M)=n < m: System has a solution.

>> [1,1;2,1;1,2]\[2;3;3]
ans =
    1.0000
    1.0000

% Case 4: Rank(M)=n < m: System has no solution, so the best least squares
%          fit is returned.

>> [1,1;2,1;1,2]\[1;1;1]
ans =
    0.3636
    0.3636

>> [1,1;1,1;2,1]\[1;2;2]
ans =
    0.5000
    1.0000

% Case 5: Rank(M) < min(m,n), and no solution exists.
% MATLAB gives a warning and returns an answer full of NaN's and Inf's

>> [1,1;1,1]\[1;2]
Warning: Matrix is singular to working precision.
ans =
   -Inf
    Inf

% Case 6: Rank(M) < min(m,n), and a solution exists.
% MATLAB will certainly give a warning; it may or may not find a solution.

>> [1,1;1,1]\[2;2]
Warning: Matrix is singular to working precision.
ans =
    NaN
    NaN

>> [1,1,1;1,1,1]\[1;1]
Warning: Rank deficient, rank = 1,  tol =    9.4206e-16.
ans =
    1.0000
         0
         0

% LU decomposition. The examples are the coefficient matrices from
% the systems at the start of the chapter. Note that in both cases the
% value of U is the row-echelon decomposition that we computed above.

>> a=[2,-4,2;2,2,1;1,1,-2]
a =
     2    -4     2
     2     2     1
     1     1    -2
```

```
>> [l,u]=lu(a)
l =
    1.0000         0         0
    1.0000    1.0000         0
    0.5000    0.5000    1.0000
u =
    2.0000   -4.0000    2.0000
         0    6.0000   -1.0000
         0         0   -2.5000

>> l*u
ans =
    2    -4     2
    2     2     1
    1     1    -2

>> b=[1,1,1,1;1,1,2,2;0,1,-1,1;1,0,2,0]
b =
    1    1    1    1
    1    1    2    2
    0    1   -1    1
    1    0    2    0

>> [l,u] = lu(b)
l =
    1    0    0    0
    1    0    1    0
    0    1    0    0
    1   -1    0    1
u =
    1    1    1    1
    0    1   -1    1
    0    0    1    1
    0    0    0    0

>> l*u
ans =
    1    1    1    1
    1    1    2    2
    0    1   -1    1
    1    0    2    0
```

5.5   Ill-Conditioned Matrices

The Gaussian elimination algorithm, although unpleasant to work through by
hand for any but very small systems, is easily understood, easily programmed,
and generally quite effective and efficient. It is one of the most frequently executed algorithms on computers. (The algorithm for computing the inverse is
substantially less important in practice.) The Gaussian elimination algorithm

has been known in Western mathematics since Gauss published it in 1809 and in Chinese mathematics since about A.D. 200 or earlier. However, in rare cases, it gets into serious trouble.

Without going deeply into this theory, we can illustrate the problem with a (contrived) example. Consider the following system of equations:

$$(1 + 10^{-9})x + y = 1, \tag{5.29}$$

$$x + (1 - 10^{-9})y = 1. \tag{5.30}$$

It is easily verified that the solution is $x = 10^9, y = -10^9$. Now let's see what MATLAB does.

```
>> format long
>> d=10^-9
d =
   1.0000e-09
>> a=[1+d, 1; 1, 1-d]
a =
   1.000000001000000    1.000000000000000
   1.000000000000000    0.999999999000000
>> a\[1;1]
Warning: Matrix is close to singular or badly scaled.
         Results may be inaccurate. RCOND = 2.775558e-17.
ans =
   1.0e+06 *
  -9.007198990992798
   9.007199999999997
>>% Note that, as warned, the answer is nowhere near right.
```

Why is MATLAB freaking out here? The answer appears when we apply Gaussian elimination. Reducing the system to row echelon form involves two steps, with $d = 10^{-9}$:

Step 1. Divide Equation (5.29) by $1 + d$, giving

$$x + \frac{1}{1+d} y = \frac{1}{1+d}, \tag{5.31}$$

$$x + (1 - d)y = 1. \tag{5.32}$$

Step 2. Subtract Equation (5.31) from Equation (5.32), giving the system of equations

$$x + \frac{1}{1+d} y = \frac{1}{1+d}, \tag{5.33}$$

$$\left(1 - d - \frac{1}{1+d}\right)y = 1 - \frac{1}{1+d}. \tag{5.34}$$

Now, note that $1 - d = 0.999999999$ (exactly) but that

$$\frac{1}{1+d} = 1 - d + d^2 + d^3 - \ldots = 0.999999999000000000999\ldots.$$

Therefore, if we compute the coefficient of y in Equation (5.34) in the natural way, by first computing the fraction $\frac{1}{1+d}$ and then subtracting this from $1 - d$, the calculation must be carried to 18 digits = 60 bits of precision, in order to detect that the coefficient is not actually 0. This is more than MATLAB ordinarily uses.

Another, related problem is that the solution of the system of equations is extremely sensitive to the exact value of the constant terms (and equally sensitive to the exact values of the coefficients, although we do not illustrate that here.) Suppose that the constant term in Equation (5.29) is changed by 1 part in a billion, from 1 to $1 + 10^{-9}$. Then the solution is obviously $x = 1, y = 0$. So changing one constant term from 1 to 1.000000001 changes the answer by 1,000,000,000!

In larger matrices, this sensitivity can turn up even with quite innocuous-looking coefficients:

```
>> m=[1:7,9; 2:8,10; 3:8,10:11; 4:8,10:12; 5:8,10:13; 6:8,10:14; 7:8,10:15 ...
      8,10:16]
m =
     1     2     3     4     5     6     7     9
     2     3     4     5     6     7     8    10
     3     4     5     6     7     8    10    11
     4     5     6     7     8    10    11    12
     5     6     7     8    10    11    12    13
     6     7     8    10    11    12    13    14
     7     8    10    11    12    13    14    15
     8    10    11    12    13    14    15    16

>> b=1+(m/1000)
b =
   1.001   1.002   1.003   1.004   1.005   1.006   1.007   1.009
   1.002   1.003   1.004   1.005   1.006   1.007   1.008   1.010
   1.003   1.004   1.005   1.006   1.007   1.008   1.010   1.011
   1.004   1.005   1.006   1.007   1.008   1.010   1.011   1.012
   1.005   1.006   1.007   1.008   1.010   1.011   1.012   1.013
   1.006   1.007   1.008   1.010   1.011   1.012   1.013   1.014
   1.007   1.008   1.010   1.011   1.012   1.013   1.014   1.015
   1.008   1.010   1.011   1.012   1.013   1.014   1.015   1.016

>> (b\[1;1;1;1;1;1;1;1])'
ans =
 -125.0000  0.0000  0.0000  0.0000  0.0000  0.0000  0.0000  125.0000

>> % Now we change one constant by 1 part in 1000 and ask the same question.
```

```
>> (b\[0.999;1;1;1;1;1;1;1])'

ans =
    4.6250  -1.0000  -1.0000  -1.0000  -1.0000  -1.0000  -1.0000  2.3750
```

Changing the first constant term from 1 to 0.999 changes the last component of the solution from 125 to 2.375. (We can get even more extreme results than this with the same matrix but different constant vectors. See Chapter 7, Exercise 7.1(b)).

Systems of equations of this kind, in which the answers are extremely sensitive to the exact values of the constant terms and the coefficients, and therefore calculations must be carried out to extreme precision, are said to be *ill-conditioned*. In principle, ill-conditioned systems of equations should be rare; a system of equations constructed by choosing coefficients at random have a very small probability of being ill-conditioned. Unfortunately, though, among problems that actually arise in practice, ill-conditioned problems are quite common.

Dealing with ill-conditioned systems of equations raises two different kinds of issues. The first is an algorithmic problem; it is important to structure an algorithm so as to avoid the buildup of roundoff error. Ill-conditioning greatly magnifies the inaccuracy due to roundoff error; in dealing with ill-conditioned matrices, it is therefore especially important to use *numerically stable* algorithms, which hold the roundoff error to a minimum. This is why the maximum element row interchange rule in row-echelon reduction is important (Section 5.2.2).

The second issue is a modeling problem. Suppose we are trying to solve a problem, and we formulate it as a system of linear equations. And suppose it turns out that we get Equations (5.29) and (5.30). Then, unless we are quite sure that the equations, the coefficients, and the constant terms are all accurate to within a factor of better than one part in a billion, we may as well throw the equations in the garbage. Even if we carry out the computation exactly and get an answer of $x = 10^9$, $y = -10^9$, we have no reason to suppose that the true values are not $x = 1$, $y = 0$.

There is, of course, no algorithmic fix to the latter problem; there is no way to get answers that are fundamentally better than your data. Nonetheless, it is important to know that when dealing with a ill-conditioned problem:

- Know not to trust the answer.

- It is probably a good idea to look for some alternative way to formulate the problem.

We discuss this issue further in Section 7.7. Trefethen and Bau (1997) offer an extensive discussion.

5.6   Computational Complexity

We conclude the chapter with a brief discussion of the computational complexity of the algorithms and problems discussed.[3] At the outset, there is a conceptual issue to address. There are actually three different ways to conceptualize numerical computations (e.g., solving systems of linear equations), each with its own drawbacks.

5.6.1   Viewpoints on Numerical Computation

In the first viewpoint, that of pure mathematics, the computation is idealized as exact arithmetic over real numbers. This is, in fact, the view taken throughout this book, and it is the usual starting point, although not the ending point, for any discussion of numerical algorithms. (If an algorithm doesn't make sense in this idealized view, it is unlikely to make sense in any other view.) Computation time is viewed as the number of arithmetic operations plus the number of other algorithmic operations, but most computations of this kind are in any case dominated by the arithmetic operations. Theorems 5.6–5.11 and Corollary 5.12 are all true in this viewpoint, and the running times discussed in Section 5.6.2 are all valid. The problem with this viewpoint is that the machine it runs on is imaginary; no actual computer can carry out arithmetic on real values to arbitrary precision in unit time per operation.

The second viewpoint, that of computation theory, is that the computation uses exact arithmetic over a class of numbers that is exactly representable in a finite data structure. For solving linear equations, the natural class of numbers would be the rational numbers; given any problem posed in terms of rational numbers, the exact answer and all the intermediate values calculated are also rational numbers. For more complex problems, such as those involving distance, it would be necessary to use an exact representation of algebraic numbers. Theorems 5.6–5.11 and Corollary 5.12 are valid in this viewpoint as well. Computation time is viewed in terms of bit operations; the time to add two integers is linear in the number of bits, and the time to multiply two integers is slightly more than linear.

However, for algorithms such as Gaussian elimination, exact arithmetic on rational numbers is entirely impractical because the number of bits required to maintain exactness increases exponentially with the size of the matrix. Likewise, then, time and space requirements increase exponentially. More sophisticated algorithms are known that are polynomial in the number of bits (Yap 2000), but these involve complex techniques in algorithmic design and analysis and are in any case impractical and not used. The computation time requirements given in Section 5.6.2 are *not* valid on this viewpoint.

---

[3]In this section, we assume familiarity with the basic concepts of algorithmic analysis.

	Pure math	Exact computation	Floating point
Realizable	×	√	√
Exact answers	√	√	×
Reasonable computation time	√	×	√

Table 5.2. Viewpoints on numerical computation.

The third viewpoint is that of numerical analysis. Here the computation is viewed as involving floating point operation using some particular precision. Computation time is given in terms of number of machine instructions; one floating point operation is one machine instruction. This is, in fact, what is *always* actually done in useful numerical programs. The time requirements given below are valid in this viewpoint.

However, floating point operations inevitably involve roundoff error, so Theorems 5.6–5.11 and Corollary 5.12 are no longer exactly true. Indeed, the whole standard notion of program correctness—that the answer obtained is always the exact answer to the question—goes out the window in this viewpoint. It is replaced by the notion of *numerical stability.* A floating-point algorithm $A(x)$ is a numerically stable algorithm for the mathematical function $f(x)$ if, for every x, there exists some $\bar{x}$ such that $\bar{x}$ is close to x (as compared to the precision of the floating-point system) and $A(\bar{x})$ is close to $f(\bar{x})$. In an implementation with roundoff, this is the best that one can hope for. The Gaussian elimination algorithm is numerically stable in all but a small number of pathological cases; with further improvements, it can be made numerically stable in all cases. Table 5.2 summarizes the pros and cons of the three viewpoints.

5.6.2   Running Times

We proceed now to the discussion of running times; these are valid in either the pure mathematical viewpoint or the floating-point viewpoint. If $\vec{u}$ and $\vec{v}$ are n-dimensional vectors and a is a scalar, then computing $a\vec{u}$, $\vec{v} + \vec{u}$, and $\vec{v} \bullet \vec{u}$ obviously requires time $O(n)$.

For a sparse vector $\vec{v}$, we can write $NZ(\vec{v})$ for the number of nonzero components of $\vec{v}$. Let $\vec{u}$ and $\vec{v}$ be sparse vectors and a be a scalar. Then, using the obvious algorithms, computing $a\vec{u}$ requires time $NZ(\vec{u})$. Computing $\vec{v} + \vec{u}$ and $\vec{v} \bullet \vec{u}$ requires time $NZ(\vec{v}) + NZ(\vec{u})$. As discussed in Section 2.7, the time requirement of indexing (that is, retrieving or setting $\vec{v}[i]$) depends on the details of the implementation; if a hash table is used, it requires expected constant time.

Multiplying an $m \times n$ matrix A times an n-dimensional vector $\vec{v}$ requires time $O(mn)$. If A and $\vec{v}$ are sparse, then computing $A\vec{v}$ requires time $O(NZ(A) + NZ(\vec{v}))$.

Let A be an $m \times p$ matrix and let B be a $p \times n$ matrix. Computing the product $A \cdot B$ in the obvious way—that is, computing the dot product of each row of A with each column of B—involves computing mn dot products of p-dimensional vectors; thus, time $O(mpn)$. If $m = p = n$ (multiplying two square matrices) the time requirement[4] is $O(n^3)$.

It should also be noted that the multiplication algorithm is highly parallelizable; in the extreme case, if you have n^3 parallel processors, you can compute the product of two $n \times n$ matrices in time $O(\log n)$.

Surprisingly, methods are actually known that are, in principle, faster for large matrices. As of the time of this writing, the fastest known algorithm (see Coppersmith & Winograd, 1990) for multiplying two $n \times n$ matrices requires time $O(n^{2.376})$. However, this algorithm and other algorithms that run in time faster than $O(n^3)$ are not used for two reasons. First, the constant factor associated with them is so large that they would pay off over the simple algorithm only when the matrices are too large to be effectively multiplied in any case. Second, they are numerically unstable.

A simple analysis of the algorithms presented in this chapter shows that, for n equations in n variables, SolveRowEchelon runs in time $O(n^2)$, and GaussianEliminate, SolveLinearEqns, and Inverse run in time $O(n^3)$. One can prove that, in principle, the faster of these matrix multiplication algorithms can be used to construct algorithms for solving linear equations and for inverting a matrix that runs in time $O(n^{2.376})$; these are not used in practice for the same reasons that the fast matrix multiplication algorithms are not used.

Exercises

Exercise 5.1. Solve the following systems of linear equations, both by hand and in MATLAB. (*Note:* In solving systems of equations by hand, there is no reason to use the "maximum element row interchange" step. If anything, one should swap with the row that has the *minimal* nonzero value in the column, in order to keep the denominators of the fractions involved as small as possible.)

(a)
$$x + y + z = 0,$$
$$2x - y - z = 2,$$
$$-x + y - z = 2.$$

---

[4]Note that there is a discrepancy here from the usual way of describing algorithmic time requirements. Usually, time requirements are described as a function of the size of the input to the algorithm; however, the size of the input in this problem is $O(n^2)$. If q is the size of the input, then the time requirement of this algorithm is $O(q^{3/2})$.

(b)
$$2x + y + 2z = 3,$$
$$x - y + 3z = -2,$$
$$x + 3y - 3z = 2.$$

(c)
$$w + x + y - z = 4,$$
$$2w + 2x - y = 4,$$
$$w - x - y - z = -2,$$
$$x + y + z = 4.$$

Exercise 5.2. The space of solutions to an underconstrained system of equations $M\vec{X} = \vec{C}$ can be characterized in an expression of the form

$$\{\vec{v} + a_1\vec{b}_1 + \ldots + a_k\vec{b}_k\},$$

where $\vec{v}$ is one solution to the system; $\vec{b}_1, \ldots, \vec{b}_k$ are a basis for Null(M); and $a_1, \ldots, a_k$ are scalar variables. For example, the space of solutions to the single equation $x + y + z = 3$ can be characterized as

$$\langle 1, 1, 1 \rangle + a \cdot \langle 1, -1, 0 \rangle + b \cdot \langle 1, 0, -1 \rangle.$$

By hand and by using MATLAB, characterize the space of solutions to the following systems of linear equations.

(a)
$$x + y - z = 1,$$
$$x + y + z = 3.$$

(b)
$$w + x + y + z = 5,$$
$$2w - x + y - z = 0.$$

(c)
$$w + x + y + z = 4,$$
$$2w - x + y - z = 0,$$
$$w - x - 2y + z = -4.$$

Programming Assignments

Assignment 5.1 (Temperature distribution). In Section 3.6.1, Application 3.8, we discuss finding steady-state temperature distribution inside an object by taking a collection of evenly spaced points, and asserting that the temperature at each point is the average of the temperature at the nearby points. It is assumed that the temperature at the border is fixed by some external constraint.

(a) Write a MATLAB function `TempDist(TL,TR,TT,TB,N)` that sets up and solves this system of equations. Assume that TL is the temperature on the left side, TR is the temperature on the right side, TT is the temperature on top, and TB is the temperature on the bottom. The function uses an $N \times N$ array of internal points. The function should return a $N \times N$ array T, where $T[I, J]$ is the temperature at point I, J (I measured from the left, J measured from the top).

For example, for the particular example presented in Application 3.8, the function call would be `TempDist(20,20,20,100,3)`, and the value returned would be

$$\begin{bmatrix} 25.71 & 27.86 & 25.71 \\ 35.00 & 40.00 & 35.00 \\ 54.29 & 62.14 & 54.29 \end{bmatrix}.$$

Note that the system of equations has N^2 equations in N^2 unknowns, and thus has a matrix of N^4 coefficients. Do not confuse the $N \times N$ array of temperatures with the $N^2 \times N^2$ array of coefficients.

(b) Use MATLAB's plotting routines to plot the temperature versus height over the two-dimensional region.

Assignment 5.2 (Curve interpolation). This assignment refers to Application 3.9 in Section 3.6.1.

(a) Using the technique described in the application, write a function `Poly-Interpolate(M)`. The input parameter M is a $2 \times n$ matrix, where each column holds the x- and y-coordinates of one point. You may assume that no two points have the same x-coordinate. The function should return the coefficients $a_{n-1}, \ldots, a_0$ of the $(n-1)$-degree polynomial $y = a_{n-1}x^{n-1} + \ldots + a_1 x_1 + a_0$. The function should also generate a plot of the curve with the input points, as shown in Figure 3.4.

For instance, for the example in Application 3.9, the function call `Poly-Interpolate([-3,-1,0,2,4; 1,0,5,0,1])` should return the vector `[0.2167, -0.4333, -2.7167, 2.9333, 5.0000]` and generate the solid curve of Figure 3.4.

(b) Write a function `SineInterpolate(M)` that interpolates a curve that is the sum of sinusoidal curves between input points. As in part (a), the input parameter M is a $2 \times n$, where each column holds the x- and y-coordinates of one point. You may assume that n is an even number. The function should return the vector of coefficients $a_0, a_1, \ldots, a_k, b_1, \ldots, b_{k-1}$ such that the function

$$y(x) = a_0 + a_1 \sin(\pi x/2) + a_2 \sin(\pi x/4) + \ldots$$
$$+ a_k \sin(\pi x/2k) + b_1 \cos(\pi x/2)$$
$$+ b_2 \cos(\pi x/4) + \ldots + b_{k-1} \cos(\pi x/2(k-1))$$

interpolates the points in M. It should also plot the curve $y(x)$.

For instance, for the example in Application 3.9, the function call `Sine-Interpolate([-3,-1,0,2,4; 1,0,5,0,1])` should return the vector `[0.0217, 1.6547, 0.9566, 2.3116, 2.6667]` and generate the dashed curve of Figure 3.4.

Assignment 5.3 (Circuit analysis). Write a MATLAB program `AnalyzeCircuit(C, R,V)` that carries out circuit analysis as described in Section 3.6.1, Application 3.7. The program should operate as follows. Let n be the number of nodes in the circuit, let b be the number of connections, and let $q = n + b$. Assign indices $1, \ldots, n$ to the nodes, where node 1 is ground (0 volts). Assign indices $1, \ldots, b$. Then the arguments are as follows:

- `C` is a $2 \times b$ array. `C[1,I]` and `C[2,I]` are the indices of the node at the tail and the head of branch `I`.

- `R` is a row vector of length b, where `R[I]` is the resistance of a resistor along branch `I`, and 0 if there is no resistor.

- `V` is a row vector of length b, where `V[I]` is a voltage source along branch `I`, and 0 if there is no voltage source.

You may assume that for every index `I` either `R[I]` or `V[I]` is nonzero, but not both.

In the example in Figure 3.2, with nodes A, B, C associated with indices $1, 2, 3$, respectively, and the branches numbered as in the diagram, the input parameters are

$$C = \begin{bmatrix} 1 & 2 & 2 & 3 \\ 2 & 1 & 3 & 1 \end{bmatrix}, \quad R = \begin{bmatrix} 0 & 100 & 75 & 25 \end{bmatrix}, \quad V = \begin{bmatrix} 100 & 0 & 0 & 0 \end{bmatrix}.$$

The program returns a vector $\vec{x}$ of dimension q, where

- for $i = 1, \ldots, n$, $\vec{x}[i]$ is the voltage at node i,

- for $i = 1, \ldots, q$, $\vec{x}[n+i]$ is the current through branch i.

Set up a $(q+1) \times q$ matrix M and a $(q+1)$-dimensional column vector c such that the system of linear equations $M\vec{x} = \vec{c}$ corresponds to the following constraints:

(a) *Ground.* The voltage at the ground is equal to 0.

(b) *Voltage source.* If $S[I, J] \neq 0$, then the voltage at J minus the voltage at I is equal $S[I, J]$.

(c) *Resistor.* If $R[I, J] \neq 0$, then the voltage at I minus the voltage at J is equal to $R[I, J]$ times the current from I to J.

(d) *Kirchoff's current law.* For each node, the sum of the currents on branches entering the node is equal to the sum of the currents on branches exiting the node. One of the equations is actually redundant (any one of them is redundant with respect to all the rest), but since the rank of the matrix is q, that doesn't do any harm.

Note that there is one equation of type (c); b equations of types (a) and (b) combined; and n equations of type (d), so in total there are $q + 1$ equations in q unknowns. Now solve the system $M \cdot \vec{x} = \vec{c}$.

Chapter 6

Geometry

Many kinds of geometric concepts can be related to vectors and linear algebra. Therefore, linear algebra is important for computer applications that have to negotiate with physical space, including graphics, image processing, computer-aided design and manufacturing, robotics, scientific computation, geographic information systems, and so on.

As we shall see, the fit between linear algebra and geometry is not perfect: central categories in linear algebra, such as the space of all linear transformations, are only moderately important geometrically; also, central categories in geometry, such as polygons, do not correspond to anything at all natural in linear algebra. But where the two theories mesh together well, they can be extremely synergistic.

6.1 Arrows

The connection between geometry and linear algebra involves a somewhat abstract geometric entity, which we will call an *arrow*.[1] An arrow has a specified length and direction, but not a specified position in the plane; it floats around without changing its length or rotating.[2] Here we use boldface letters, such as **p**, for points, and letters with thick arrows over them, such as $\vec{x}$, for arrows.

Points and arrows have four geometric operations defined on them (Figure 6.1). Let **p, q, r**, and **s** be points; let $\vec{a}$, $\vec{b}$, and $\vec{c}$ be arrows, and let t and w be numbers. Then

---

[1] An arrow is a kind of vector, under the general definition given in Section 4.3.1; but there does not seem to be standard term for this specific geometric category.

[2] The problem in explaining geometric vector theory is that there is no really good way to draw or visualize an arrow. We can draw all arrows with their tails at the origin, but then it is difficult to explain the difference between an arrow and the point at the head of the arrow. Or we can draw "floating" arrows, with their tails at some other points, but then it is difficult to explain that the arrow from, say, $\langle 1,3 \rangle$ to $\langle 2,5 \rangle$ is the same as the arrow from $\langle 3,2 \rangle$ to $\langle 4,4 \rangle$, since, pictorially, it is obviously a different thing. It is also difficult to visualize a geometric arrow as a physical arrow that can be moved around the plane because it is difficult to imagine why we aren't allowed to rotate it.

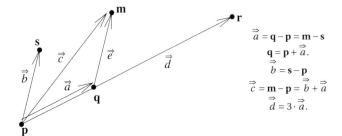

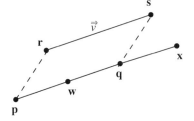

Figure 6.1. Operations on points and arrows.

Figure 6.2. The differences $\mathbf{q} - \mathbf{p}$, $\mathbf{s} - \mathbf{r}$, and $\mathbf{x} - \mathbf{w}$ are the same arrow $\vec{v}$.

- The difference $\mathbf{q} - \mathbf{p}$ is an arrow; namely, the arrow whose head is at $\mathbf{q}$ when its tail is at $\mathbf{p}$.

- The arrow $\mathbf{q} - \mathbf{p}$ is considered *the same* as the arrow $\mathbf{s} - \mathbf{r}$ if the ray from $\mathbf{r}$ to $\mathbf{s}$ and the ray from $\mathbf{p}$ to $\mathbf{q}$ have the same length and are parallel (Figure 6.2).

- The sum $\mathbf{p} + \vec{a}$ of a point plus an arrow is equal to the point $\mathbf{q}$ satisfying $\mathbf{q} - \mathbf{p} = \vec{a}$.

- The sum of two arrows $\vec{a} + \vec{b}$ is an arrow, equal to $((\mathbf{p} + \vec{a}) + \vec{b}) - \mathbf{p}$.

- The product $t \cdot \vec{a}$ is an arrow. Let $\vec{a} = \mathbf{q} - \mathbf{p}$. Then $t \cdot \vec{a} = \mathbf{r} - \mathbf{p}$, where $\mathbf{p}, \mathbf{q}$, and $\mathbf{r}$ are colinear; $d(\mathbf{p}, \mathbf{r}) = |t| \cdot d(\mathbf{p}, \mathbf{q})$ and if $t > 0$, then $\mathbf{r}$ and $\mathbf{q}$ are on the same side of $\mathbf{p}$; if $t < 0$, then $\mathbf{r}$ and $\mathbf{q}$ are on opposite sides of $\mathbf{p}$.

The following geometric rules establish that arrows are vectors, in the abstract sense (see Section 4.3.1):

$$\vec{a} + \vec{b} = \vec{b} + \vec{a}, \tag{6.1}$$

$$(\vec{a} + \vec{b}) + \vec{c} = \vec{a} + (\vec{b} + \vec{c}), \tag{6.2}$$

$$(\mathbf{p} + \vec{a}) + \vec{b} = \mathbf{p} + (\vec{a} + \vec{b}), \tag{6.3}$$

$$t \cdot (\vec{a} + \vec{b}) = t \cdot \vec{a} + t \cdot \vec{b}, \tag{6.4}$$

$$(t + w) \cdot \vec{a} = t \cdot \vec{a} + w \cdot \vec{a}, \tag{6.5}$$

$$(t \cdot w) \cdot \vec{a} = t \cdot (w \cdot \vec{a}). \tag{6.6}$$

A point $\mathbf{r}$ lies on the line $\mathbf{pq}$ if the arrows $\mathbf{q} - \mathbf{p} = t(\mathbf{r} - \mathbf{p})$ for some value of t. Therefore, the set of all points on the line connecting $\mathbf{p}$ and $\mathbf{q}$ is given by $\{\mathbf{p} + t \cdot (\mathbf{q} - \mathbf{p}) \mid t \in \mathbb{R}\}$.

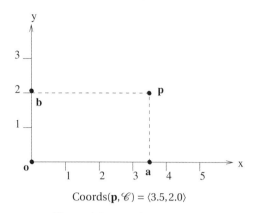

$$\text{Coords}(\mathbf{p}, \mathscr{C}) = \langle 3.5, 2.0 \rangle$$

Figure 6.3. Coordinate system.

6.2   Coordinate Systems

Suppose that we have a collection of points in the plane; think of these as relatively small, fixed, physical things, such as thumbtacks in a bulletin board or lampposts in a city. We want to record the position of these in a database or a gazeteer. The standard solution, of course, is to use coordinates in a coordinate system. We have already discussed this in Application 2.1 in Chapter 2, but it will be helpful to go over it again, very carefully this time.

A coordinate system for the plane consists of a point $\mathbf{o}$, the origin; a unit distance d_1 (e.g., a meter or a mile); and two orthogonal arrows $\vec{x}$ and $\vec{y}$ of length d_1 (Figure 6.3). The coordinate vector of an arrow $\vec{a}$ with respect to this coordinate system is the pair of numbers $\langle s, t \rangle$ such that $\vec{a} = s \cdot \vec{x} + t \cdot \vec{y}$. The coordinate vector of a point $\mathbf{p}$ with respect to this coordinate system is the pair of numbers $\langle s, t \rangle$ such that $\mathbf{p} = \mathbf{o} + s \cdot \vec{x} + t \cdot \vec{y}$. If $\mathscr{C}$ is a coordinate system, $\mathbf{p}$ is a point, and $\vec{v}$ is an arrow, then we write "Coords$(\mathbf{p}, \mathscr{C})$" and "Coords$(\vec{v}, \mathscr{C})$" to mean the coordinates of $\mathbf{p}$ and $\vec{v}$ in system $\mathscr{C}$.

The x-axis of the coordinate system is the line $\{\mathbf{o} + t\vec{x} \mid t \in \mathbb{R}\}$, and the y-axis is the line $\{\mathbf{o} + t\vec{y} \mid t \in \mathbb{R}\}$.

The coordinates of $\mathbf{p}$ can be measured by dropping perpendiculars from $\mathbf{p}$ to point $\mathbf{a}$ on the x-axis and to point $\mathbf{b}$ on the y-axis. The coordinate $\mathbf{p}[1]$ is then $\pm d(\mathbf{a}, \mathbf{o})/d_1$, with a positive sign if $\mathbf{a}$ and $\mathbf{o} + \vec{x}$ are on the same side of $\mathbf{o}$ and a negative sign if they are on opposite sides. Likewise, the coordinate $\mathbf{p}[2]$ is $\pm d(\mathbf{b}, \mathbf{o})/d_1$, with the corresponding rule for sign.

If L is any distance, then the measure of L in $\mathscr{C}$, Coords$(L, \mathscr{C}) = L/d_1$. In three dimensions, a coordinate system includes a third arrow $\vec{z}$, which is orthogonal to both $\vec{x}$ and $\vec{y}$. The coordinates of an arrow and coordinates of a point are defined and measured analogously.

It is then a geometric theorem, easily proven, that the operations on points and arrows correspond, in the obvious way, to the operations of the same name on their coordinate vectors; for example, $\text{Coords}(\mathbf{p} + \vec{v}, \mathscr{C}) = \text{Coords}(\mathbf{p}, \mathscr{C}) + \text{Coords}(\vec{v}, \mathscr{C})$.

I have belabored this at such length to emphasize the distinction between, on the one hand, the geometric domain of points and arrows and the geometric operations on them and, on the other hand, the numeric domain of coordinate vectors and the arithmetic operations on *them*, and to emphasize that the geometric operations do what they do on the geometric objects regardless of how we choose to represent them as numeric vectors. The following text uses the practice, convenient though imprecise, of conflating the geometric objects with their coordinate system representation; for example, I write $\vec{v} \bullet \vec{u}$ where what I really mean is $\text{Coords}(\vec{v}, \mathscr{C}) \bullet \text{Coords}(\vec{u}, \mathscr{C})$. I explicitly mention the coordinate system only in cases for which there is more than one coordinate system under consideration, particularly in Chapter 7.

6.3   Simple Geometric Calculations

The power of the coordinate system approach to geometry is that all kinds of geometric concepts and calculations can now be expressed in terms of the coordinates in the entities involved. In this book, we confine ourselves exclusively to those geometric concepts and operations that turn out to be closely related to concepts and operations of linear algebra.

6.3.1   Distance and Angle

Chapter 2 already discussed two geometric formulas that use the dot product:

- The length of arrow $\vec{v}$ is $\sqrt{(\vec{v} \bullet \vec{v})} = |\vec{v}|$. The distance between points $\mathbf{p}$ and $\mathbf{q}$ is $\sqrt{(\mathbf{p} - \mathbf{q}) \bullet (\mathbf{p} - \mathbf{q})} = |\mathbf{p} - \mathbf{q}|$. As examples, the length of arrow $\langle 1, 1 \rangle$ is $\sqrt{1^2 + 1^2} = \sqrt{2}$, and the distance from $\langle 0, 4, 3 \rangle$ to $\langle -1, 2, 4 \rangle$ is

$$\sqrt{(-1 - 0)^2 + (2 - 4)^2 + (4 - 3)^2} = \sqrt{6}.$$

- The angle ϕ between arrows $\vec{u}$ and $\vec{v}$ satisfies $\cos(\phi) = \vec{u} \bullet \vec{v} / |\vec{u}| \cdot |\vec{v}|$. In particular, if $\vec{v}$ is orthogonal to $\vec{u}$, then $\phi = \pi/2$, $\cos(\phi) = 0$, so $\vec{u} \bullet \vec{v} = 0$. For example, if $\vec{u} = \langle 1, 3 \rangle$ and $\vec{v} = \langle 4, 1 \rangle$, then $\cos(\phi) = \vec{u} \bullet \vec{v} / |\vec{u}| \cdot |\vec{v}| = 7/\sqrt{10}\sqrt{17} = 0.5369$, so $\phi = \cos^{-1}(0.5369) = 1.0041$ radians $= 57.53°$.

The set of points **p** that are at a distance r from a center point **c** is the circle, in two dimensions, or the sphere, in three dimensions, with radius r and center **c**. It can be represented by the equation $|\mathbf{p} - \mathbf{c}| = r$, or equivalently, $(\mathbf{p} - \mathbf{c}) \bullet (\mathbf{p} - \mathbf{c}) = r^2$. A point **p** is inside this circle if $(\mathbf{p} - \mathbf{c}) \bullet (\mathbf{p} - \mathbf{c}) < r^2$. It is outside if $(\mathbf{p} - \mathbf{c}) \bullet (\mathbf{p} - \mathbf{c}) > r^2$.

6.3.2 Direction

A *direction* in two-space or three-space can be taken to be a *unit* arrow; that is, an arrow of length d_0. These are often written with a hat over them, such as $\hat{u}$. The *direction* of nonzero arrow $\vec{u}$, denoted $\text{Dir}(\vec{u})$ or $\hat{u}$, is defined as $(d_0/|\vec{u}|) \cdot \vec{u}$; thus, in coordinate system $\mathscr{C}$ with unit length d_0, we have $\text{Coords}(\vec{u}, \mathscr{C}) = \text{Coords}(|\vec{u}|, \mathscr{C}) \cdot \text{Coords}(\hat{u}, \mathscr{C})$ (Figure 6.4).

Arrows $\vec{u}$ and $\vec{v}$ are *parallel* if $\vec{v} = c\vec{u}$ for some $c \neq 0$; equivalently $\text{Dir}(\vec{u}) = \pm\text{Dir}(\vec{v})$. For example, if $\vec{u} = \langle 3, 4 \rangle$, then, since $|\vec{u}| = 5$, we have $\hat{u} = \text{Dir}(\vec{u}) = \langle 3/5, 4/5 \rangle$. If $\vec{v} = \langle 1, 1 \rangle$ then, since $|\vec{v}| = \sqrt{2}$, we have $\hat{v} = \text{Dir}(\vec{v}) = \langle 1/\sqrt{2}, 1/\sqrt{2} \rangle$. The vector $\langle 9, 12 \rangle = 3\vec{u} = 15\hat{u}$ and the vector $\langle -6, -8 \rangle = -2\vec{u} = -10\hat{u}$ are parallel to $\vec{u}$.

If $\vec{v} = c\vec{u}$ then $\vec{v}$ and $\vec{u}$ are linearly dependent, so the matrix whose rows are $\vec{v}, \vec{u}$ has rank 1.

In two dimensions, let $\vec{v}$ be the arrow with coordinates $\langle V_x, V_y \rangle$. The vector that results from rotating $\vec{v}$ by $90°$ is $\langle -V_y, V_x \rangle$ for a counterclockwise rotation (assuming the usual layout of the x- and y-axes) and $\langle V_y, -V_x \rangle$ for a clockwise rotation. It is trivial to check that these have the same length as $\vec{v}$ and are orthogonal to $\vec{v}$.

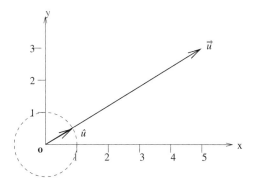

Figure 6.4. Arrows and their directions.

6.3.3   Lines in Two-Dimensional Space

As stated before, the line from **p** to **q** is the set of points $\{\mathbf{p} + t \cdot (\mathbf{q} - \mathbf{p}) \mid t \in \mathbb{R}\}$. In two dimensions, another way to characterize the line from **p** to **q** is as follows. Let $\vec{v} = \mathbf{q} - \mathbf{p}$, and let $\vec{w}$ be the arrow that is orthogonal to $\vec{v}$. Let **s** be any point in two-space; then there exist a and b such that $\mathbf{s} = \mathbf{p} + a\vec{v} + b\vec{w}$.

Point **s** is on the line **pq** if and only if $b = 0$. Note that

$$\mathbf{s} \bullet \vec{w} = (\mathbf{p} + a\vec{v} + b\vec{w}) \bullet \vec{w} = \mathbf{p} \bullet \vec{w} + a\vec{v} \bullet \vec{w} + b\vec{w} \bullet \vec{w} = \mathbf{p} \bullet \vec{w} + b\vec{w} \bullet \vec{w}.$$

So, if $b = 0$, then $\mathbf{s} \bullet \vec{w} = \mathbf{p} \bullet \vec{w}$ and conversely. Thus, a point **s** is on the line **p, q** if and only if it satisfies $\mathbf{s} \bullet \vec{w} = \mathbf{p} \bullet \vec{w}$, where $\vec{w}$ is a vector orthogonal to $\mathbf{q} - \mathbf{p}$. Note that $\mathbf{p} \bullet \vec{w}$ is a constant, independent of **s**. Note also that if x and y are variables representing the coordinates of **s**, then this is a single linear equation in x and y: $\vec{w}[1] \cdot x + \vec{w}[2] \cdot y = \mathbf{p} \bullet \vec{w}$.

The line $\mathbf{L} = \{\mathbf{s} \mid \mathbf{s} \bullet \vec{w} = c\}$ divides two-dimensional space into three regions: the line itself, points on one side, and points on the other side. A point **a** on the $\vec{w}$ side of **L** satisfies $\mathbf{a} \bullet \vec{w} > c$; a point **b** on the opposite side satisfies $\mathbf{b} \bullet \vec{w} < c$. The two sides of the line are called the *half-spaces* defined by the line. A *closed half-space* includes the line itself, whereas an *open half-space* excludes the line.

Note that we now have three ways to represent a line:

(a) as a pair of points **p, q**;

(b) in parameterized form $\{\mathbf{p} + t\vec{v} \mid t \in \mathbb{R}\}$, where $\vec{v} = \mathbf{q} - \mathbf{p}$;

(c) as the solutions to the linear equation $\{\mathbf{s} \mid \mathbf{s} \bullet \vec{w} = c\}$, where $\vec{w}$ is orthogonal to $\vec{v}$ and where c is the constant $\mathbf{p} \bullet \vec{w}$.

With each of these, there is a method to carry out a simple operation, such as checking whether a point **s** is on the line:

- **s** is on line **pq** if $\mathbf{s} - \mathbf{p}$ is parallel to $\mathbf{s} - \mathbf{q}$, which holds if the matrix with these two vectors as rows has rank 1.

- **s** is on the line $\{\mathbf{p} + t\vec{v} \mid t \in \mathbb{R}\}$ if the pair of linear equations

$$\vec{v}[1]\,t = \mathbf{s}[1] - \mathbf{p}[1],$$
$$\vec{v}[2]\,t = \mathbf{s}[2] - \mathbf{p}[2]$$

has a solution. (This is a pair of linear equations in the single unknown t.)

- **s** is on the line $\{\mathbf{s} \mid \mathbf{s} \bullet \vec{w} = c\}$ if $\mathbf{s} \bullet \vec{w} = c$.

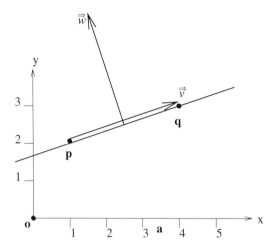

Figure 6.5. Representations of a line in two-space

Example 6.1. Taking Figure 6.5 for an example, we let **p** be the point $\langle 1,2 \rangle$ and **q** be the point $\langle 4,3 \rangle$. Then $\vec{v} = \mathbf{q} - \mathbf{p} = \langle 3,1 \rangle$. So the line **pq** can be expressed as the set $\{\mathbf{p} + t\vec{v} \mid t \in \mathbb{R}\} = \{\langle 1+3t, 2+t \rangle \mid t \in \mathbb{R}\}$. The arrow $\vec{w} = \langle -1,3 \rangle$ is orthogonal to $\vec{v}$. So the line **pq** can also be expressed as the set $\{\mathbf{s} \mid \mathbf{s} \bullet \vec{w} = \mathbf{p} \bullet \vec{w}\}$; that is $\{\langle x,y \rangle \mid -x + 3y = 5\}$.

We can check that the point $\mathbf{s} = \langle 7,4 \rangle$ is on the line by checking one of these operations:

(a) The matrix

$$\left[\begin{array}{c} \mathbf{s} - \mathbf{p} \\ \mathbf{s} - \mathbf{q} \end{array} \right] = \left[\begin{array}{cc} 6 & 2 \\ 3 & 1 \end{array} \right]$$

has rank 1.

(b) The pair of equations $\vec{v}\,t = \mathbf{s} - \mathbf{p}$; that is,

$$3t = (7-1),$$
$$t = (4-2)$$

has a solution (namely, $t = 2$).

(c) The dot product $\mathbf{s} \bullet \vec{w} = \mathbf{p} \bullet \vec{w}$. That is, $\langle 7,4 \rangle \bullet \langle -1,3 \rangle = \langle 1,2 \rangle \bullet \langle -1,3 \rangle = 5$.

Note that the point $\langle 2,4 \rangle$ is on the $\vec{w}$ side of **pq**; thus, $-1 \cdot 2 + 3 \cdot 4 = 10 > 5$. The point $\langle 2,2 \rangle$ is on the $-\vec{w}$ side of **pq**; thus, $-1 \cdot 2 + 3 \cdot 2 = 4 < 5$.

6.3.4   Lines and Planes in Three-Dimensional Space

The situation in three-dimensional space is similar to that in two dimensions, but it is a little more complicated. Some of the properties of lines in two-space carry over to lines in three-space, some properties carry over to planes in three-space, and some things are simply different.

Planes in three-space.   In three-dimensions, a point $\mathbf{s}$ is in the plane containing points $\mathbf{p}, \mathbf{q}$, and $\mathbf{r}$ if $\mathbf{s}-\mathbf{p}$ is a linear sum of $\mathbf{q}-\mathbf{p}$ and $\mathbf{r}-\mathbf{p}$. That is, let $\vec{u} = \mathbf{q}-\mathbf{p}$ and $\vec{v} = \mathbf{r}-\mathbf{p}$; then the plane is the set of points $\{\mathbf{p} + a \cdot \vec{u} + b \cdot \vec{v} \mid a, b \in \mathbb{R}\}$.

Again, there is an alternative representation using the dot product. Let $\vec{w}$ be a vector that is orthogonal to both $\vec{u}$ and $\vec{v}$. Let $\mathbf{s}$ be any point in three-space; then there exist a, b, and c such that $\mathbf{s} = \mathbf{p} + a\vec{u} + b\vec{v} + c\vec{w}$. Point $\mathbf{s}$ is in the plane $\mathbf{pqr}$ if and only if $c = 0$. But

$$\mathbf{s} \bullet \vec{w} = \mathbf{p} \bullet \vec{w} + a\vec{u} \bullet \vec{w} + b\vec{v} \bullet \vec{w} + c\vec{w} \bullet \vec{w} = \mathbf{p} \bullet \vec{w} + c\vec{w} \bullet \vec{w}.$$

So $c = 0$ if and only if $\mathbf{s} \bullet \vec{w} = \mathbf{p} \bullet \vec{w}$. Thus, the plane is the set of points $\{\mathbf{s} \mid \mathbf{s} \bullet \vec{w} = \mathbf{p} \bullet \vec{w}\}$. Note that this is a single linear equation in three variables: $\vec{w}[1] \cdot \mathbf{s}[1] + \vec{w}[2] \cdot \mathbf{s}[2] + \vec{w}[3] \cdot \mathbf{s}[3] = \vec{w} \cdot \mathbf{p}$. (The values of $\vec{w}$ and $\mathbf{p}$ are known; the coordinates of $\mathbf{s}$ are variables.)

This plane divides the rest of three-space into the two open half-spaces $\{\mathbf{s} \mid \mathbf{s} \bullet \vec{w} > \mathbf{p} \bullet \vec{w}\}$ and $\{\mathbf{s} \mid \mathbf{s} \bullet \vec{w} < \mathbf{p} \bullet \vec{w}\}$.

Finding a vector $\vec{w}$ perpendicular to $\vec{u}$ and $\vec{v}$ is trickier in three-space than in two-space. A simple formula for this is the *cross-product* $\vec{u} \times \vec{v}$, which is computed as follows. Let the coordinates of $\vec{u}$ and $\vec{v}$ be $\langle u_x, u_y, u_z \rangle$ and $\langle v_x, v_y, v_z \rangle$. Then the coordinates of the cross-product $\vec{w}$ are given by

$$\vec{w} = \vec{u} \times \vec{v} = \langle u_y v_z - u_z v_y, u_z v_x - u_x v_z, u_x v_y - u_y v_x \rangle.$$

It is straightforward to check that $\vec{w} \bullet \vec{u} = 0$ and $\vec{w} \bullet \vec{v} = 0$, and that if $\vec{u}$ and $\vec{v}$ are not parallel, then $\vec{u} \times \vec{v} \neq \vec{0}$.

As an example, let $\mathbf{p} = \langle 0, 1, 1 \rangle$, $\mathbf{q} = \langle 1, 2, 1 \rangle$, and $\mathbf{r} = \langle 2, 0, 0 \rangle$. Then $\vec{u} = \mathbf{q} - \mathbf{p} = \langle 1, 1, 0 \rangle$ and $\vec{v} = \mathbf{r} - \mathbf{p} = \langle 2, -1, -1 \rangle$. Therefore, any point of the form $\mathbf{p} + a\vec{u} + b\vec{v}$ is in the plane; for $a = 2, b = 2$, this is the point $\langle 6, 1, -1 \rangle$, and for $a = 0, b = -1$, this is $\langle -2, 2, 2 \rangle$. The cross-product $\vec{w} = \vec{u} \times \vec{v} = \langle -1, 1, -3 \rangle$. So a point $\mathbf{s}\langle x, y, z \rangle$ on the plane satisfies the equation $\mathbf{s} \bullet \vec{w} = \mathbf{p} \bullet \vec{w}$; that is, $-x + y - 3z = -2$. The open half-spaces are given by $-x + y - 3z > 2$ and $-x + y - 3z < 2$.

Another way to determine whether point $\mathbf{s}$ lies in the same plane as $\mathbf{p}, \mathbf{q}, \mathbf{r}$ is to observe that, if so, $\mathbf{s} - \mathbf{p}$, $\mathbf{s} - \mathbf{q}$, and $\mathbf{s} - \mathbf{r}$ are linearly dependent. Therefore, the matrix with these three vectors as rows has rank at most 2.

Lines in three-space. As in two-space, the line **pq** is the set of points {**p** + $t\vec{v} \mid t \in \mathbb{R}$} where $\vec{v} = \mathbf{q} - \mathbf{p}$. A line in three-space is the solution space of a *pair* of linear equations. Let $\vec{w}$ and $\vec{u}$ be two nonparallel arrows that are both orthogonal to $\vec{v}$. Let **s** be any point in three-space; then there exist scalars a, b, and c such that $\mathbf{s} = \mathbf{p} + a\vec{v} + b\vec{u} + c\vec{w}$. Now we have

$$\mathbf{s} \bullet \vec{u} = (\mathbf{p} + a\vec{v} + b\vec{u} + c\vec{w}) \bullet \vec{u} = \mathbf{p} \bullet \vec{u} + (b\vec{u} + c\vec{w}) \bullet \vec{u}. \tag{6.7}$$

Likewise,

$$\mathbf{s} \bullet \vec{w} = \mathbf{p} \bullet \vec{w} + (b\vec{u} + c\vec{w}) \bullet \vec{w}. \tag{6.8}$$

Clearly, if $b = c = 0$, then we have the pair of equations

$$\mathbf{s} \bullet \vec{u} = \mathbf{p} \bullet \vec{u} \tag{6.9}$$

and

$$\mathbf{s} \bullet \vec{w} = \mathbf{p} \bullet \vec{w}. \tag{6.10}$$

Less obviously, the converse also holds. We know that Equation (6.7) and (6.8) hold for all **s**. Suppose now that Equations (6.9) and (6.10) also hold for some particular **s**. Then, subtracting Equation (6.9) from Equation (6.7) and Equation (6.10) from Equation (6.8), we get the pair of equations $(b\vec{u} + c\vec{w}) \bullet \vec{u} = 0$ and $(b\vec{u} + c\vec{w}) \bullet \vec{w} = 0$. But since the arrow $b\vec{u} + c\vec{w}$ is in the same plane as $\vec{u}$ and $\vec{w}$, it can be orthogonal to both $\vec{u}$ and $\vec{w}$ only if $b\vec{u} + c\vec{w} = \vec{0}$; since $\vec{u}$ and $\vec{w}$ are nonparallel, that means that $b = c = 0$.

We have shown that point **s** is on line **pq** if and only $\mathbf{s} \bullet \vec{u} = \mathbf{p} \bullet \vec{u}$ and $\mathbf{s} \bullet \vec{w} = \mathbf{p} \bullet \vec{w}$. Let the coordinates of **s** be $\langle x, y, z \rangle$; then **s** is on the line from **p** to **q** only if x, y, z satisfy the *two* linear equations

$$\vec{u}[1]x + \vec{u}[2]y + \vec{u}[3]z = \vec{u} \bullet \mathbf{p},$$
$$\vec{w}[1]x + \vec{w}[2]y + \vec{w}[3]z = \vec{w} \bullet \mathbf{p}.$$

To find two nonparallel vectors $\vec{u}$ and $\vec{w}$ that are orthogonal to $\vec{v}$, we can proceed as follows. Let i be an index such that $\vec{v}[i] \neq 0$, and let j and k be the other two indices. Then we can define $\vec{u}$ so that $\vec{u}[i] = \vec{v}[j]$, $\vec{u}[j] = -\vec{v}[i]$, and $\vec{u}[k] = 0$; and define $\vec{w}$ so that $\vec{w}[i] = \vec{v}[k]$, $\vec{w}[k] = -\vec{v}[i]$, and $\vec{w}[j] = 0$.

In some cases, it is useful to require also that $\vec{u}$ and $\vec{w}$ be perpendicular to one another; in that case, define $\vec{u}$ as above and $\vec{w}$ as the cross-product $\vec{v} \times \vec{u}$.

As an example, let $\mathbf{p} = \langle 1, 1, 1 \rangle$ and let $\mathbf{q} = \langle 3, 4, 5 \rangle$. Then $\vec{v} = \mathbf{q} - \mathbf{p} = \langle 2, 3, 4 \rangle$. Choosing $i = 3, j = 1, k = 2$, we have $\vec{u} = \langle -4, 0, 2 \rangle$ and $\vec{w} = \langle 0, -4, 3 \rangle$. So the line **pq** is defined by the two equations $\mathbf{s} \bullet \vec{u} = \mathbf{p} \bullet \vec{u}$, or $-4x + 2z = -2$; and $\mathbf{s} \bullet \vec{w} = \mathbf{p} \bullet \vec{w}$, or $-4y + 3z = -1$.

Again, we have three formats for representing a line, and each format has a different associated method for checking whether the point is on the line.

- A line is determined by two points **pq**. Point **s** is on the line if the 2×3 matrix with rows $\mathbf{s} - \mathbf{p}$, $\mathbf{s} - \mathbf{q}$ has rank 1.

- A line is expressed in parameterized form $\{\mathbf{p} + t\vec{v} \mid t \in \mathbb{R}\}$, where $\vec{v} = \mathbf{q} - \mathbf{p}$. Point **s** is on the line if the system of three equations in one variable $\vec{v}t = \mathbf{s} - \mathbf{p}$ has a solution.

- A line is expressed as a pair of linear equations $\mathbf{s} \bullet \vec{u} = a$ and $\mathbf{s} \bullet \vec{w} = b$. Point **s** is on the line if it satisfies the two equations.

6.3.5   Identity, Incidence, Parallelism, and Intersection

Section 6.3.4 presented the following three kinds of representations for lines in two-space and lines and planes in three-space:

Representation in terms of points. A line is represented in terms of two points on the line; a plane is represented in terms of three noncolinear points in the plane.

Parameterized representation. A line is represented in the form $\mathbf{p} + t\vec{v}$. A plane is represented in the form $\mathbf{p} + a\vec{v} + b\vec{w}$, where $\vec{v}$ and $\vec{w}$ are nonparallel vectors parallel to the plane.

Linear equations. A line in two-space or a plane in three-space is represented in terms of a single linear equation $\mathbf{s} \bullet \vec{w} = \mathbf{p} \bullet \vec{w}$, where $\vec{w}$ is the normal to the plane. A line in three-space is represented in terms of a pair of linear equations $\mathbf{s} \bullet \vec{w} = \mathbf{p} \bullet \vec{w}$, $\mathbf{s} \bullet \vec{u} = \mathbf{p} \bullet \vec{u}$, where $\vec{w}$ and $\vec{u}$ are not parallel and are orthogonal to the line.

None of these representations is unique; that is, for any geometric object (line or plane), and any style of representations, there are multiple ways to represent the same object. For instance, the equations $x + 2y = 2$ and $2x + 4y = 4$ represent the same line in the plane, and similarly for the other representational styles. This is to some extent unavoidable; there is no method for representing lines or planes in terms of numerical parameters that is entirely non-problematic. For instance, lines in the plane are often represented in the form $y = mx + b$, which does give a unique representation for lines not parallel to the y-axis, but requires that lines that are parallel to the y-axis be given a different kind of representation $x = a$. (Also, this representation is unstable for lines that are nearly parallel to the y-axis.)

With each kind of these objects, and with each of these representations, a number of basic problems need to be solved:

Identity. Do two representations represent the same object?

Conversion. Convert the representation of an object in one format to another format.

Incidence. Does a point lie on a line or plane? Does a line lie in a plane?

Generation. Generate a point that lies on a given line or plane, or a line that lies in a given plane.

Parallelism. Are two objects parallel, or do they intersect, or (for two lines in three-space) are they skew?

Intersection. If two objects do intersect, find their intersection.

Section 6.3.4 discussed many of the problems of the incidence of a point on a line or plane. The remaining problems are likewise straightforward. In general, the methods involved fall into one of four categories, in increasing order of complexity, presented here with an example of each.

Trivial. If plane $P1$ is represented by equation $E1$ and plane $P2$ is represented by equation $E2$ then their intersection is the line represented by the pair of equations $\{E1, E2\}$.

Simple arithmetic. If line L is represented in terms of the coordinates of two points $\mathbf{p}$ and $\mathbf{q}$ that lie on L, then the conversion to parameterized form is $\mathbf{p} + t(\mathbf{q} - \mathbf{p})$.

Computing rank. Two planes characterized parametrically as $\mathbf{p} + a\vec{v} + b\vec{w}$ and $\mathbf{q} + c\vec{u} + d\vec{x}$ are parallel or identical if the matrix with rows $\vec{v}, \vec{w}, \vec{u}, \vec{x}$ has rank 2.

Solving systems of linear equations (one, two, or three equations in one, two, or three variables). Finding the intersection of two lines, or finding the intersection of a line and a plane, in any representation, involves solving a system of equations (assuming that the point answer is to be represented in terms of coordinates).

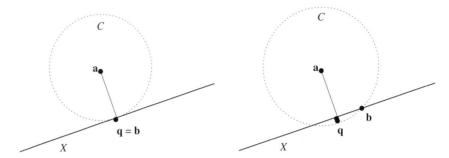

Figure 6.6. The line from **a** to Proj(**a**, X) is perpendicular to X.

6.3.6   Projections

The *projection* of a point **a** onto a plane or line X, denoted Proj(**a**, X), is the point **q** in X that is closest to **a**. The distance from **a** to X is, by definition, the distance from **a** to **q**. Point **q** is the only point on X such that the line from **a** to **q** is orthogonal to X. To prove this, let **b** be any point on **x**, and draw the circle C centered at **a** with radius **ab** (Figure 6.6). Then the following statements are equivalent:

- **b** =Proj(**a**, X).

- **b** is the closest point in X to **a**.

- **b** is on the boundary of C, and every other point on X is outside C.

- X is tangent to C at **b**.

- X is orthogonal to the radius of C, **ab**.

Suppose X is the line $\{\mathbf{p} + t\vec{v} \mid t \in \mathbb{R}\}$. Let $\hat{v} = \mathrm{Dir}(\vec{v})$. Then **q**= Proj(**a**, X) = $\mathbf{p} + ((\mathbf{a} - \mathbf{p}) \bullet \hat{v}) \cdot \hat{v}$. The proof of this is given by the equation

$$(\mathbf{a} - \mathbf{q}) \bullet \hat{v} = (\mathbf{a} - (\mathbf{p} + ((\mathbf{a} - \mathbf{p}) \bullet \hat{v}) \cdot \hat{v})) \bullet \hat{v} = (\mathbf{a} - \mathbf{p}) \bullet \hat{v} - (\mathbf{a} - \mathbf{p}) \bullet \hat{v} = 0,$$

so $\mathbf{a} - \mathbf{q}$ is orthogonal to X, and, by the above proof, **q**=Proj(**a**, X).

In three-space, suppose X is the plane $\{\mathbf{p} + s\vec{u} + t\vec{v} \mid s, t \in \mathbb{R}\}$, where $\vec{u}$ and $\vec{v}$ are orthogonal. Let $\hat{u} = \mathrm{Dir}(\vec{u})$ and $\hat{v} = \mathrm{Dir}(\vec{v})$. Then

$$\mathrm{Proj}(\mathbf{a}, X) = \mathbf{p} + ((\mathbf{a} - \mathbf{p}) \bullet \hat{v}) \cdot \hat{v} + ((\mathbf{a} - \mathbf{p}) \bullet \hat{u}) \cdot \hat{u}.$$

The proof is essentially identical to the two-space proof.

6.4   Geometric Transformations

One of the most important applications of linear algebra to geometry is the use of linear transformation to represent some of the ways in which objects or images can be moved around in space. Transformations have many computer science applications, such as

- in two-dimensional graphics, moving an image in the picture plane,

- in computations involving solid objects, such as robotics or CAD/CAM, calculating the motion of a solid object through space,

- in graphics or computer vision, determining how an object moving through space appears in an image.

What linear algebra gives us is a language for characterizing the relation of one particular position of an object to another particular position. Such a relation is called a *geometric transformation*. We are not concerned here with describing the continuous motion of an object over time, in which it moves through a whole range of positions; that motion requires calculus or real analysis, and is beyond the scope of this book.

Geometrical transformations fall into categories such as translations, rotations, rigid motions, and so on. The categories are organized in a hierarchy; for example, the category of rigid motions includes the category of translations.

The categories of geometric transformations that we present here are (naturally) those that can be expressed by simple operations on coordinate vectors, particularly linear transformations. With many kinds of geometric transformations, doing this elegantly requires introducing a new method of representing points as numeric vectors. So far we have used the *natural* representation, in which a geometric point in two-space is represented by a two-dimensional vector, and a point in three-space is represented by a three-dimensional vector. In the new representation, called the *homogeneous* representation, a point in two-space is represented by a three-dimensional vector, and a point in three-space is represented by a four-dimensional vector. The homogeneous representation is introduced in Section 6.4.3.

An important—indeed, a defining—characteristic of each category of geometric transformations is the class of *invariants* of the category, meaning geometric features or relations that are left unchanged by transformations in the category. For instance, a transformation Γ is a translation if and only if point subtraction $\mathbf{q}-\mathbf{p}$ is an invariant; that is, for all points $\mathbf{p}$ and $\mathbf{q}$, $\Gamma(\mathbf{q})-\Gamma(\mathbf{p}) = \mathbf{q}-\mathbf{p}$. A transformation Γ is in the class of rigid mappings with reflection if distance is an invariant; that is, if $d(\Gamma(\mathbf{p}),\Gamma(\mathbf{q})) = d(\mathbf{p},\mathbf{q})$ for all points $\mathbf{p},\mathbf{q}$.

Figure 6.7 shows the hierarchy of categories of transformations that we consider here.

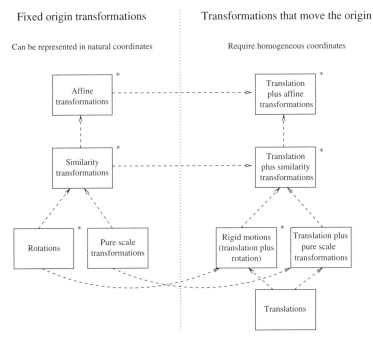

Figure 6.7. Hierarchy of categories of transformations. A dashed arrow from any category C to any category D means that D includes C. Any category C with an asterisk is actually two categories: C1, the transformations in C excluding reflections; and C2, all transformations in C, including reflections. Category C2 includes C1.

6.4.1   Translations

For any arrow $\vec{v}$, *translation* by $\vec{v}$ is the function that moves every point in parallel by $\vec{v}$; that is, $\Gamma(\mathbf{p}) = \mathbf{p} + \vec{v}$ for every point $\mathbf{p}$. Thus, a translated figure is moved in the two-space or three-space changing only position, and keeping the shape and the orientation constant (Figure 6.8).

The fundamental invariant of a translation is point subtraction; if Γ is a translation, then $\Gamma(\mathbf{q}) - \Gamma(\mathbf{p}) = \mathbf{q} - \mathbf{p}$ for all points $\mathbf{p}, \mathbf{q}$. It follows, then, that distances, angles, and orientations are all invariants of a translation.

Translation is a simple but important category in computer graphics; moving a window or an image in a display is (usually) a pure translation. In natural coordinates, the coordinates of the translation of point $\mathbf{p}$ by vector $\vec{v}$ are just the sum of the coordinates of $\mathbf{p}$ plus the coordinates of $\vec{v}$. Translations are easily composed and inverted. If $\mathbf{p}$ is translated first by $\vec{v}$ and then by $\vec{u}$, the result is the same as a single translation by $\vec{v} + \vec{u}$. The inverse of translating $\mathbf{p}$ by $\vec{v}$ is translating it by $-\vec{v}$.

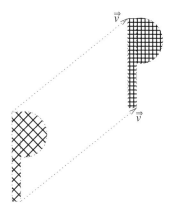

Figure 6.8. Translation.

6.4.2   Rotation around the Origin

In two-space, a rotation around the origin is carried out by drawing the figure on a piece of paper and sticking a thumbtack through the paper into a fixed board at the origin, and then turning the paper while leaving the origin fixed. In three-space, a rotation around the origin is carried out by taking a solid object, marking one point on the object as the origin, and then twisting the object in space while keeping the origin fixed in place. Mechanically, this can be done by attaching the object to a fixed frame by using a ball joint, where the center of the joint is at the origin.

Rotations around the origin in two-space. The two-dimensional case is easier to visualize and illustrate than the three-space. Rotations can be characterized in terms of coordinate vectors as follows. For a paper rotated around a thumbtack attached to a board at the origin, we want to know how points drawn on the paper move. We measure this motion by using a fixed coordinate system drawn on the board. Put the paper in its starting position. On the paper, draw the unit x and y arrows, $\vec{x}$ and $\vec{y}$, placing their tails at the origin. Draw a dot at any point $\mathbf{p}$ with coordinates $\langle a, b \rangle$. Thus $\mathbf{p} = \mathbf{o} + a\vec{x} + b\vec{y}$. Now rotate the paper. Let $\Gamma(\vec{x})$, $\Gamma(\vec{y})$, and $\Gamma(\mathbf{p})$ be the new positions of the arrows and the dots. Clearly, the relations on the paper among $\mathbf{o}, \mathbf{p}, \vec{x}$, and $\vec{y}$ have not changed, so we have $\Gamma(\mathbf{p}) = \mathbf{o} + a\Gamma(\vec{x}) + b\Gamma(\vec{y})$. Let $\mathscr{C}$ be the fixed coordinate system attached to the board, and let $\vec{p} = \text{Coords}(\mathbf{p}, \mathscr{C})$, $\vec{p}\,' = \text{Coords}(\Gamma(\mathbf{p}), \mathscr{C})$, $\vec{x}\,' = \text{Coords}(\Gamma(\mathbf{x}), \mathscr{C})$, and $\vec{y}\,' = \text{Coords}(\Gamma(\mathbf{y}), \mathscr{C})$. Then we have $\vec{p}\,' = a\vec{x}\,' + b\vec{y}\,'$.

Now let R be a matrix whose columns are $\vec{x}\,'$ and $\vec{y}\,'$. Then we can write this equation as $\vec{p}\,' = R\vec{p}$. In other words, the result of the rotation Γ corresponds to matrix multiplication by the matrix R.

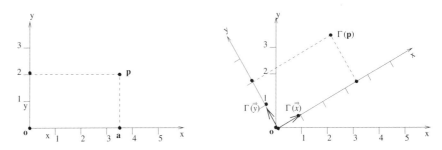

Figure 6.9. Rotation around the origin.

For example, Figure 6.9 shows the case for $\vec{p} = \langle 3.5, 2 \rangle$ and Γ is a rotation by 30°. Then $\Gamma(\vec{x}) = \langle \cos(30°), \sin(30°) \rangle = \langle .866, 0.5 \rangle$ and $\Gamma(\vec{y}) = \langle -\sin(30°), \cos(30°) \rangle = \langle -0.5, 8.66 \rangle$ so

$$\Gamma(\mathbf{p}) = \begin{bmatrix} 0.866 & -0.5 \\ 0.5 & 0.866 \end{bmatrix} \cdot \begin{bmatrix} 3.5 \\ 2 \end{bmatrix} = \begin{bmatrix} 2.03 \\ 3.48 \end{bmatrix}.$$

What can we say about the matrix R? One thing we can do is to calculate it. If Γ is a positive rotation by angle θ, then, by trigonometry, the coordinates of $\Gamma(\vec{x})$ are $\langle \cos(\theta), \sin(\theta) \rangle$ and the coordinates of $\Gamma(\vec{y})$ are $\langle -\sin(\theta), \cos(\theta) \rangle$, so R is the matrix

$$R = \begin{bmatrix} \cos(\theta) & -\sin(\theta) \\ \sin(\theta) & \cos(\theta) \end{bmatrix}.$$

But that is trigonometry, not linear algebra (hence the nonlinear functions). From the standpoint of linear algebra, we can observe that, since we are not folding or bending the paper, the rotated arrows $\Gamma(\vec{x})$ and $\Gamma(\vec{y})$ still have length 1 and are still at right angles. Since the coordinates of these two arrows are the columns of R, that means that the columns of R have the following elegant properties:

1. $R[:, i] \bullet R[:, i] = 1$.

2. If $i \neq j$, then $R[:, i] \bullet R[:, j] = 0$.

Thus, R is an orthonormal matrix, so $R^T \cdot R = I$.

It is easy to show that, for any orthonormal matrix R and vectors $\vec{v}$ and $\vec{u}$, the length of $\vec{v}$ and the angle between $\vec{v}$ and $\vec{u}$ is invariant under multiplication by R by showing that the dot product is invariant, as follows:

$$(R\vec{u}) \bullet (R\vec{v}) = (R\vec{u})^T R\vec{v} = \vec{u}^T R^T R\vec{u} = \vec{u}^T \vec{v} = \vec{u} \bullet \vec{v}.$$

We have shown that any rotation around the origin corresponds to multiplication by an orthonormal matrix. Is the converse true? Unfortunately, not

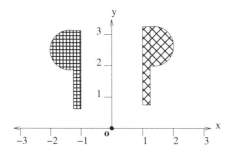

Figure 6.10. Reflection.

quite. Consider the matrix

$$R = \begin{bmatrix} -1 & 0 \\ 0 & 1 \end{bmatrix};$$

then the product

$$R \cdot \begin{bmatrix} x \\ y \end{bmatrix} = \begin{bmatrix} -x \\ y \end{bmatrix}.$$

The effect of multiplication by R on a P shape is shown in Figure 6.10. The left-hand P is not a rotation of the right-hand P; there is no way of turning the paper to make the right-hand P line up with the left-hand P. The only way is to hold the paper up to a mirror, or to draw the right-hand P on thin paper and look at it through the paper from the other side. In short, the left-hand P is a *reflection* of the right-hand P; in this case, the reflection is across the y-axis.

A reflection across a line through the origin also corresponds to an orthonormal matrix. Conversely, if R is an orthonormal matrix, then multiplication by R carries out either a rotation around the origin or a reflection across a line through the origin.

Note that a reflection also changes the sign of the angles between directions. For instance, the signed angle from $\vec{x}$ to $\vec{y}$ is 90° counterclockwise. The reflection matrix R maps $\vec{x}$ to $-\vec{x}$ and maps $\vec{y}$ to itself, and the signed angle from $-\vec{x}$ to $\vec{y}$ is −90° counterclockwise. However, since the dot product just gives the cosine of the angle, and $\cos(-\theta) = \cos(\theta)$, there is no way to detect this from the dot product.

So how can we tell a matrix that carries out a reflection from one for a rotation? The solution is to use the *determinant*.[3] The determinant of a 2×2 matrix

---

[3]Determinants are, in general, not extremely important for the kinds of applications we are considering in this book, so we are treating them at much less length than most linear algebra texts do.

is given by the formula,

$$\text{Det}\left(\left[\begin{array}{cc} a & b \\ c & d \end{array}\right]\right) = ad - bc.$$

For any orthonormal matrix R, either $\text{Det}(R) = 1$, in which case R is a rotation, or $\text{Det}(R) = -1$, in which case R is a reflection. We discuss determinants further in Section 6.4.7.

The invariants associated with rotations and reflections around the origin are

- the position of the origin,

- the distance between any two points,

- the unsigned angle between any two directions.

If we exclude reflections, then the invariants associated purely with rotations include also the signed angle between any two directions.

Rotations around the origin in three-space The linear algebra associated with rotations around the origin in three-space is almost the same as in two-space, but the trigonometry is considerably more difficult.

The argument that proves that a rotation around the origin corresponds to multiplication by an orthonormal matrix works in three dimensions in exactly the same way as in two dimensions. All we need to do is to add a third coordinate vector $\vec{z}$. In particular, we let Γ be any rotation around the origin in three-space, let $\mathbf{p}$ be any point, and let $\mathscr{C}$ be a coordinate system with origin $\mathbf{o}$ and unit direction arrows $\vec{x}, \vec{y}, \vec{z}$. We then let $\vec{p} = \text{Coords}(\mathbf{p}, \mathscr{C})$, $\vec{p}\,' = \text{Coords}(\Gamma(\mathbf{p}), \mathscr{C})$, $\vec{x}\,' = \text{Coords}(\Gamma(\mathbf{x}), \mathscr{C})$, $\vec{y}\,' = \text{Coords}(\Gamma(\mathbf{y}), \mathscr{C})$, $\vec{z}\,' = \text{Coords}(\Gamma(\mathbf{z}), \mathscr{C})$; and let R be the 3×3 matrix whose columns are $\vec{x}\,', \vec{y}\,'$, and $\vec{z}\,'$. Then

- $\vec{p}\,' = R \cdot \vec{p}$,

- R is an orthonormal matrix; that is, $R^T \cdot R = I$.

Conversely, if R is an orthonormal matrix, then multiplication by R corresponds either to a rotation around the origin or to reflection across a plane through the origin. In three-space, a reflection transforms a left-hand glove into a right-hand glove and vice versa, or a left-hand screw into a right-hand screw; luckily for the manufacturers of gloves, and unluckily for those of us who wear and lose gloves, there is no way to achieve this by using rotation. An orthonormal matrix R is rotation without reflection if its determinant is 1; it is a reflection if its determinant is -1. The determinant of a 3×3 matrix is defined in Section 6.4.7.

The trigonometry of three-space rotations is much more complicated than the trigonometry of two-space rotations. Just as a rotation around the origin in two-space can be represented in terms of a single angle θ, a rotation around the origin in three-space can be represented in terms of three angles, θ, ϕ, and ψ. However, the following differences introduce some complications.

- Whereas two-space has only one reasonable way to represent a rotation as an angle, three-space has many different ways; in fact, several different ways are used in practice.

- Any method of representing three-space rotations in terms of three angles necessarily suffers from "topological singularities," for which the representation becomes severely awkward.

- In two-space, the same angle representation can be used for both directions and rotations. In three-space, directions are represented by *two* angles (e.g., latitude and longitude), whereas rotations are represented by *three* angles.

- In two-space, the composition of a rotation by θ followed by a rotation by ϕ is just a rotation by $(\theta + \phi) \bmod 2\pi$. No representation of three-space rotations in terms of angles has so simple a rule for composition; in fact, to compute a composition, it is generally easiest to convert to matrix notation, do matrix multiplication, and convert back.

We discuss one particular simple case of three-dimensional rotations in Section 7.1.3.

6.4.3   Rigid Motions and the Homogeneous Representation

At this point, we know how to calculate the results of translating a figure and of rotating a figure around the origin. But suppose that we want to move an object freely around space, combining rotations and translations as the spirit moves us. Or suppose that we want carry out a rotation around some point that is not the origin? Such a transformation is called a *rigid motion*. A rigid motion preserves distances and angles. Again, there is a distinction between rigid transformations that are not reflections, which actually can be physically carried out by turning the object around, and those that are reflections, which cannot be achieved by any kind of turning.

One way to do combined translations and rotations is just to carry out a sequence of vector additions and matrix multiplications on the natural coordinates. We can also perform rotations around points other than the origin in this way. Figure 6.11 shows an example of this. Let **c** and **d** be the points with coordinates $\langle 1,1 \rangle$ and $\langle 3,1 \rangle$, and suppose we want to compute the location of

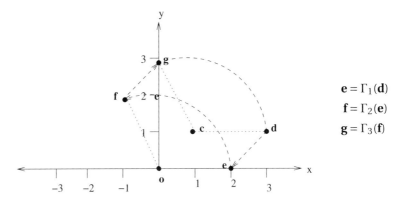

Figure 6.11. Rotation around a point that is not the origin.

d after performing a rotation of $120°$ around **c**. This can be done as a sequence of three transformations:

- Γ_1 is a translation by $\mathbf{o} - \mathbf{c}$, computed by adding the vector $\langle -1, -1 \rangle$. Thus, $\Gamma_1(\mathbf{c}) = \langle 0, 0 \rangle$; $\Gamma_1(\mathbf{d}) = \langle 2, 0 \rangle$.

- Γ_2 is a rotation around **o** by $120°$, computed by multiplying by the matrix

$$\begin{bmatrix} -0.5 & 0.866 \\ -0.866 & -0.5 \end{bmatrix}.$$

  Thus, $\Gamma_2(\Gamma_1(\mathbf{c})) = \langle 0, 0 \rangle$; $\Gamma_2(\Gamma_1(\mathbf{d})) = \langle -1.0, -1.732 \rangle$.

- Γ_3 is a translation by $\mathbf{c} - \mathbf{o}$, computed by adding the vector $\langle 1, 1 \rangle$. Thus, $\Gamma_3(\Gamma_2(\Gamma_1(\mathbf{c}))) = \langle 1, 1 \rangle$; $\Gamma_3(\Gamma_2(\Gamma_1(\mathbf{d}))) = \langle 0, -0.732 \rangle$.

The composition $\Gamma_3 \circ \Gamma_2 \circ \Gamma_1$ is the desired transformation.

If we are doing a one-time calculation, this may actually be the easiest approach, but it becomes awkward if we want to do many such calculations, particularly if we want to compose rigid motions.

There is really no way around this awkwardness using natural coordinates. Obviously, rotation does not correspond to any kind of vector addition. Almost equally obviously, translation does not correspond to any kind of matrix multiplication because translation moves all points and matrix multiplication leaves the zero vector unchanged.

The solution is to use a different coordinate representation, called *homogeneous coordinates*.[4] In homogeneous coordinates, as in natural coordinates, a

---

[4]Strictly speaking, the representation we discuss here is known as *normalized homogeneous coordinates*. There is a more general version of homogeneous coordinates, in which a point with natural coordinates $\langle a, b \rangle$ is represented by a vector $\langle ra, rb, r \rangle$ for any $r \neq 0$.

coordinate system is still defined in terms of an origin $\mathbf{o}$ and two or three co-ordinate arrows, $\vec{x}, \vec{y}$ in two-space, plus $\vec{z}$ in three-space. However, a point in two-space is represented by a three-dimensional numeric vector; the first two coordinates are the natural coordinates, and the third coordinate is always 1. Likewise, a point in three-space is represented by a four-dimensional numeric vector; the first three coordinates are the natural coordinates, and the fourth coordinate is always 1. Arrows in two-space and three-space are likewise represented by using three- and four-dimensional vectors; the first two (or three) coordinates are the natural coordinates, and the last coordinate is 0.

We denote homogeneous coordinates of point $\mathbf{p}$ or arrow $\vec{v}$ with respect to coordinate system $\mathscr{C}$ as $\mathrm{Hc}(\mathbf{p}, \mathscr{C})$ and $\mathrm{Hc}(\vec{v}, \mathscr{C})$, respectively. For example, in Figure 6.11, $\mathrm{Hc}(\mathbf{c}, \mathscr{C}) = \langle 1, 1, 1 \rangle$; $\mathrm{Hc}(\mathbf{d}, \mathscr{C}) = \langle 3, 1, 1 \rangle$; and $\mathrm{Hc}(\mathbf{d} - \mathbf{c}, \mathscr{C}) = \langle 2, 0, 0 \rangle$. We often omit the argument $\mathscr{C}$ in cases where the coordinate system is fixed.

The basic arithmetical operations on points and arrows in terms of coordinates still work as before; $\mathrm{Hc}(\mathbf{p} + \vec{v}) = \mathrm{Hc}(\mathbf{p}) + \mathrm{Hc}(\vec{v})$ and so on. Likewise, the length of $\vec{v}$ is equal to $|\mathrm{Hc}(\vec{v})|$, and the angle θ between $\vec{u}$ and $\vec{v}$ satisfies

$$\cos(\theta) = \frac{\mathrm{Hc}(\vec{u}) \bullet \mathrm{Hc}(\vec{v})}{|\mathrm{Hc}(\vec{u})| \cdot \mathrm{Hc}(\vec{v})|}.$$

We can visualize this as follows, at least in the case of the three-dimensional representation of two-space.[5] We represent geometric two-space as the plane $z = 1$ in three-dimensional vector space (Figure 6.12). The homogeneous coordinates of a point $\mathbf{p}$ relative to the plane are the natural coordinates of the vector $\mathbf{p} - \mathbf{o}_3$ where $\mathbf{o}_3$ is the origin of the embedding three-space. A two-space arrow $\vec{v}$ is represented by the corresponding point in the plane $z = 0$. (Note also that this eliminates the confusion, discussed earlier, between visualizing points and visualizing arrows.)

The advantage of using homogeneous coordinates is that now both rotation and translation correspond to multiplying the coordinate vector by a transformation matrix, so the composition of two transformations is just the matrix product of the corresponding transformation matrices. Specifically, a translation by arrow $\vec{v}$, which has natural coordinates $\vec{v}$, corresponds to multiplication of the homogeneous coordinates by the matrix

$$\left[\begin{array}{c|c} I & \vec{v} \\ \hline \vec{0}^T & 1 \end{array} \right].$$

---

[5]This also works in principle for the four-dimensional representation of three-space, but that is more difficult to visualize.

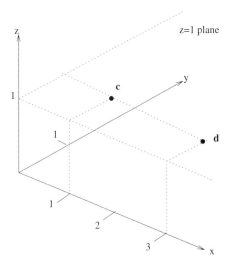

Figure 6.12. Homogeneous coordinates.

If rotation Γ around the origin is represented in natural coordinates as matrix R, then it is represented in homogeneous coordinates by the matrix

$$\left[\begin{array}{c|c} R & \vec{0} \\ \hline \vec{0}^T & 1 \end{array}\right].$$

This notation, with horizontal and vertical lines, describes the division of the matrix into subrectangles. For instance,

$$\text{if } \vec{v} = \left[\begin{array}{c} 5 \\ 6 \end{array}\right], \text{ then } \left[\begin{array}{c|c} I & \vec{v} \\ \hline \vec{0}^T & 1 \end{array}\right] \text{ denotes the matrix } \left[\begin{array}{ccc} 1 & 0 & 5 \\ 0 & 1 & 6 \\ 0 & 0 & 1 \end{array}\right].$$

Applying this notation to the former example of rotating point $\mathbf{d}$ around point $\mathbf{c}$, again we consider the overall transformation Γ as the composition of the three transformations $\Gamma_3 \circ \Gamma_2 \circ \Gamma_1$, but since all these transformations are now matrix multiplication, their composition corresponds to the product of the matrices. Specifically,

$$\Gamma_1 = \left[\begin{array}{ccc} 1 & 0 & -1 \\ 0 & 1 & -1 \\ 0 & 0 & 1 \end{array}\right], \quad \Gamma_2 = \left[\begin{array}{ccc} -0.5 & -0.866 & 0 \\ 0.866 & -0.5 & 0 \\ 0 & 0 & 1 \end{array}\right], \quad \Gamma_3 = \left[\begin{array}{ccc} 1 & 0 & 1 \\ 0 & 1 & 1 \\ 0 & 0 & 1 \end{array}\right],$$

so

$$\Gamma = \Gamma_3 \cdot \Gamma_2 \cdot \Gamma_1 = \left[\begin{array}{ccc} -0.5 & -0.866 & 0.6340 \\ 0.866 & -0.5 & 2.3660 \\ 0 & 0 & 1 \end{array}\right] \text{ and } \Gamma \cdot \mathbf{d} = \left[\begin{array}{c} -1 \\ -1.732 \\ 1 \end{array}\right].$$

Figure 6.13. Rigid motion.

In general, for any rigid transformation Γ, the corresponding matrix in homogeneous coordinates has the form

$$\left[\begin{array}{c|c} R & \vec{v} \\ \hline \vec{0}^T & 1 \end{array} \right],$$

where R is an orthonormal matrix. Since this matrix is equal to

$$\left[\begin{array}{c|c} I & \vec{v} \\ \hline \vec{0}^T & 1 \end{array} \right] \cdot \left[\begin{array}{c|c} R & \vec{0} \\ \hline \vec{0}^T & 1 \end{array} \right],$$

it expresses Γ as the composition of the rotation around the origin described by R in natural coordinates, followed by a translation by $\vec{v}$.

Conversely, if R is an orthonormal matrix and $\vec{v}$ is a vector, then the matrix

$$A = \left[\begin{array}{c|c} R & \vec{v} \\ \hline \vec{0}^T & 1 \end{array} \right]$$

corresponds to a rigid motion (Figure 6.13) in homogeneous coordinates. A is not a reflection if $\mathrm{Det}(R) = 1$, and A is a reflection if $\mathrm{Det}(R) = -1$.

The invariants of a rigid motion are

- the distance between two points and the length of an arrow,

- the angle between two arrows.

6.4.4   Similarity Transformations

A *similarity* or *scale* transformation expands or contracts everything by a constant factor in all directions (Figure 6.14). This transformation is important in

Figure 6.14. Scale transformation.

imaging applications; in graphics, it is often necessary to expand or contract an image without otherwise distorting it; in vision, the image of an object expands or contracts as the distance from the eye to the object increases or decreases. It is not very important in physical applications since physical objects rarely expand and contract uniformly.

The simplest form of scale transformation is the change of scale without rotation, reflection, or translation. This corresponds to a simple scalar multiplication of natural coordinates:

$$\text{Coords}(\Gamma(\mathbf{p}), \mathscr{C}) = c \cdot \text{Coords}(\mathbf{p}, \mathscr{C}),$$

where $c > 0$.

In homogeneous coordinates, this corresponds to multiplication by the matrix

$$\left[\begin{array}{c|c} c \cdot I & \vec{0} \\ \hline \vec{0}^T & 1 \end{array} \right], \text{ where } c > 0.$$

For example, if $\mathbf{p}$ is the point $\langle 3, 5 \rangle$ and Γ is a scale expansion by a factor of 2, then the natural coordinates of $\Gamma(\mathbf{p})$ are $2 \cdot \langle 3, 5 \rangle = \langle 6, 10 \rangle$. The homogeneous coordinates of $\Gamma(\mathbf{p})$ are

$$\left[\begin{array}{ccc} 2 & 0 & 0 \\ 0 & 2 & 0 \\ 0 & 0 & 1 \end{array} \right] \cdot \left[\begin{array}{c} 3 \\ 5 \\ 1 \end{array} \right] = \left[\begin{array}{c} 6 \\ 10 \\ 1 \end{array} \right].$$

The invariants of a pure scale transformation of this kind are the position of the origin, directions of arrows, angles between arrows, and ratios between distances (that is, for any points $\mathbf{p}, \mathbf{q}, \mathbf{r}, \mathbf{s}, d(\mathbf{p}, \mathbf{q})/d(\mathbf{r}, \mathbf{s})$ is unchanged).

The most general form of scale transformation combines it with a rotation, reflection, and/or translation. In homogeneous coordinates, this corresponds to multiplication by a matrix of the form

$$
\left[
\begin{array}{c|c}
c \cdot R & \vec{v} \\
\hline
\vec{0}^T & 1
\end{array}
\right],
$$

where $c \neq 0$ and R is a orthonormal matrix.

Conversely, a matrix M corresponds to a general scale transformation if the following conditions hold. Let A be the upper left-hand corner of such a matrix (that is, all but the last row and column). Then

- the last row of M has the form $\langle 0, ..., 0, 1 \rangle$,

- $A^T A$ is a diagonal matrix with the same value (c^2) all along the main diagonal, and 0 elsewhere.

The transformation is not a reflection if $\text{Det}(A) > 0$. It is a reflection if $\text{Det}(A) < 0$.

The invariants of a general scale transformation are the angles between arrows and the ratios of distances.

6.4.5 Affine Transformations

For the final class of transformations, it is easiest to go in the opposite direction, from matrices to geometry. Let M be any matrix of the form

$$
M = \left[
\begin{array}{c|c}
A & \vec{v} \\
\hline
\vec{0}^T & 1
\end{array}
\right].
$$

Then multiplication by M transforms one vector of homogeneous coordinates (i.e., the vector with final component 1) to another. What is the geometric significance of this operation?

To answer this question, it is easiest to consider the case where $\vec{v} = \vec{0}$, so that the origin remains fixed. In this case, the transformation can also be viewed as matrix multiplication of the natural coordinates of a point by the matrix A; that is,

$$
\text{Coords}(\Gamma(\mathbf{p}), \mathscr{C}) = A \cdot \text{Coords}(\mathbf{p}, \mathscr{C}).
$$

In two-space, let $\vec{x}$ and $\vec{y}$ be the x and y coordinate arrows. Let $\mathbf{p}$ be a point with coordinates $\langle a, b \rangle$. Then we have

$$
\text{Coords}(\Gamma(\vec{x}), \mathscr{C}) = A \begin{bmatrix} 1 \\ 0 \end{bmatrix} = A[:, 1] \text{ and } \text{Coords}(\Gamma(\vec{y}), \mathscr{C}) = A \begin{bmatrix} 0 \\ 1 \end{bmatrix} = A[:, 2],
$$

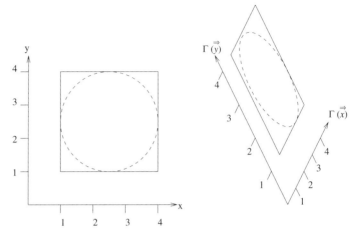

Figure 6.15. Affine transformation.

so

$$\text{Coords}(\Gamma(\mathbf{p}), \mathscr{C}) = A \begin{bmatrix} a \\ b \end{bmatrix} = aA[:,1] + bA[:,2]$$

$$= a \cdot \text{Coords}(\Gamma(\vec{x}), \mathscr{C}) + b \cdot \text{Coords}(\Gamma(\vec{y}), \mathscr{C}).$$

In other words, what Γ does is to map the coordinate directions $\vec{x}$ and $\vec{y}$ to two other vectors $\Gamma(\vec{x}), \Gamma(\vec{y})$, and then to map any point with coordinates $\langle a, b \rangle$ to the point $a\Gamma(\vec{x}) + b\Gamma(\vec{y})$. The result is in general a rotation and a skewing of the figure. For example, Figure 6.15 shows the affine transformation corresponding to multiplication by the transformation

$$M = \begin{bmatrix} 1/4 & -1/2 \\ 1/2 & 1 \end{bmatrix}.$$

Here, $\Gamma(\vec{x}) = \langle 1/4, 1/2 \rangle$ and $\Gamma(\vec{y}) = \langle -1/2, 1 \rangle$.

If $\Gamma(\vec{x}), \Gamma(\vec{y})$ are not parallel, then Rank(A)=2, and Γ is a bijection of the plane to itself. If $\Gamma(\vec{x})$ and $\Gamma(\vec{y})$ are parallel, then Rank(A)=1, and any point $\mathbf{p}$ is mapped onto a point on the line $\{\mathbf{o} + t\Gamma(\vec{x}) \mid t \in \mathbb{R}\}$; that is, the plane is collapsed down into a line. If $\Gamma(\vec{x}) = \Gamma(\vec{y}) = \vec{0}$, then Rank($A$)=0 and Γ collapses the plane down to a point. A transformation that maps the plane to a line or a point is said to be *degenerate*; a transformation that maps the plane to the plane is *invertable*.

The general affine transformation corresponds to multiplication of homogeneous coordinates by a matrix M of the form

$$\left[\begin{array}{c|c} A & \vec{v} \\ \hline \vec{0}^T & 1 \end{array}\right].$$

This is the composition of an affine transformation A around the origin followed by a translation of $\vec{v}$.

Affine transformations in three-space are similar. Let $\vec{x}, \vec{y}, \vec{z}$ be the coordinate directions and let point $\mathbf{p} = a\vec{x} + b\vec{y} + c\vec{z}$. Then an affine transformation maps $\mathbf{p}$ to the point $\Gamma(\mathbf{p}) = \vec{v} + a\Gamma(\vec{x}) + b\Gamma(\vec{y}) + c\Gamma(\vec{z})$, where $\vec{v}$ is the translation part of the mapping, independent of $\mathbf{p}$. If $\text{Rank}(A)=3$, then Γ maps three-space to itself; that is, Γ is invertable. If $\text{Rank}(A)=2$, 1, or 0, then Γ collapses three-space to a plane, a line, or a point, respectively; that is, Γ is degenerate.

The case of $\Gamma(\vec{x}) = a\vec{x}, \Gamma(\vec{y}) = b\vec{y}$ corresponds to a change in the aspect ratio, an operation available in most image editors. Otherwise, the primary application for geometric affine transformation is in images of objects rotating in three-space, as described in Section 6.4.6.

The fundamental invariant of an affine transformation Γ is identity of point subtraction; that is, if $\mathbf{b} - \mathbf{a} = \mathbf{d} - \mathbf{c}$, then $\Gamma(\mathbf{b}) - \Gamma(\mathbf{a}) = \Gamma(\mathbf{d}) - \Gamma(\mathbf{c})$. It follows from this that Γ is defined over arrows and that addition of arrows to one another and to points and scalar multiplication of arrows are likewise invariants; that is,

$$\Gamma(\mathbf{a} + \vec{u}) = \Gamma(\mathbf{a}) + \Gamma(\vec{u}),$$

$$\Gamma(\vec{u} + \vec{v}) = \Gamma(\vec{u}) + \Gamma(\vec{v}),$$

$$\Gamma(c \cdot \vec{u}) = c \cdot \Gamma(\vec{u}).$$

6.4.6 Image of a Distant Object

One application of linear transformations is to describe the changing image of a distant object moving in space.

In image formation, there is a lens with focal point $\mathbf{f}$ that projects an image onto a image plane $\mathbf{I}$ (Figure 6.16). Any point in space $\mathbf{p}$ is projected onto a point $\Gamma(\mathbf{p})$, which is the intersection of the line $\mathbf{pf}$ with the plane $\mathbf{I}$. To describe this, we will use two coordinate systems. Let $\mathbf{o}$ be the projection of $\mathbf{f}$ onto $\mathbf{I}$. In the coordinate system $\mathscr{C}$ for three-space, the origin will be the focus $\mathbf{f}$, the unit distance will be $d(\mathbf{f}, \mathbf{o})$, the z-axis $\vec{z} = \mathbf{f} - \mathbf{o}$ (note that this is perpendicular to $\mathbf{I}$) and the other two axes $\vec{x}$ and $\vec{y}$ will be two other orthogonal vectors. Thus, in the coordinate system $\mathscr{C}$, $\mathbf{I}$ is the plane $z = -1$. In the coordinate system $\mathscr{B}$ for $\mathbf{I}$, the origin $\mathbf{o}$ is the projection of $\mathbf{f}$ onto $\mathbf{I}$, and the two coordinate axes are $-\vec{x}$ and $-\vec{y}$. Then, for any point $\mathbf{p}$ in three space, if $\text{Coords}(\mathbf{p}, \mathscr{C}) = \langle x, y, z \rangle$, then

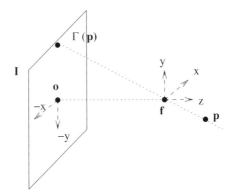

Figure 6.16. Projection onto an image plane.

Coords$(\Gamma(\mathbf{p}), \mathscr{B}) = \langle x/z, y/z \rangle$. Note that for any fixed z, this is linear in x and y but it is not linear in z.

Now, suppose we have a planar piece of lucite $\mathbf{P}$ that we can move around in space, and we have drawn a region Q on $\mathbf{P}$. We are interested in how the image of Q in $\mathbf{I}$ changes as we move $\mathbf{P}$. In some important cases, this is a linear transformation or nearly so.

Case 1. (Rigid motion.) Suppose that $\mathbf{P}$ is parallel to $\mathbf{I}$ and is kept fixed. The piece of lucite is first rotated by rotation R within $\mathbf{P}$ and then moved by translation $\vec{v}$ within $\mathbf{P}$.

(This may seem counterintuitive; you might suppose that if we move $\mathbf{P}$ very far up or down, the image of Q becomes small and foreshortened. That, however, is because we are thinking of the case where we turn our eye or head to keep Q in sight. If we are projecting onto a plane $\mathbf{I}$, which is kept constant, this doesn't happen because the projection is likewise far from the point $\mathbf{o}$. This is only an adequate model of ocular vision in the case where $\mathbf{P}$ remains quite close to a fixed line of sight.)

Case 2. (Scale transformation.) Suppose $\mathbf{P}$ is kept parallel to $\mathbf{I}$, $\mathbf{P}$ is translated by arrow $\vec{v}$, which is parallel to $\vec{z}$; and the z-coordinate of P moves from z_0 to $z_1 = z_0 + \vec{v} \bullet \vec{z}$. Then a point $\mathbf{p}$ in $\mathbf{P}$ moves from coordinates $\langle x, y, z_0 \rangle$ to coordinates $\langle x, y, z_1 \rangle$, so $\Gamma(\mathbf{p})$ moves from $\langle x/z_0, y/z_0 \rangle$ to $\langle x/z_1, y/z_1 \rangle$. That is, the image undergoes a scale transformation around the origin by a factor of z_0/z_1.

Case 3. (Affine transformation.) Let $\mathbf{P}$ be at any plane in space. Let $\mathbf{q}$ be a point in the region Q, and let $\vec{u}$ and $\vec{v}$ be two orthogonal unit arrows parallel to $\mathbf{P}$. Let z_0 be the z-coordinate of some point in Q, and assume that the diameter of Q is small as compared to z_0. Let $\mathbf{p}$ be a point in region Q whose coordinates in $\mathscr{D}$ are $\langle a, b \rangle$; thus, $\mathbf{p} = \mathbf{q} + a\vec{u} + b\vec{v}$, where a and b are small compared to z_0. The coordinate vector of $\Gamma(\mathbf{p})$ in $\mathscr{B}$ is then

$$\text{Coords}(\Gamma(\mathbf{p}), \mathscr{B}) = \left\langle \frac{\vec{q}[x] + a\vec{u}[x] + b\vec{v}[x]}{\vec{p}[z]}, \frac{\vec{q}[y] + a\vec{u}[x] + b\vec{v}[x]}{\vec{p}[z]} \right\rangle.$$

However, since

$$\vec{q}[z] - (a + b) \le \vec{p}[z] \le \vec{q}[z] + a + b,$$
$$\vec{q}[z] - (a + b) = z_0(1 - (a + b)/z_0),$$
$$\vec{q}[z] + (a + b) = z_0(1 + (a + b)/z_0),$$

we have $1/\vec{p}[z] \approx 1/z_0$. So

$$\text{Coords}(\Gamma(\mathbf{p}), \mathscr{B}) \approx \left\langle \frac{\vec{q}[x]}{z_0} + \frac{a\vec{u}[x] + b\vec{v}[x]}{z_0}, \frac{\vec{q}[y] + a\vec{u}[x] + b\vec{v}[x]}{z_0} \right\rangle$$
$$= \begin{bmatrix} \vec{q}[x]/z_0 \\ \vec{q}[y]/z_0 \end{bmatrix} + \begin{bmatrix} \vec{u}[x]/z_0 & \vec{v}[x]/z_0 \\ \vec{u}[y]/z_0 & \vec{v}[y]/z_0 \end{bmatrix} \cdot \begin{bmatrix} a \\ b \end{bmatrix}.$$

Thus, the restriction of Γ to the region Q is approximately an affine transformation. It is easy to show that if $\mathbf{P}$ is not perpendicular to $\mathbf{I}$, then there is an inverse Δ mapping $\Gamma(Q)$ to Q, which is also an affine transformation.

Now, suppose that Q has image $\Gamma(Q)$, and Q undergoes a rigid transformation Φ, still satisfying the condition that the distance from Q to $\mathbf{I}$ is much greater than the diameter of Q. Then the image of the new position of Q is $\Gamma(\Phi(Q))$. Let W be the first image and Y be the second image; then we have $Y = \Gamma(\Phi(\Delta(W)))$. But then W and Y are related by the composition $\Gamma \circ \Phi \circ \Delta$; since all these are approximately affine transformations, their composition is also approximately an affine transformation. In short, two images of a moving planar figure are related by a transformation that is approximately an affine transformation, as long as the distance from the figure to the image plane is always much greater than the diameter of the figure.

Reflections in this scenario correspond to flipping the plane $\mathbf{P}$ around so that the eye is now looking at the back rather than the front. (This is why we imagine the figure as being drawn on lucite.)

But suppose we want to move a planar figure around in space freely, without restrictive assumptions, and we want an exact description of how the image changes, not an approximation. The transformations involved then are *projective* transformations. Projective geometry is in many ways more interesting than affine geometry; however, as the transformation is nonlinear, it is beyond the scope of this book.

6.4.7 Determinants

We have defined the *determinant* of a 2×2 matrix, and we alluded to the determinants of other matrices. The determinant function is very important in more advanced study of linear algebra; however, in the kinds of applications discussed in this book, it is not critical and rarely computed. This section briefly defines the determinant of an $n \times n$ matrix and enumerates some of its properties without proof or discussion.

To define the determinant of an $n \times n$ matrix, we need the notion of the *minor* to an element of a matrix. If A is an $n \times n$ matrix and i and j are indices between 1 and n, then the minor to $A[i, j]$, denoted $C^{i,j}$, is the $(n-1) \times (n-1)$ matrix consisting of all the rows of A except the ith and all the columns of A except the jth.

We can now define the determinant recursively as follows:

- If A is the 1×1 matrix $[a]$ then $\mathrm{Det}(A) = a$.

- If A is an $n \times n$ matrix, with $n > 1$, and i is any index between 1 and n, then

$$\mathrm{Det}(A) = \sum_{j=1}^{n} (-1)^{i+j} A[i, j] \cdot \mathrm{Det}(C^{i,j}).$$

  That is, we go across the ith row, multiply $A[i, j]$ by the determinant of its minor, and add these products up, alternating signs at each step. The answer at the end is the same whatever row we choose.

  Alternatively, we can carry out the same operation by going down the jth column:

$$\mathrm{Det}(A) = \sum_{i=1}^{n} (-1)^{i+j} A[i, j] \cdot \mathrm{Det}(C^{i,j}).$$

  Again, we get the same answer, whichever column we choose.

As an example, let

$$A = \begin{bmatrix} 1 & 2 & 3 \\ 4 & 0 & -1 \\ -3 & -2 & -4 \end{bmatrix}.$$

Then, multiplying across the first row,

$$\mathrm{Det}(A) = 1 \cdot \mathrm{Det}\left(\begin{bmatrix} 0 & -1 \\ -2 & -4 \end{bmatrix}\right) - 2 \cdot \mathrm{Det}\left(\begin{bmatrix} 4 & -1 \\ -3 & -4 \end{bmatrix}\right) + 3 \cdot \mathrm{Det}\left(\begin{bmatrix} 4 & 0 \\ -3 & -2 \end{bmatrix}\right)$$

$$= 1 \cdot ((0 \cdot -4) - (-2 \cdot -1)) - 2 \cdot ((4 \cdot -4) - (-1 \cdot -3)) + 3 \cdot ((4 \cdot -2) - (0 \cdot -3))$$

$$= (1 \cdot -2) - (2 \cdot -19) + (3 \cdot -8) = 12,$$

or multiplying down the second column,

$$\mathrm{Det}(A) = -2 \cdot \mathrm{Det}\left(\begin{bmatrix} 4 & -1 \\ -3 & -4 \end{bmatrix}\right) + 0 \cdot \mathrm{Det}\left(\begin{bmatrix} 1 & 3 \\ -3 & -4 \end{bmatrix}\right) - (-2) \cdot \mathrm{Det}\left(\begin{bmatrix} 1 & 3 \\ 4 & -1 \end{bmatrix}\right)$$

$$= -2 \cdot ((4 \cdot -4) - (-3 \cdot -1)) + 0 - 2 \cdot ((1 \cdot -1) - (3 \cdot 4))$$

$$= (-2 \cdot -19) + (2 \cdot -13) = 12.$$

This formula is elegant, but it leads to an $O(n!)$ time algorithm. A more efficient algorithm uses row-echelon reduction. To compute $\mathrm{Det}(A)$ more efficiently, carry out the row-echelon reduction of A (Section 5.2), adding the following steps:

1. At the start, set variable $D \leftarrow 1$.

2. Whenever you divide a row by a constant, $A[i,:] \leftarrow A[i,:]/c$, set $D \leftarrow c \cdot D$.

3. Whenever you swap two rows, set $D \leftarrow -D$.

4. Return D times the product of the elements in the main diagonal.

Note that D is not changed when the row-echelon operation $A[j,:] \leftarrow A[j,:] - cA[i,:]$ is executed.

The following properties of the determinant should be noted, where A is an $n \times n$ matrix, and Γ is the transformation corresponding to multiplication by A.

- A is a singular matrix if and only if $\mathrm{Det}(A) = 0$.

- If $\mathrm{Det}(A) < 0$, then Γ is a reflection. If $\mathrm{Det}(A) > 0$, then Γ is not a reflection.

- If $n = 2$ and R is a region in two-space, then $\mathrm{area}(\Gamma(R)) = |\mathrm{Det}(A)| \cdot \mathrm{area}(R)$. If $n \geq 3$ and R is a region in three-space, then $\mathrm{volume}(\Gamma(R)) = |\mathrm{Det}(A)| \cdot \mathrm{volume}(R)$. The corresponding formula holds for $n > 3$, where "volume" is interpreted as n-dimensional volume. These formulas hold whether natural or homogeneous coordinates are used.

- $$\mathrm{Det}(A^T) = \mathrm{Det}(A).$$
$$\mathrm{Det}(A^{-1}) = 1/\mathrm{Det}(A).$$
$$\mathrm{Det}(A \cdot B) = \mathrm{Det}(A) \cdot \mathrm{Det}(B).$$

The MATLAB function for computing the determinant of matrix M is, not surprisingly, det(m).

```
>> det ([1,2,3;  4,0,-1;  -3,-2,-4])
ans  =
    12
```

6.4.8   Coordinate Transformation on Image Arrays

If A is an image array so that $A[i, j]$ is the grayscale at the pixel at location i, j, then it is often desirable to apply the geometric transformation being discussed to the image. The same formulas apply, of course, but they have to be applied to the *indices* i, j rather than to the values. (Changes to the values give image operations such as brightening, changing color, or heightening contrast.)

Suppose we have an $m \times n$ image array A of square pixels. Let us consider the coordinate system with the origin is at the top left-hand corner, unit length is the size of a pixel, x-coordinate is horizontal, and y-coordinate is vertically down (top to bottom because that is the way MATLAB prints matrices). The image consists of a nonwhite figure against a white ground. We wish to apply to the figure a linear transformation, whose matrix in homogeneous coordinates is M, filling in any gaps with white, and cropping any part of the figure that lies outside the canvas. To a very crude first approximation, the following MATLAB code will work:

```
function B = TransformImage(M,A);
  white = 255; % gray level for white.
  [n,m] = size(A);
  for i = 1:n
    for j = 1:m
      B(i,j) = white;
    end
  end
  for i = 1:n
    for j = 1:m
      v = floor(M*[i;j;1])
      if (1 <= v(1) & v(1) <= n & 1 <= v(2) & v(2) <= m & A(i,j) ~= white)
        B(v(1),v(2)) = A(i,j);
      end
    end
  end
end
```

The problem with this treatment, however, is that except in very specific cases, a pixel in B does not correspond to a single pixel in A; it overlaps with the image of several pixels in A. The only cases for which there is a one-to-one correspondence are when M is a translation by a pair of integers, M is a rotation by a multiple of $90°$, M is a reflection around the coordinate axes, or M is a composition of these. In any other case, the result of this simple algorithm will probably look seriously wrong. The fixes to this are beyond the scope of this book; see, for example Foley, van Dam, Feiner, & Hughes (1990, Section 17.4).

Exercises

Use MATLAB as needed.

Exercise 6.1. Represent the plane containing the points $\langle 2,0,1 \rangle$, $\langle -1,1,1 \rangle$, $\langle 1,1,0 \rangle$ in the form $\{\vec{p} \mid \vec{p} \cdot \vec{w} = c\}$.

Exercise 6.2. Find the projection from the point $\langle 3,3,3 \rangle$ on the plane in Exercise 6.1 and find the distance from the point to the plane.

Exercise 6.3. Find an orthonormal basis for the plane in Exercise 6.1.

Exercise 6.4. Find the intersection of the plane in Exercise 6.1 with the line $\langle 1,1,0 \rangle + t \cdot \langle 2,1,1 \rangle$.

Exercise 6.5. The intersection of the plane in Exercise 6.1 with the plane $x + y + 2z = 4$ is a line. Characterize this line in the form $\mathbf{p} + t\vec{w}$.

Exercise 6.6. For each matrix M listed below, consider the product $M \cdot \vec{v}$, where $\vec{v}$ has the homogeneous coordinates of a point in two-space. State whether this operation carries out a translation, rigid motion, scale transformation, invertable affine transformation, or degenerate affine transformation; whether it leaves the origin fixed; and whether it is a reflection. Note that $|\langle 0.28, 0.96 \rangle| = 1$. Given that fact, you should be able to do these by inspection, without putting pencil to paper, let alone running MATLAB.
  Draw a sketch of what the operations look like.

$$\begin{bmatrix} 0.28 & -0.96 & 0 \\ 0.96 & 0.28 & 0 \\ 0 & 0 & 1 \end{bmatrix}, \quad \begin{bmatrix} 0.28 & 0.96 & 1 \\ -0.96 & 0.28 & 3 \\ 0 & 0 & 1 \end{bmatrix}, \quad \begin{bmatrix} 0.96 & 0.28 & 3 \\ 0.28 & -0.96 & 2 \\ 0 & 0 & 1 \end{bmatrix},$$

$$\begin{bmatrix} 1 & 0 & 3 \\ 0 & 1 & 2 \\ 0 & 0 & 1 \end{bmatrix}, \quad \begin{bmatrix} 2.8 & -9.6 & 0 \\ 9.6 & 2.8 & 0 \\ 0 & 0 & 1 \end{bmatrix}, \quad \begin{bmatrix} 0.96 & 0.28 & 0 \\ 0.28 & 0.96 & 0 \\ 0 & 0 & 1 \end{bmatrix},$$

$$\begin{bmatrix} 1 & 2 & 3 \\ 4 & 5 & 6 \\ 0 & 0 & 1 \end{bmatrix}, \quad \begin{bmatrix} 1 & 2 & 0 \\ 4 & 8 & 0 \\ 0 & 0 & 1 \end{bmatrix}.$$

Problems

Problem 6.1. Give geometric proofs of the equations for geometric rules given at the end of Section 6.1.

Problem 6.2. Let $\vec{u}$ and $\vec{v}$ be two three-dimensional vectors, and let $\vec{w} = \vec{u} \times \vec{v}$ be their cross-product, as defined in Section 6.3.4.

(a) Prove that $\vec{w} = \vec{0}$ if and only if $\vec{u}$ and $\vec{v}$ are parallel.

(b) Prove that $\vec{w}$ is orthogonal to both $\vec{u}$ and $\vec{v}$.

Problem 6.3. In Figure 6.15, assuming that the coordinate system on the left has unit length, what is the area of the parallelogram on the right? What is the area of the ellipse? Recall that, as stated in Section 6.4.5, the transformation matrix is

$$M = \begin{bmatrix} 1/4 & -1/2 \\ 1/2 & 1 \end{bmatrix}.$$

Programming Assignments

Assignment 6.1 (Pappus's theorem).

(a) Write a function QuadIntersect(A,B,C,D) that takes as arguments four points A,B,C,D and returns the coordinates of the intersection of the line containing A and B with the line containing C and D. Don't worry about checking for special cases (e.g., there is no line because A == B or the two lines are parallel).[6] For example, QuadIntersect([0,0], [1,1], [1,0], [0,1]) should return [0.5,0.5], and QuadIntersect([0,0], [10,0], [0,5], [-1,4]) should return [0,-5].

(b) Pappus's theorem states the following: Suppose that points **a,b,c** lie on one line and points **x,y,z** lie on another. Construct the following points:

point **m**, the intersection of lines **ay** and **bx**; point **n**, the intersection of lines **az** and **cx**; and point **p**, the intersection of lines **bz** and **cy**.

Then **m,n,p** lie on a line (Figure 6.17).

------

[6]Here I am encouraging utterly irresponsible behavior. Not checking for special cases is very bad programming style; one of the hallmarks of production-quality software as opposed to immature software is "failing gracefully"—that is, returning meaningful values and generating useful error/warning messages for erroneous or otherwise improper inputs. However, in geometric programming especially, there tend to be lots of special cases, and it can be very difficult to find them all and to decide what should be done. The code for handling special cases can easily be several times longer than the code for handling the standard case. The object of the assignments in this course is to teach the mathematics and to give students practice in rapid prototyping, not to teach high-quality software engineering.

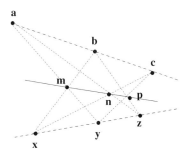

Figure 6.17. Pappus's theorem.

Write a function `RandomPappus` with no arguments that generates a random diagram illustrating Pappus's theorem. That is, construct random points **a**, **b**, **x**, **y**; construct a random point **c** on **ab** and a random point **z** on **xy**; and then construct the corresponding diagram. *Note:* in rare cases the calculation may fail; more commonly, the diagram may end up so ugly as to be useless. Do not worry about these.)

Curiously, if you use the MATLAB rank function to check whether **m**, **n**, and **p** are colinear, using the default tolerance—that is, if you set up the matrix with row **p** − **m** and **n** − **m** and check whether it has rank 1—it fails 16% of the time. We discuss the issues involved here in Section 7.9.2.

Assignment 6.2. Write a function `CircumscribeTriangle(A,B,C)` that takes as input the coordinates of three points in the plane, `A`,`B`,`C`, and draws the triangle connecting them and the circumscribing circle (the circle that goes through all three points).

The center of the circumscribing circle is the intersection of the perpendicular bisectors to the sides. Therefore, it can be computed as follows:

- Find the midpoint **x** of side **ab**.

- Find the midpoint **y** of side **bc**.

- Find the line L through **x** perpendicular to **ab**.

- Find the line M through **y** perpendicular to **bc**.

- Find the intersection **o** of L with M. (If **a**, **b**, and **c** are collinear, then there is no intersection and MATLAB will give an error at this point.)

- Let r be the distance from **o** to **a**. Then the circle centered at **o** of radius r is the circumscribing circle.

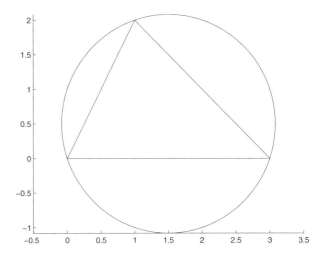

Figure 6.18. CircumscribeTriangle([0,0], [3,0], [1,2]).

To draw a circle of center **o** and radius r, compute the points $\mathbf{o} + r \cdot \langle \sin(2\pi t/N),$ $\cos(2\pi t/N)\rangle$ for $t = 1,\ldots,N$ for some large N (e.g., $N = 100$). Figure 6.18 shows the drawing for the function call CircumscribeTriangle([0,0], [3,0], [1,2]).

Assignment 6.3 (Rotating a polyhedron). Use MATLAB to show how the appearance of a three-dimensional object (a convex polyhedron) changes as a result of rotation (assuming that the distance to the object is constant, and large compared to the size of the object).

This problem is greatly simplified by restricting it to *convex polyhedra*. The advantage of using a convex polyhedron is that it is easy to determine what parts of the surface are visible; a face of a convex polyhedron is visible if the normal to the face points toward the viewer.

The assignment is to write a function DrawRotatedPolyhedron(M,P). The input parameters are M, a 3×3 rotation (orthogonal) matrix, and P, a data structure representing a convex polyhedron. What the function does is to draw a two-dimensional picture of P after rotation by M in a form described below.

The input data structure representing an n-face polyhedron P is a cellular array of size n, where each cell is a face of the polyhedron. A face of the polyhedron with k vertices is a $3 \times k$ array in which each column of the array contains the coordinates of the vertices. The columns are in counterclockwise order as viewed by someone outside the solid and looking at the face.

For example, the unit cube is represented as a cellular array {X1,X2,Y1,Y2,Z1, Z2}, where X1 is the face at the low end of the x-axis, X2 is the face at the high

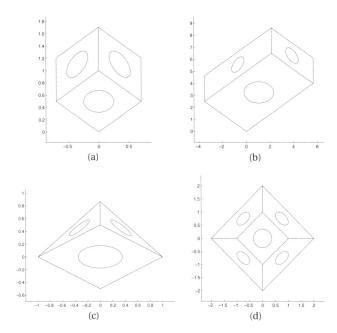

Figure 6.19. (a) DrawRotatedPolyhedron(EulerRotation(pi/4, pi/4, 0), cube), (b) DrawRotatedPolyhedron(EulerRotation(pi/4, pi/4, 0), box), (c) DrawRotatedPolyhedron(EulerRotation(0, pi/3, 0), pyramid), (d) DrawRotatedPolyhedron(EulerRotation(0, pi, 0), frustum).

end of the x-axis, and so on. Specifically,

$$
X1 = \begin{bmatrix} 0 & 0 & 0 & 0 \\ 0 & 0 & 1 & 1 \\ 0 & 1 & 1 & 0 \end{bmatrix}, \quad
X2 = \begin{bmatrix} 1 & 1 & 1 & 1 \\ 0 & 1 & 1 & 0 \\ 0 & 0 & 1 & 1 \end{bmatrix}, \quad
Y1 = \begin{bmatrix} 0 & 1 & 1 & 0 \\ 0 & 0 & 0 & 0 \\ 0 & 0 & 1 & 1 \end{bmatrix},
$$

$$
Y2 = \begin{bmatrix} 0 & 0 & 1 & 1 \\ 1 & 1 & 1 & 1 \\ 0 & 1 & 1 & 0 \end{bmatrix}, \quad
Z1 = \begin{bmatrix} 0 & 0 & 1 & 1 \\ 0 & 1 & 1 & 0 \\ 0 & 0 & 0 & 0 \end{bmatrix}, \quad
Z2 = \begin{bmatrix} 0 & 1 & 1 & 0 \\ 0 & 0 & 1 & 1 \\ 1 & 1 & 1 & 1 \end{bmatrix}.
$$

This and other shapes are defined as scripts in the file *polyhedra.m*. The function EulerRotation(Psi,Phi,Theta) generates the three-dimensional rotation with Euler angles Psi, Phi, and Theta.

The picture of the shape will reflect the appearance of the rotated shape as seen from below; thus, it will be the projection of the faces that are visible below onto the $x - y$-plane. The picture will show (1) the visible vertices and edges (2) in the center of each face, the projection of a small circle around the center (which will project as an ellipse). Some sample outputs are shown in Figure 6.19.

Constructing this picture involves the following steps:

- Apply the matrix M to each of the points in P, generating a new poly-hedron P'. (Since the absolute position of of the polyhedron does not matter, you can use natural coordinates and a 3×3 rotation matrix rather than homogeneous coordinates and a 4×4 matrix.)

- Exclude faces that are not visible from below. To do this, compute the outward normal to each face, and exclude any face whose normal has a positive z-component. To find the outward normal to a face, choose any three consecutive vertices of the face $\mathbf{a}, \mathbf{b}, \mathbf{c}$; the cross-product $(\mathbf{c} - \mathbf{b}) \times (\mathbf{a} - \mathbf{b})$ is the outward normal.

- Project the vertices and edges onto the $x-y$-plane simply by ignoring the z-coordinate. Draw these in the picture.

- Compute the central circles on each visible face as follows:

  - Compute the center $\mathbf{o}$ of the face as the average of the vertices of the face.

  - Choose a radius r as half the distance from $\mathbf{o}$ to the nearest of the edges. The distance from point $\mathbf{p}$ to line $\mathbf{L}$ is equal to the distance from $\mathbf{p}$ to $\text{Proj}(\mathbf{p}, \mathbf{L})$ (Section 6.3.6).

  - Find two orthonormal arrows $\vec{u}, \vec{v}$ in the plane of the face. If $\mathbf{a}, \mathbf{b}, \mathbf{c}$ are vertices in the face, then you can choose $\vec{u} = \text{Dir}(\mathbf{b} - \mathbf{a})$ as one and $\vec{v} = \text{Dir}(\mathbf{c} - \mathbf{p})$ as the other, where $\mathbf{p}$ is the projection of $\mathbf{c}$ on line $\mathbf{ab}$.

  - Compute points on the circle in 3D as

$$\mathbf{o} + r \cos(2\pi t / N)\vec{u} + r \sin(2\pi t / N)\vec{v}$$

    for $t = 0, \ldots, N$.

  - Project these points onto the $x-y$-plane by ignoring the z-coordinate, and connect the points to plot the ellipse.

Chapter 7

Change of Basis, DFT, and SVD

7.1 Change of Coordinate System

Suppose that we have the job of creating a gazetteer that records the positions of all the lampposts in a city. We chose a coordinate system $\mathscr{C}$ with origin $\mathbf{o}$ and coordinate arrows $\vec{x}, \vec{y}$; did lots of measurements; and recorded the positions of all the thousands of lampposts in our city gazetteer.

Now suppose the day after we finish this, the boss calls us into his office, and tells us he wants to use a different coordinate system $\mathscr{D}$; this one will have origin $\mathbf{q}$, unit length m, and coordinate arrows $\vec{i}$ and $\vec{j}$. Is there a simple way to convert the $\mathscr{C}$ coordinates to $\mathscr{D}$ coordinates?

The solution is simple. All we have to do is to measure the coordinates of $\mathbf{o}, \vec{x}$, and $\vec{y}$ in $\mathscr{D}$. Suppose that the coordinates of $\mathbf{o}$ are $\langle a, b \rangle$; the coordinates of $\vec{x}$ are $\langle c, d \rangle$, and the coordinates of $\vec{y}$ are $\langle e, f \rangle$. Let $\mathbf{p}$ be a point whose coordinates in the $\mathbf{o}, \vec{x}, \vec{y}$ system have been measured to be $\langle s, t \rangle$. Then

$$\mathbf{p} = \mathbf{o} + s\vec{x} + t\vec{y}$$

$$= (\mathbf{q} + a\vec{i} + b\vec{j}) + s(c\vec{i} + d\vec{j}) + t(e\vec{i} + f\vec{j})$$

$$= \mathbf{q} + (a + sc + te)\vec{i} + (b + sd + tf)\vec{j}.$$

Using homogeneous coordinates, we can write this as

$$\mathrm{Hc}(\mathbf{p}, \mathscr{D}) = \begin{bmatrix} c & e & a \\ d & f & b \\ 0 & 0 & 1 \end{bmatrix} \cdot \mathrm{Hc}(\mathbf{p}, \mathscr{C}).$$

That is, the conversion from homogeneous coordinates in $\mathscr{C}$ to $\mathscr{D}$ coordinates is just a linear transformation.

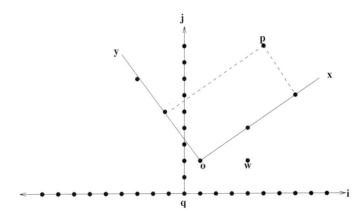

Figure 7.1. Coordinate transformation.

For example, suppose $\mathrm{Coords}(\mathbf{o}, \mathscr{D}) = \langle 1, 2 \rangle$, $\mathrm{Coords}(\vec{x}, \mathscr{D}) = \langle 3, 2 \rangle$, and $\mathrm{Coords}(\vec{y}, \mathscr{D}) = \langle -2, 3 \rangle$, and suppose $\mathrm{Coords}(\mathbf{p}, \mathscr{C}) = \langle 2, 1 \rangle$ (Figure 7.1). Then

$$\mathrm{Hc}(\mathbf{p}, \mathscr{D}) = \begin{bmatrix} 3 & -2 & 1 \\ 2 & 3 & 2 \\ 0 & 0 & 1 \end{bmatrix} \cdot \begin{bmatrix} 2 \\ 1 \\ 1 \end{bmatrix} = \begin{bmatrix} 5 \\ 9 \\ 1 \end{bmatrix}.$$

Conversely, to convert from coordinates in $\mathscr{D}$ to coordinates in $\mathscr{C}$, we can solve the corresponding system of linear equations. For instance, with the above coordinate systems, if $\mathrm{Coords}(\mathbf{w}, \mathscr{D}) = \langle 4, 2 \rangle$, we have

$$\begin{bmatrix} 3 & -2 & 1 \\ 2 & 3 & 2 \\ 0 & 0 & 1 \end{bmatrix} \cdot \begin{bmatrix} \mathbf{w}[x] \\ \mathbf{w}[y] \\ 1 \end{bmatrix} = \begin{bmatrix} 4 \\ 2 \\ 1 \end{bmatrix}.$$

So $\mathrm{Coords}(\mathbf{w}, \mathscr{C}) = \langle 9/13, -6/13 \rangle$.

If the two coordinate systems have the same origin, then natural coordinates can be used instead of homogeneous coordinates. The analysis is the same.

7.1.1 Affine Coordinate Systems

We just skipped a step. In chapter 6, we required that the coordinate arrows $\vec{x}$ and $\vec{y}$ have the same (unit) length and be perpendicular. But in the calculation we just did, we didn't require that to hold for either $\vec{x}, \vec{y}$ or for $\vec{i}, \vec{j}$. What is the meaning of a coordinate system in which the coordinate vectors are not the same length and are not perpendicular?

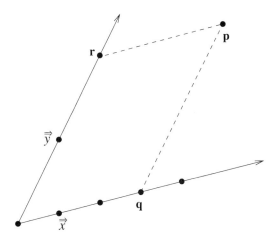

Figure 7.2. Affine coordinates.

Such a coordinate system is known as an *affine* coordinate system, and can be defined as long as the two coordinate vectors are not parallel. Let $\mathscr{C}$ be an affine coordinate system with origin $\mathbf{o}$ and coordinate vectors $\vec{x}, \vec{y}$. Then the coordinates of a point $\mathbf{p}$ are the pair $\langle a, b \rangle$ such that $\mathbf{p} = \mathbf{o} + a\vec{x} + b\vec{y}$. These can be found by using the following procedure. Let L and M be the two coordinate axes $L = \{\mathbf{o} + t\vec{x} \mid t \in \mathbb{R}\}$ and $M = \{\mathbf{o} + t\vec{y} \mid t \in \mathbb{R}\}$. Draw the line parallel to M through $\mathbf{p}$ and let $\mathbf{q}$ be the point where this line intersects L, and draw the line parallel to L through $\mathbf{p}$ and let $\mathbf{r}$ be the point where this line intersects L. Then $a = (\mathbf{q} - \mathbf{p})/\vec{x}$ and $b = (\mathbf{r} - \mathbf{p})/\vec{y}$. For instance, in Figure 7.2, $a = 3$ and $b = 2$. Note that if $\vec{x}$ and $\vec{y}$ are orthogonal, then the line parallel to M is the perpendicular to L, which is how we defined this construction in Section 6.2—but in the general case, we use the parallels rather than the perpendiculars.

The analogous construction applies to affine coordinate systems in three-space, except that the condition that $\vec{x}$ and $\vec{y}$ are not parallel is replaced by the condition that $\vec{x}, \vec{y},$ and $\vec{z}$ are not coplanar.

Addition of vectors in affine coordinate systems works as usual:

$$\text{Coords}(\mathbf{p} + \vec{v}, \mathscr{C}) = \text{Coords}(\mathbf{p}, \mathscr{C}) + \text{Coords}(\vec{v}, \mathscr{C}).$$
$$\text{Coords}(\vec{u} + \vec{v}, \mathscr{C}) = \text{Coords}(\vec{u}, \mathscr{C}) + \text{Coords}(\vec{v}, \mathscr{C}).$$
$$\text{Coords}(c \cdot \vec{v}, \mathscr{C}) = c \cdot \text{Coords}(\vec{v}, \mathscr{C}).$$

However, the dot product formulas no longer work properly for computing lengths, distances, and angles.

7.1.2   Duality of Transformation and Coordinate Change; Handedness

The changes of coordinate system discussed thus far in this chapter and the geometric transformations discussed in Section 6.4 create the same kind of changes to coordinates in ways that are dual to one another, in the following sense. Suppose that on Sunday, Ed records that the corners of the dining room table are located at coordinates $\langle 2,3 \rangle$, $\langle 5,3 \rangle$, $\langle 5,4 \rangle$, and $\langle 2,4 \rangle$. On Monday, Dora records that the corners are at $\langle 8,6 \rangle$, $\langle 11,6 \rangle$, $\langle 11,7 \rangle$, and $\langle 8,7 \rangle$. One possibility is that the table has undergone a translation by $\langle 6,3 \rangle$. Another possibility is that Dora is using a different coordinate system, one with the same coordinate directions and unit length, but with an origin whose coordinates in Ed's system are $\langle -6,-3 \rangle$. (Or, of course, there may have been both a transformation and a change of coordinate system; but we are interested here in comparing only the two pure cases.)

So there is a correspondence between transformations and changes of coordinate systems and between categories of transformations and categories of coordinate system changes. In particular:

- As in this example, a translation by $\langle x, y \rangle$ corresponds to moving the origin by $\langle -x, -y \rangle$. The directions and unit length of the coordinate system remain unchanged.

- A pure rotation corresponds to rotating the coordinate system in the opposite direction. In particular, an orthonormal coordinate system is changed to another orthonormal coordinate system.

- A expansion by c corresponds to a contraction of the unit length of the coordinate system by c. The origin and coordinate directions remain unchanged.

Reflections are more interesting. A reflection corresponds to changing the handedness of the coordinate system—changing a left-handed coordinate system into a right-handed one, or vice versa (Figure 7.3). In the plane, a right-handed coordinate system places the y-axis $90°$ counterclockwise from the x-axis; a left-handed coordinate system places the y-axis clockwise from the x-axis. In a right-handed coordinate system in three-space, the axes are aligned so that if you place your right hand along the x-axis, pointing your fingers in the positive x direction, and your palm is facing in the positive y direction, then your thumb is pointing in the positive z-direction. In a left-handed coordinate system, the same holds with your left hand.

Unlike other properties of bases that we have considered, the handedness of a coordinate system depends on the *order* of the basis elements and not just on the set of coordinate vectors. If $\langle \hat{x}, \hat{y} \rangle$ is a right-handed coordinate system,

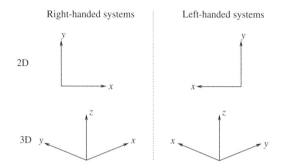

Figure 7.3. Right- and left-handed coordinate systems.

then $\langle \hat{y}, \hat{x} \rangle$ is a left-handed coordinate system; it corresponds (in the above sense) to the reflection across the line $x = y$. In three dimensions, if $\langle \hat{x}, \hat{y}, \hat{z} \rangle$ is a right-handed coordinate system, then $\langle \hat{y}, \hat{z}, \hat{x} \rangle$ and $\langle \hat{z}, \hat{x}, \hat{y} \rangle$ are also right-handed, whereas $\langle \hat{y}, \hat{x}, \hat{z} \rangle$, $\langle \hat{x}, \hat{z}, \hat{y} \rangle$, and $\langle \hat{z}, \hat{y}, \hat{x} \rangle$ are left-handed.

The distinction between right- and left-handed coordinate systems—or, more precisely, between a pair of coordinate systems with the same handedness and a pair with opposite handedness—carries over into higher dimensions. In any dimension, any coordinate system has one of two handednesses. Swapping the order of two vectors in the basis flips the handedness; doing a second swap restores it. For instance, let $\mathcal{B} = \langle \vec{b}_1, \vec{b}_2, \vec{b}_3, \vec{b}_4, \vec{b}_5 \rangle$ be a basis in $\mathbb{R}^5$. Let $\mathcal{C} = \langle \vec{b}_1, \vec{b}_4, \vec{b}_3, \vec{b}_2, \vec{b}_5 \rangle$ be the result of swapping the second and fourth vector in $\mathcal{B}$. Then $\mathcal{C}$ has the opposite handedness to $\mathcal{B}$. Let $\mathcal{D}$ be the result of swapping the second and fifth vector in $\mathcal{C}$: $\mathcal{D} = \langle \vec{b}_1, \vec{b}_5, \vec{b}_3, \vec{b}_2, \vec{b}_4 \rangle$. Then $\mathcal{D}$ has the opposite handedness to $\mathcal{C}$, and the same handedness as $\mathcal{B}$.

7.1.3   Application: Robotic Arm

Simple robotic arms can be analyzed by using coordinate transformations. Here we model a robotic arm as a sequence of k links, each of a fixed length, with a pin joint between successive links that allows rotation in the plane perpendicular to the pin. The first link is attached to the origin by a pin joint. The robot directly controls the angles $\theta_1, \ldots, \theta_k$, where θ_1 is the angle in the x-y plane of the first link, and for $i > 1$, θ_i is the angle between the forward directions on the $(i-1)$th link and the ith link (Figure 7.4). The question is, for a given sequence of angles $\theta_1, \ldots, \theta_k$, what is the position of the end of the last link, and what is the direction of the last link? (The inverse problem—given a target position and direction, find a sequence of angles that achieves it—is known as the "inverse kinematics" problem, and is a difficult problem, beyond the scope of this book.)

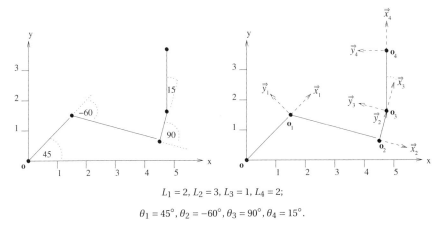

$$L_1 = 2,\ L_2 = 3,\ L_3 = 1,\ L_4 = 2;$$

$$\theta_1 = 45°,\ \theta_2 = -60°,\ \theta_3 = 90°,\ \theta_4 = 15°.$$

Figure 7.4. Two-dimensional robotic arm.

Let us consider first the two-dimensional case, in which all the pins are vertical, and hence all the links rotate in the x-y plane like a folding yardstick. In this case, we can ignore the vertical dimension altogether. We solve the problem by attaching a little coordinate system $\mathscr{C}_i$ with origin $\mathbf{o}_i$ and coordinate directions $\vec{x}_i$ and $\vec{y}_i$ to the end of ith link where $\vec{x}_i$ points in the direction of the link. $\mathscr{C}_0$ is the external, absolute coordinate system. Let L_i be the length of the ith link. Then for $i = 1, \ldots, k$, we have the following relations:

- $\vec{x}_i, \vec{y}_i$ are rotated by θ_i from $\vec{x}_{i-1}, \vec{y}_{i-1}$,

- $\mathbf{o}_i$ is located at $\mathbf{o}_{i-1} + L_i \cdot \vec{x}_i$.

It would seem at first[1] that one could view this problem in terms of a sequence of geometric transformations from $\mathscr{C}_0$ to $\mathscr{C}_1$ to $\mathscr{C}_2$. But that doesn't work because geometric transformations are measured with respect to the absolute coordinate system $\mathscr{C}_0$, whereas here we are given the coordinate of $\mathscr{C}_2$ in terms of $\mathscr{C}_1$, and of $\mathscr{C}_3$ in terms of $\mathscr{C}_2$. So we have to combine coordinate transformations: to find the coordinate transformation from $\mathscr{C}_n$ to $\mathscr{C}_{n-1}$, from $\mathscr{C}_{n-1}$ to $\mathscr{C}_{n-2}, \ldots$, from $\mathscr{C}_1$ to $\mathscr{C}_0$; and compose them to get the coordinate transformation from $\mathscr{C}_n$ to $\mathscr{C}_0$. Since the coordinate transformation is the inverse of the geometric transformation, this composition will give the coordinates of $\mathscr{C}_n$ in $\mathscr{C}_0$.

---

[1] To be precise, I originally wrote this section using geometric transformations, and was surprised to find that I was getting the wrong answers. It took me 15 minutes of thought to work out what the problem was; a fine example of how the kind of confusion discussed in Section 7.3 can arise even in very concrete, geometric applications.

We will represent the coordinate transformation from $\mathscr{C}_n$ to $\mathscr{C}_i$ in homogeneous coordinates in terms of a matrix

$$T_i = \left[\begin{array}{c|c} R_i & \vec{v}_i \\ \hline \vec{0}^T & 1 \end{array}\right].$$

The coordinate transformation from $\mathscr{C}_i$ to $\mathscr{C}_{i-1}$ consists of two parts in sequence.

1. A translation of the origin by $-L_i\vec{x}$, relative to the coordinate system $\mathscr{C}_i$. We call this intermediate coordinate system $\mathscr{D}_i$. The corresponding coordinate transformation matrix is

$$\begin{bmatrix} 1 & 0 & L_i \\ 0 & 1 & 0 \\ 0 & 0 & 1 \end{bmatrix}.$$

2. A rotation by $-\theta_i$ around the origin of $\mathscr{D}_i$. The corresponding coordinate transformation matrix is

$$\begin{bmatrix} \cos(\theta_i) & -\sin(\theta_i) & 0 \\ \sin(\theta_i) & \cos(\theta_i) & 0 \\ 0 & 0 & 1 \end{bmatrix}.$$

Therefore, the combined transformation from $\mathscr{C}_i$ to $\mathscr{C}_{i-1}$ is the composition of these two matrices, or

$$T_i = \begin{bmatrix} \cos(\theta_i) & -\sin(\theta_i) & 0 \\ \sin(\theta_i) & \cos(\theta_i) & 0 \\ 0 & 0 & 1 \end{bmatrix} \cdot \begin{bmatrix} 1 & 0 & L_i \\ 0 & 1 & 0 \\ 0 & 0 & 1 \end{bmatrix} = \begin{bmatrix} \cos(\theta_i) & -\sin(\theta_i) & L_i\cos(\theta_i) \\ \sin(\theta_i) & \cos(\theta_i) & L_i\sin(\theta_i) \\ 0 & 0 & 1 \end{bmatrix}.$$

The coordinate transformation from $\mathscr{C}_n$ to $\mathscr{C}_0$ is then the composition of all of these $T_1 \cdot T_2 \cdot \ldots \cdot T_n$. The upper left 2×2 square gives the rotation of $\mathscr{C}_n$ from $\mathscr{C}_0$, and the first two elements in the third column give the coordinates of $\mathbf{o}_n$ in terms of $\mathscr{C}_0$.

Actually, in the two-dimensional case, we could have done this more simply without matrix algebra because in two dimensions, rotational angles simply add. Define $\phi_1 = \theta_1$, $\phi_2 = \theta_1 + \theta_2$, $\phi_3 = \theta_1 + \theta_2 + \theta_3$, and so on up to $\phi_n = \theta_1 + \ldots + \theta_n$. Then ϕ_i is the total rotation between $\mathscr{C}_0$ and $\mathscr{C}_i$. The arrow, in absolute coordinates, from $\mathbf{o}_{i-1}$ to $\mathbf{o}_i$ is $\langle L_i\cos(\phi_i), L_i\sin(\phi_i)\rangle$, so the coordinates of $\mathbf{o}_n$ in $\mathscr{C}_0$ are

$$\langle L_1\cos(\phi_1) + L_2\cos(\phi_2) + \ldots + L_n\cos(\phi_n), L_1\sin(\phi_1) + L_2\sin(\phi_2) + \ldots + L_n\sin(\phi_n)\rangle.$$

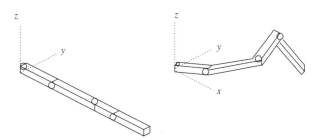

Figure 7.5. Three-dimensional robotic arm.

Moving on to the three-dimensional case, however, there is no alternative to using matrix algebra. Let us consider the following model of an arm.[2] Again, the arm consists of n links of length $L_1, \ldots, L_n$ connected in sequence by pin joints. The angle at the connection between the $(i-1)$th and ith links is θ_i, and when all the θ_i are zero, then the links lie stretched out along the positive x-axis. However, the direction of the pins alternates. Specifically, the pin at the origin is always vertical, so that the first link rotates in the x-y plane. When $\theta_1 = 0$, the second pin is parallel to the y-axis, so the second link rotates in the x-z plane. When all the θs are 0, the third pin is again vertical, the fourth pin is again parallel to the y-axis, the fifth is vertical, the sixth is parallel to the y-axis, and so on (Figure 7.5).

The linear algebra for the three-dimensional case is essentially the same as for the two-dimensional case. Again, we attach a little coordinate system $\mathscr{C}_i$ to the far end of the ith link, and we compute the coordinate transformation from $\mathscr{C}_n$ to $\mathscr{C}_0$ by using matrix multiplication using homogeneous coordinates. The only difference is in the rotations.

As in the two-dimensional case, the transformation from $\mathscr{C}_i$ to $\mathscr{C}_{i-1}$ consists of two parts:

1. Translating the origin by $-L_i \vec{x}$; the corresponding matrix is

$$\begin{bmatrix} 1 & 0 & 0 & L_i \\ 0 & 1 & 0 & 0 \\ 0 & 0 & 1 & 0 \\ 0 & 0 & 0 & 1 \end{bmatrix}.$$

2. If i is even, rotating about the $\vec{z}$ axis of $\mathscr{D}_i$; the corresponding matrix is

$$\begin{bmatrix} \cos(\theta_i) & -\sin(\theta_i) & 0 & 0 \\ \sin(\theta_i) & \cos(\theta_i) & 0 & 0 \\ 0 & 0 & 1 & 0 \\ 0 & 0 & 0 & 1 \end{bmatrix}.$$

---
[2]I do not know of any actual arm built like this, but it is a convenient model.

If i is odd, rotating about the $\vec{y}$ axis of $\mathcal{D}_i$; the corresponding matrix is

$$
\begin{bmatrix}
\cos(\theta_i) & 0 & -\sin(\theta_i) & 0 \\
0 & 1 & 0 & 0 \\
\sin(\theta_i) & 0 & \cos(\theta_i) & 0 \\
0 & 0 & 0 & 1
\end{bmatrix}.
$$

If i is even, then the composition of the two transformations is

$$
T_i =
\begin{bmatrix}
\cos(\theta_i) & -\sin(\theta_i) & 0 & 0 \\
\sin(\theta_i) & \cos(\theta_i) & 0 & 0 \\
0 & 0 & 1 & 0 \\
0 & 0 & 0 & 1
\end{bmatrix}
\cdot
\begin{bmatrix}
1 & 0 & 0 & L_i \\
0 & 1 & 0 & 0 \\
0 & 0 & 1 & 0 \\
0 & 0 & 0 & 1
\end{bmatrix}
$$

$$
=
\begin{bmatrix}
\cos(\theta_i) & -\sin(\theta_i) & 0 & L_i\cos(\theta_i) \\
\sin(\theta_i) & \cos(\theta_i) & 0 & L_i\sin(\theta_i) \\
0 & 0 & 1 & 0 \\
0 & 0 & 0 & 1
\end{bmatrix}.
$$

If i is odd, then the composition of the two transformations is

$$
T_i =
\begin{bmatrix}
\cos(\theta_i) & 0 & -\sin(\theta_i) & 0 \\
0 & 1 & 0 & 0 \\
\sin(\theta_i) & 0 & \cos(\theta_i) & 0 \\
0 & 0 & 0 & 1
\end{bmatrix}
\cdot
\begin{bmatrix}
1 & 0 & 0 & L_i \\
0 & 1 & 0 & 0 \\
0 & 0 & 1 & 0 \\
0 & 0 & 0 & 1
\end{bmatrix}
$$

$$
=
\begin{bmatrix}
\cos(\theta_i) & 0 & -\sin(\theta_i) & L_i\cos(\theta_i) \\
0 & 1 & 0 & 0 \\
\sin(\theta_i) & 0 & \cos(\theta_i) & L_i\sin(\theta_i) \\
0 & 0 & 0 & 1
\end{bmatrix}.
$$

The net transformation from $\mathcal{C}_n$ to $\mathcal{C}_0$ is the product $T = T_1 \cdot T_2 \cdot \ldots \cdot T_n$. The upper 3×3 block of T is the net rotation from $\mathcal{C}_0$ to $\mathcal{C}_n$. The first three elements of the right-hand column of T are the absolute coordinates of $\mathbf{o}_n$.

7.2   The Formula for Basis Change

We now move from the geometric case to the general case of n-dimensional vectors. Suppose that we have two bases, $\mathcal{B}$ and $\mathcal{C}$, for $\mathbb{R}^n$ and we are given the coordinates of vector $\vec{v}$ in $\mathcal{B}$. How can we calculate the coordinates of $\vec{v}$ in $\mathcal{C}$? Theorem 7.2 gives the answer in the case where either $\mathcal{B}$ or $\mathcal{C}$ is the standard basis $\{e^1, \ldots, e^n\}$; Corollary 7.3 gives the answer in general. First, we prove Lemma 7.1, which gives a general answer but not in terms that can immediately be computed.

Lemma 7.1. *Let V be an n-dimensional vector space. Let $\mathscr{B} = \vec{b}_1, \ldots, \vec{b}_n$ and $\mathscr{C} = \vec{c}_1, \ldots, \vec{c}_n$ be two ordered bases for V. Define the $n \times n$ matrix M such that $M[:, j] = \mathrm{Coords}(\vec{b}_j, \mathscr{C})$. Then, for any vector $\vec{v}$ in V,*

$$\mathrm{Coords}(\vec{v}, \mathscr{C}) = M \cdot \mathrm{Coords}(\vec{v}, \mathscr{B}).$$

Proof: Let $\mathrm{Coords}(\vec{v}, \mathscr{B}) = \langle a_1, \ldots, a_n \rangle$. Then

$$\begin{aligned}
\vec{v} &= a_1 \vec{b}_1 + \ldots + a_n \vec{b}_n \\
&= a_1 \cdot (M[1,1] \cdot \vec{c}_1 + M[2,1] \cdot \vec{c}_2 + \ldots + M[n,1] \cdot \vec{c}_n) \\
&\quad + a_2 \cdot (M[1,2] \cdot \vec{c}_1 + M[2,2] \cdot \vec{c}_2 + \ldots + M[n,2] \cdot \vec{c}_n) \\
&\quad + \ldots + a_n \cdot (M[1,n] \cdot \vec{c}_1 + M[2,n] \cdot \vec{c}_2 + \ldots + M[n,n] \cdot \vec{c}_n) \\
&= (M[1,1] \cdot a_1 + M[1,2] \cdot a_2 + \ldots + M[1,n] \cdot a_n) \cdot \vec{c}_1 \\
&\quad + (M[2,1] \cdot a_1 + M[2,2] \cdot a_2 + \ldots + M[2,n] \cdot a_n) \cdot \vec{c}_2 \\
&\quad + \ldots + (M[n,1] \cdot a_1 + M[n,2] \cdot a_2 + \ldots + M[n,n] \cdot a_n) \cdot \vec{c}_n. \quad \square
\end{aligned}$$

So

$$\begin{aligned}
\mathrm{Coords}(\vec{v}, \mathscr{C}) = \langle\ & M[1,1] \cdot a_1 + M[1,2] \cdot a_2 + \ldots + M[1,n] \cdot a_n, \\
& M[2,1] \cdot a_1 + M[2,2] \cdot a_2 + \ldots + M[2,n] \cdot a_n, \ldots, \\
& M[n,1] \cdot a_1 + M[n,2] \cdot a_2 + \ldots + M[n,n] \cdot a_n\ \rangle \\
= &\ M \cdot \mathrm{Coords}(\vec{v}, \mathscr{B}).
\end{aligned}$$

Theorem 7.2. *Let $\mathscr{C} = \vec{c}_1, \ldots, \vec{c}_n$ be a basis for $\mathbb{R}^n$. Let $\vec{u}$ be a vector and let $\vec{v} = \mathrm{Coords}(\vec{u}, \mathscr{C})$. Let M be the $n \times n$ matrix whose columns are the vectors in $\mathscr{C}$, or $M[:, j] = \vec{c}_j$. Then*

(a) $\vec{u} = M\vec{v}$,

(b) $\vec{v} = M^{-1}\vec{u}$.

Proof: The proof of result (a) is immediate from Lemma 7.1 with $\mathscr{B}$ being the standard basis for $\mathbb{R}^n$, since $\mathrm{Coords}(\vec{c}_j, \mathscr{B}) = \vec{c}_j$. The proof of result (b) is immediate from (a). $\quad \square$

Corollary 7.3. *Let $\mathscr{B} = \vec{b}_1, \ldots, \vec{b}_n$ and $\mathscr{C} = \vec{c}_1, \ldots, \vec{c}_n$ be two bases for $\mathbb{R}^n$. Let $\vec{w}$ be a vector. Let B be the matrix whose columns are $\vec{b}_1, \ldots, \vec{b}_n$; let C be the matrix with columns $\vec{c}_1, \ldots, \vec{c}_n$. Then $\mathrm{Coords}(\vec{w}, \mathscr{C}) = C^{-1}\vec{w} = C^{-1} \cdot B \cdot \mathrm{Coords}(\vec{w}, \mathscr{B})$.*

Proof: Immediate from Theorem 7.2.

7.3   Confusion and How to Avoid It

Despite the simple form of Theorem 7.2, it can be quite confusing, for several reasons. In the first place, the whole idea of vectors can seem to be floating off into the ether. We started with a vector $\vec{v}$ being an n-tuple of numbers, which is nice and concrete. But now the same (?) vector $\vec{v}$ is being represented by a n-tuple of coordinates relative to a basis. Moreover, (1) by choosing the vector space and the basis properly, we can get $\vec{v}$ to have any coordinate vector we want (other than the zero vector); and (2) the basis vectors themselves are just tuples of numbers. So in what sense is this the same vector? And what is the difference between two different vectors, if we can use any vector to represent any other vector? Moreover, how do we even pin down what the basis vectors mean, when they also are just tuples of numbers and likewise can turn into anything else?

Abstract mathematics of various kinds can trigger this kind of vertigo, although this particular example is an unusually severe one for math that is considered comparatively elementary and concrete. When hit by these kinds of feelings, the best thing to do is to stop working at the abstract level and go back to concrete instances; geometric examples are good, because we can draw pictures. As we did in the previous section, we want to be sure to distinguish between the concrete things on the one hand and their coordinate vectors on the other.

Really, there is nothing more in this indeterminacy than the fact that a person's height may be 6 when measured in feet and 72 when measured in inches, but it seems more confusing in this abstract setting.

The second source of confusion is that we have now introduced a second form of linear transformation. Previous to this section, a linear transformation was an operation that turned one thing into a different thing, such as a basket vector into a price vector, or a population distribution at one time into a different population distribution at a different time. In Theorem 7.2, by contrast, the "thing" remains the same; what is changing is the way we are representing the thing. The same matrix multiplication is used for both. Again, this is basically the same as the fact that you multiply a height measurement by 12 either when an object gets 12 times taller or when you change from feet to inches.

The third source of confusion is that it is easy to get the direction of Corollary 7.3 backward. To change a coordinate vector in $\mathscr{B}$ to a coordinate vector in $\mathscr{C}$, we multiply by the matrix of coordinates of $\mathscr{B}$ in $\mathscr{C}$, and not vice versa. Again, if we want to change from a measurement in feet to one in inches, we multiply by 12, the measure of one foot in inches.

7.4   Nongeometric Change of Basis

A final source of discomfort about these changes of bases, in nongeometric applications, is that it is not obvious what the point of nonstandard bases is in the first place. In geometry, the choice of coordinate system is largely arbitrary anyway, so it makes sense to go from one to another. But in the kinds of nongeometric applications we have considered, it is much less clear why we would want to do this.

Consider a specific example of shopping baskets; to make it easy, suppose that there are three products: a gallon of milk, a loaf of bread, and a pound of butter. The natural representation for a basket with x gallons of milk, y loaves of bread, and z pounds of butter uses the vector $\langle x, y, z \rangle$. Of course, we *could* use any three linearly independent baskets we want as a basis, say,

$$\vec{b}_1 = 2.5 \text{ gallons of milk}, -4 \text{ loaves of bread, and } 3 \text{ pounds of butter},$$

$$\vec{b}_2 = -3 \text{ gallons of milk}, \quad 2 \text{ loaves of bread, and } 1 \text{ pound of butter},$$

$$\vec{b}_3 = \quad 0 \text{ gallons of milk}, -6 \text{ loaves of bread, and } 7 \text{ pounds of butter}.$$

Then we could represent a basket with 1 gallon of milk as the coordinate vector $\langle 1, 1/2, -1/2 \rangle$. But why is this anything other than perverse?

In fact, there are many different reasons to consider alternative bases; a change of basis is as common in applications of linear algebra as actual change, if not more so. We consider a number of these in the rest of this chapter.

7.5   Color Graphics

There are domains other than geometric for which there can be different equally plausible bases for the vector space. One example is color graphics. Color is essentially a three-dimensional vector space;[3] a single color is the sum of three primary colors, each weighted by an intensity. However, different display systems use different sets of primary colors; for instance, a color printer uses a set that is different from a color terminal. Each triple of primary colors can be considered as a basis for the space of all colors; conversion of the intensity vector for a particular color from one system of primary colors to another is thus a matter of basis change. (This is an idealization; in practice, things are more complicated.)

---

[3]The dimension is a function of the human eye, which has three different kinds of cones. Pigeons are believed to have five primary colors. Stomatopods, a kind of shrimp, have twelve.

7.6   Discrete Fourier Transform (Optional)

The discrete[4] Fourier transform (DFT) is a basis change that is used throughout signal processing and in many other applications. This will require a little work, but it is well worth the time.

A simple signal processor has an input $\mathbf{a}(t)$ and an output $\mathbf{g}(t)$, both of which are functions of time. (In this section, we use boldface letters for functions of time.) The output depends on the input, so $\mathbf{g} = F(\mathbf{a})$ for some function F. Note that $\mathbf{g}(t)$ may depend on the entire signal $\mathbf{a}$ (or at least all of $\mathbf{a}$ prior to t) and not just on the single value $\mathbf{a}(t)$. Many signal processors have the following elegant properties, or conditions:

1. F is linear. That is, for any inputs $\mathbf{a}$ and $\mathbf{b}$, $F(\mathbf{a}+\mathbf{b}) = F(\mathbf{a}) + F(\mathbf{b})$. For any constant c, $F(c \cdot \mathbf{a}) = c \cdot F(\mathbf{a})$.

2. F is time-invariant. If input $\mathbf{b}$ is the same as $\mathbf{a}$ delayed by δ, then the output for $\mathbf{b}$ is the same as the output for $\mathbf{a}$ delayed by δ. Symbolically. if $\mathbf{b}(t) = \mathbf{a}(t-\delta)$ then $(F(\mathbf{b}))(t) = (F(\mathbf{a}))(t-\delta)$

3. The output to a sinusoidal input is a sinusoid of the same frequency, possibly amplified and time-delayed. That is, if $\mathbf{a}(t) = \sin(\omega t)$, and $\mathbf{g} = F(\mathbf{a})$ then $\mathbf{g}(t) = A(\omega) \sin(\omega t + \delta(\omega))$. As shown, the quantities $A(\omega)$ and $\delta(\omega)$ may depend on the frequency ω but are independent of time t. The quantities $A(\omega)$ and $\delta(\omega)$ are the *frequency response characteristics* of the signal processor.

Condition (3) may seem rather constrained. However, if the amplifier is characterized by a differential equation, which many physical systems are, and if it is damped, meaning that if the input becomes zero, the output eventually dies down to zero, and if it satisfies conditions (1) and (2), then it necessarily satisfies condition (3).

Using the trigonometric identity $\sin(\alpha + \beta) = \sin(\alpha)\cos(\beta) + \sin(\beta)\cos(\alpha)$, we can rewrite the equation in condition (3) as follows:

$$\mathbf{g}(t) = A(\omega) \sin(\omega t + \delta(\omega)) = A(\omega) \cos(\delta(\omega)) \sin(\omega t) + A(\omega) \sin(\delta(\omega)) \cos(\omega t).$$

Moveover, since $\cos(\alpha) = \sin(\alpha + \pi/2)$ (i.e., the cosine is just a time-advanced sine), it follows that the response to the input $\cos(\omega t)$ is just

$$A(\omega) \cos(\delta(\omega)) \cos(\omega t) - A(\omega) \sin(\delta(\omega)) \sin(\omega t).$$

Now, suppose that we take an input time signal $\mathbf{a}(t)$ and the corresponding output time signal $\mathbf{g}(t)$ from time $t = 0$ to $t = T$. We choose an even number

---

[4]The continuous Fourier transform is also a basis change, but for an infinite-dimensional space of functions.

$2K$ and we sample both signals at the $2K$ points, $t = T/2K$, $t = 2T/2K$, $t = 3T/2K, \ldots, t = 2KT/2K = T$. This gives us two vectors, $\vec{a}$ and $\vec{g}$, of dimension $2K$. Thus, for $I = 1, \ldots, 2K$, $\vec{a}[I] = \mathbf{a}(I \cdot T/2K)$ and $\vec{g}[I] = \mathbf{g}(I \cdot T/2K)$. In general for any function $\mathbf{f}(t)$ from 0 to T, we consider the sample of $2K$ points to be the vector $\vec{f}[I] = \mathbf{f}(I \cdot T/2K)$ for $I = 1, \ldots, 2K$.

We are now dealing in the space of $2K$-dimensional vectors. We define the following set of vectors $\mathcal{V} = \{\vec{v}_1, \ldots, \vec{v}_{2K}\}$:

$$\vec{v}_1 = \vec{1},$$

for $I, J = 1, \ldots, K$,

$$\vec{v}_{2J}[I] = \cos(I \cdot J\pi/K),$$

for $I, J = 1, \ldots, K-1$,

$$\vec{v}_{2J+1}[I] = \sin(I \cdot J\pi/K).$$

That is: $\vec{v}_1$ is just the 1 vector. The Jth even-numbered vector is a sample of the cosine function $\cos(J\pi t/T)$, and the Jth odd-numbered vector is a sample of the sine function $\sin(J\pi t/T)$.

For example, with $K = 2$, $2K = 4$, we have

$$\vec{v}_1 = \langle 1, 1, 1, 1 \rangle,$$
$$\vec{v}_2 = \langle \cos(\pi/2), \cos(2\pi/2), \cos(3\pi/2), \cos(4\pi/2) \rangle = \langle 0, -1, 0, 1 \rangle,$$
$$\vec{v}_3 = \langle \sin(\pi/2), \sin(2\pi/2), \sin(3\pi/2), \sin(4\pi/2) \rangle = \langle 1, 0, -1, 0 \rangle,$$
$$\vec{v}_4 = \langle \cos(2\pi/2), \cos(4\pi/2), \cos(6\pi/2), \cos(8\pi/2) \rangle = \langle -1, 1, -1, 1 \rangle.$$

We would seem to be missing the final sine function, but since that would be

$$\langle \sin(2\pi/2), \sin(4\pi/2), \sin(6\pi/2), \sin(8\pi/2) \rangle = \langle 0, 0, 0, 0 \rangle,$$

it is useless.

With $K = 3$, $2K = 6$, we have

$$\vec{v}_1 = \langle 1, 1, 1, 1, 1, 1 \rangle,$$
$$\vec{v}_2 = \langle \cos(\pi/3), \cos(2\pi/3), \cos(3\pi/3), \cos(4\pi/3), \cos(5\pi/3), \cos(6\pi/3) \rangle$$
$$\qquad = \langle 0.5, -0.5, -1, -0.5, 0.5, 1 \rangle,$$
$$\vec{v}_3 = \langle \sin(\pi/3), \sin(2\pi/3), \sin(3\pi/3), \sin(4\pi/3), \sin(5\pi/3), \sin(6\pi/3) \rangle$$
$$\qquad = \langle 0.866, 0.866, 0, -0.866, -0.866, 0 \rangle,$$
$$\vec{v}_4 = \langle \cos(2\pi/3), \cos(4\pi/3), \cos(6\pi/3), \cos(8\pi/3), \cos(10\pi/3), \cos(12\pi/3) \rangle$$
$$\qquad = \langle -0.5, -0.5, 1, -0.5, -0.5, 1 \rangle,$$
$$\vec{v}_5 = \langle \sin(2\pi/3), \sin(4\pi/3), \sin(6\pi/3), \sin(8\pi/3), \sin(10\pi/3), \sin(12\pi/3) \rangle$$
$$\qquad = \langle 0.866, -0.866, 0, 0.866, -0.866, 0 \rangle,$$
$$\vec{v}_6 = \langle \cos(3\pi/3), \cos(6\pi/3), \cos(9\pi/3), \cos(12\pi/3), \cos(15\pi/3), \cos(18\pi/3) \rangle$$
$$\qquad = \langle -1, 1, -1, 1, -1, 1 \rangle.$$

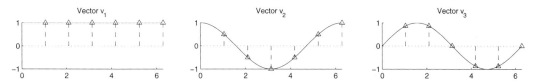

Figure 7.6. Discrete Fourier vectors as samplings of sine and cosine curves: $\vec{v}_1$, $\vec{v}_2$, and $\vec{v}_3$.

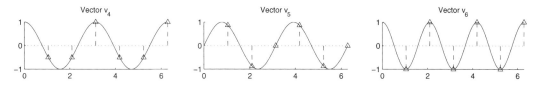

Figure 7.7. Discrete Fourier vectors as samplings of sine and cosine curves: $\vec{v}_4$, $\vec{v}_5$, and $\vec{v}_6$.

These vectors are the $2K$ *real Fourier vectors*. The corresponding sinusoidal functions are called the $2K$ *real Fourier components*. These are the functions

$$\mathbf{v}_1(t) = 1,$$

for $J = 1, \dots, K$,

$$\mathbf{v}_{2J}(t) = \cos(J\pi t/T),$$

for $J = 1, \dots, K-1$,

$$\mathbf{v}_{2J+1}(t) = \sin(J\pi t/T).$$

Thus, vector $\vec{v}_J$ is the $2K$ sample of function $\mathbf{v}_{2J}(t)$ over the interval $[0, T]$ (see Figures 7.6 and 7.7).

It is a fact that the $2K$ real Fourier vectors are linearly independent (in fact, orthogonal) and therefore form a basis for $\mathbb{R}^{2K}$, although we will not prove this. Therefore, the sample vectors $\vec{a}$ and $\vec{g}$ can be written as linear sums over the Fourier vectors:

$$\vec{a} = p_1 \cdot \vec{v}_1 + \dots + p_{2K} \cdot \vec{v}_{2K},$$
$$\vec{g} = q_1 \cdot \vec{v}_1 + \dots + q_{2K} \cdot \vec{v}_{2K}.$$

These coordinates are said to be in the *frequency domain*. The original coordinates are in the *time domain*.

It is reasonable to suppose that if these equations hold for a dense sample with large K, then they approximately hold for the actual continuous functions involved. That is,

$$\mathbf{a} \approx p_1 \cdot \mathbf{v}_1 + \dots + p_{2K} \cdot \mathbf{v}_{2K},$$
$$\mathbf{g} \approx q_1 \cdot \mathbf{v}_1 + \dots + q_{2K} \cdot \mathbf{v}_{2K}.$$

(There is a large theory of approximation, which discusses under what circumstances this is valid and how accurate you can expect the approximation to be.)

We now return to the signal processing function F that we introduced at the start of this section. It turns out that the function F does something rather simple with the Fourier vectors. Let us define the sequence of numbers $d_1,\ldots,d_{2K}$ as follows:

$$d_1 = A_1,$$

for $J = 1,\ldots,K$,

$$d_{2J} = A(\pi J/T)\cos(\delta(\pi J/T)),$$

for $J = 1,\ldots,K-1$,

$$d_{2J+1} = A(\pi J/T)\sin(\delta(\pi J/T)).$$

Then $F(\mathbf{v}_1) = d_1\mathbf{v}_1$.

For $J = 1,\ldots,K$,

$$
\begin{aligned}
F(\mathbf{v}_{2J}) &= F(\cos(\pi Jt/T)) \\
&= A(\pi J/T)\cos(\pi J/T)\cos(\pi Jt/T) - A(\pi J/T)\sin(\pi J/T)\sin(\pi Jt/T) \\
&= d_{2J}\mathbf{v}_{2J} - d_{2J+1}\mathbf{v}_{2J+1}.
\end{aligned}
$$

For $J = 1,\ldots,K-1$,

$$
\begin{aligned}
F(\mathbf{v}_{2J+1}) &= F(\sin(\pi Jt/T)) \\
&= A(\pi J/T)\cos(\pi J/T)\sin(\pi Jt/T) + A(\pi J/T)\sin(\pi J/T)\cos(\pi Jt/T) \\
&= d_{2J+1}\mathbf{v}_j + d_{2J}\mathbf{v}_{J+1}.
\end{aligned}
$$

We can combine all of this to get the following:

$$
\begin{aligned}
q_1\mathbf{v}_1 + \ldots + q_{2K}\mathbf{v}_{2K} &\approx \mathbf{g} \\
&= F(\mathbf{a}) \\
&\approx F(p_1\mathbf{v}_1 + \ldots + p_{2K}\mathbf{v}_{2K}) \\
&= p_1 d_1\mathbf{v}_1 + p_2(d_2\mathbf{v}_2 - d_3\mathbf{v}_3) + p_3(d_2\mathbf{v}_3 + d_3\mathbf{v}_2) + p_4(d_4\mathbf{v}_4 - d_5\mathbf{v}_5) \\
&\quad + p_5(d_4\mathbf{v}_5 + d_5\mathbf{v}_4) + \ldots + p_{2K}d_{2K}\mathbf{v}_{2K} \\
&= d_1 p_1\mathbf{v}_1 + (d_2 p_2 + d_3 p_3)\mathbf{v}_2 + (-d_3 p_2 + d_2 p_3)\mathbf{v}_3 + (d_4 p_4 + d_5 p_5)\mathbf{v}_4 \\
&\quad + (-d_5 p_4 + d_4 p_5)\mathbf{v}_5 + \ldots + d_{2K} p_{2K}\mathbf{v}_{2K}.
\end{aligned}
$$

Comparing the coefficients of $\mathbf{v}_J$ in the first expression above and the last, and using the fact that the $\mathbf{v}_J$ are linearly independent, we conclude that

$$q_1 = d_1 p_1,$$
$$q_2 = d_2 p_2 + d_3 p_3,$$
$$q_3 = -d_3 p_2 + d_2 p_3,$$
$$q_4 = d_4 p_4 + d_5 p_5,$$
$$q_5 = -d_5 p_4 + d_4 p_5,$$
$$\vdots$$
$$q_{2K} = d_{2K} p_{2K}.$$

So $\vec{q}$ is a linear transformation of $\vec{p}$ with a matrix D that is nearly a diagonal:

$$\vec{q} = \begin{bmatrix} d_1 & 0 & 0 & 0 & 0 & \cdots & 0 \\ 0 & d_2 & d_3 & 0 & 0 & \cdots & 0 \\ 0 & -d_3 & d_2 & 0 & 0 & \cdots & 0 \\ 0 & 0 & 0 & d_4 & d_5 & \cdots & 0 \\ 0 & 0 & 0 & -d_5 & d_4 & \cdots & 0 \\ \vdots & \vdots & \vdots & \vdots & \vdots & \vdots & \vdots \\ 0 & 0 & 0 & 0 & 0 & \cdots & d_{2K} \end{bmatrix} \cdot \vec{p}.$$

D represents the signal processing characteristics of the device. It is a sparse matrix with $4K - 2$ nonzero elements. The multiplication $D \cdot \vec{p}$ can be carried out in time $O(K)$. It is also easy to invert, if you want to go from output to input. D^{-1} is a matrix of the same structure, consisting of the inverse of each 1×1 or 2×2 square. So the inverse can be computed in time $O(K)$; it is again a sparse matrix, and multiplication by the inverse is carried out in time $O(K)$.

But, of course, this simple form applies only to vectors in the *frequency domain*, whereas we actually always start from vectors in the time domain, and generally we want to end there. That is where our theory of basis change comes in. Let M be the matrix whose columns are the Fourier vectors. Then multiplication by M^{-1} transforms vectors in the time domain to vectors in the frequency domain, and multiplication by M transforms vectors in the frequency domain back into the time domain. So $\vec{g} = M \cdot D \cdot M^{-1} \cdot \vec{a}$.

That doesn't sound very promising; M and M^{-1} are large, nonsparse matrices. However, three facts come to our rescue here:

- The Fourier matrix of size $2K$ depends only on K and not on the signal processor characteristics F and certainly not on the input $\vec{p}$. Therefore, M and its inverse can be computed once and for all and then reused for all processors and inputs.

- Since the columns of M are orthogonal, M^{-1} is just M^T with each row divided by its length squared.

- Multiplying a vector $\vec{v}$ by M and by M^{-1} does not require time $O(K^2)$; it requires only time $O(K \log K)$. The algorithm to achieve this is known as the fast Fourier transform. In its modern form, it was discovered by Cooley and Tukey, but a version was known to Gauss. The explanation of this algorithm is beyond the scope of this book; see, for example, (Cormen et al., 2009, Chapter 30).

The fast Fourier transform is executed by electronic devices all over the world millions of times every second. It is no exaggeration to say that modern electronic and communication depend critically on the fact that this basis transformation can be computed in time $O(K \log K)$ rather than time $O(K^2)$.

7.6.1   Other Applications of the Fourier Transform

Many applications of the Fourier transform exist. One important category of application is in the *recognition* of waveforms. For example, Figure 7.8 shows the waveforms for a sustained middle A note as produced by a tuning fork, a flute, a violin, and a singer, together with their Fourier transforms (see Petersen, 2004). (To be precise, the diagrams in the right column show the *energy* at each frequency, which is the sum of the squares of the amplitudes of the sine and cosine wave at that frequency.) The waveforms are obviously very different, but it is not immediately obvious how one would best distinguish

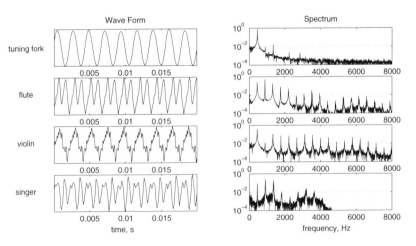

Figure 7.8. Waveforms and energy spectra. (Petersen, 2004, Figure 3; courtesy of Mark Petersen and the MAA.)

them algorithmically. However, the energy spectra can be easily distinguished, by comparing the energies at the fundamental frequencies and the harmonics (the other peaks, integer multiples of the fundamental) between the different spectra. Note that the energy spectra are drawn on a logarithmic scale, so a small difference in height corresponds to a substantial ratio in the energy. The perceived quality of the tones likewise reflects characteristics of the energy spectrum:

> The timbre, or sound quality of an instrument, is due to the relative strengths of the harmonics. A pure tone, composed of the fundamental alone, is shrill and metallic, like a tuning fork. Power in the higher harmonics add warmth and color to the tone. Figure [7.8] shows that a flute is dominated by the lower harmonics, accounting for its simple, whistle-like timbre. In contrast violins have power in the higher harmonics which gives them a warmer, more complex sound. The effect of higher harmonics is clearly seen within the wave form, where the violin has complex oscillations within each period. (Petersen, 2004)

(Energy spectrum analysis does not by any means account for all the recognizable differences in timbre between different musical instruments; another aspect is the shape of the waveform at the beginning of the note, which is quite different from the steady state.)

Another common application is in *denoising* a signal. In many situations, it can be assumed that the low-frequency components of a waveform are the signal and the high-frequency components are noise. A low-pass filter, which keeps only the low-frequency components, can easily be constructed by performing a DFT, deleting the high-frequency components, and then translating back to the time domain. (Such filters can also be built out of analog components, but software is often easier to construct, modify, and integrate.)

Innumerable other applications of the DFT exist, including image deblurring, image compression, recovering signals from samples, tomography, and finance (Marks, 2009).

7.6.2   The Complex Fourier Transform

The discrete Fourier transform is generally presented by using vectors and matrices whose components are complex numbers. In particular, code for computing the DFT, such as MATLAB's `fft` function, mostly use the complex version. The complex DFT is in many ways simpler than the real DFT presented above; in the complex DFT, the matrix D is a pure diagonal matrix. However, explaining the complex DFT would require a discussion of the theory of matrices over complex numbers; this theory has some important differences from matrices over the reals and it is beyond the scope of this book. Section 7.10 presents "black box" code to convert the complex `fft` function provided by MATLAB to the real DFT described previously.

7.7   Singular Value Decomposition

In this section, we show that *every* linear transformation takes a particularly simple form, with one particular choice of the basis in the domain and another in the range. This is known as the *singular value decomposition (SVD)*. We begin by defining the SVD, and then we describe some of its properties and applications.

Theorem 7.4. *Let M be an $m \times n$ matrix. Let $r = \text{Rank}(M)$. Then there exists an orthonormal basis $\hat{b}_1, \ldots, \hat{b}_n$ for $\mathbb{R}^n$ such that $\{M \cdot \hat{b}_1 \cdots M \cdot \hat{b}_r\}$ is an orthogonal basis for* $\text{Image}(M)$.

In Theorem 7.4, the vectors $\hat{b}_1, \ldots, \hat{b}_n$ are called the *right singular vectors* for M; these form an orthonormal basis for $\mathbb{R}^n$. The vectors $\hat{b}_{r+1}, \ldots, \hat{b}_n$ are in $\text{Null}(M)$ and are thus an orthonormal basis for $\text{Null}(M)$. For $i = 1, \ldots, n$, let $\sigma_i = |M \cdot \hat{b}_i|$. The values $\sigma_1, \ldots, \sigma_n$ are called the *singular values* for M. Note that if $r < n$, then for $i = r + 1, \ldots, n$, $\sigma_i = 0$. For $i = 1, \ldots, r$, let $\hat{c}_i = M \cdot \hat{b}_i / \sigma_i$. The vectors $\hat{c}_1, \ldots, \hat{c}_r$ are the *left singular vectors* for M. By Theorem 7.4, $\hat{c}_1, \ldots, \hat{c}_r$ are an orthonormal basis for $\text{Image}(M)$. The proof of Theorem 7.4 is not difficult; see Section 7.7.2.

The set of singular values is uniquely defined for any M, but the set of singular vectors is not quite unique. If the value σ_i occurs only once in the list of singular values, then there are two choices for the associated vector, and one is the negative of the other. If the same value occurs q times in the list of singular values, then there is a q-dimensional subspace of vectors associated with that value, and any orthonormal basis for that subspace will serve as the associated right singular vectors.

The SVD can be visualized as follows: Consider the n-dimensional sphere S of unit vectors in $\mathbb{R}^n$. Now consider $M \cdot S = \{M \cdot \hat{s} \mid \hat{s} \in S\}$. $M \cdot S$ is an *ellipsoid* within the subspace $\text{Image}(M)$. The semiaxes of the ellipsoid are the values of $M \cdot \hat{b}_i$. So the lengths of the semiaxes are the singular values. The singular vectors $\hat{b}_i$ are the inverse image of the axes under M^{-1}.

The singular values and vectors can be defined by using Algorithm 7.1.

Thus, the first singular vector $\hat{b}_1$ is the unit vector $\hat{u}$ that maximizes $M\hat{u}$; the second singular vector $\hat{b}_2$ is the unit vector $\hat{u}$ orthogonal to $\hat{b}_1$ that maximizes $M\hat{u}$; the third singular vector $\hat{b}_3$ is the unit vector $\hat{u}$ orthogonal to both $\hat{b}_1$ and $\hat{b}_2$ that maximizes $M\hat{u}$; and so on.

Alternatively, we can go from smallest to largest. That is, the last singular vector $\hat{b}_n$ is the unit vector $\hat{u}$ that minimizes $M\hat{u}$; the second to last singular vector $\hat{b}_{n-1}$ is the unit vector $\hat{u}$ orthogonal to $\hat{b}_1$ that minimizes $M\hat{u}$; and so on. We end up with the same sets of singular values and singular vectors.

As an actual algorithm, this does not work very well because calculating the maximum in step (3) of Algorithm 7.1 is difficult; it involves the maximization of the quadratic function $(M\hat{b}) \bullet (M\hat{b})$ over the quadratic surface $\hat{b} \bullet \hat{b} = 1$, and

$\mathcal{V} \leftarrow \mathbb{R}^n;$ (1)
for $(i \leftarrow 1, \ldots, n)$ (2)
    $\hat{b}_i \leftarrow$ the unit vector $\hat{b}$ in $\mathcal{V}$ for which $|M \cdot \hat{b}|$ is maximal; (3)
    $\sigma_i \leftarrow |M \cdot \hat{b}_i|.$ (4)
    if $\sigma_i \neq 0$ then $\hat{c}_i = M \cdot \vec{b}_i / \sigma_i;$ (5)
    $\mathcal{V} \leftarrow$ the orthogonal complement of $\{\hat{b}\}$ in $\mathcal{V}$ (6)
endfor

Algorithm 7.1. SVD algorithm.

over the cross section of that surface with the linear subspace $\mathcal{V}$. The algorithms that are actually used in computing the SVD are beyond the scope of this book; see Trefethen and Bau (1997).

Example 7.5. Refer to Figure 7.9. Let

$$M = \begin{bmatrix} .0752 & 1.236 \\ -0.636 & -0.048 \end{bmatrix},$$

and let $\hat{b}_1 = \langle 0.6, 0.8 \rangle$, $\hat{b}_2 = \langle -0.8, 0.6 \rangle$, $\hat{c}_1 = \langle 0.96, -0.28 \rangle$, and $\hat{c}_2 = \langle 0.28, 0.96 \rangle$. It is easily checked that $\{\hat{b}_1, \hat{b}_2\}$ and $\{\hat{c}_1, \hat{c}_2\}$ are orthonormal bases, and that $M\hat{b}_1 = 1.5 \cdot \vec{c}_1$ and $M\hat{b}_2 = 0.5 \cdot \vec{c}_2$. Thus, the right singular values are $\hat{b}_1$ and $\hat{b}_2$, the left singular values are $\hat{c}_1$ and $\hat{c}_2$, and the singular values are $\sigma_1 = 1.5$ and $\sigma_2 = 0.5$.

Example 7.6. Let

$$M = \begin{bmatrix} 1.0447 & 1.4777 & 0.9553 \\ -2.4761 & -2.2261 & -0.9880 \end{bmatrix},$$

and let $\hat{b}_1 = \langle 2/3, 2/3, 1/3 \rangle$, $\hat{b}_2 = \langle -2/3, 1/3, 2/3 \rangle$, $\hat{b}_3 = \langle 1/3, -2/3, 2/3 \rangle$, $\hat{c}_1 = \langle 0.5, -0.866 \rangle$, and $\hat{c}_2 = \langle 0.866, 0.5 \rangle$. It is easily checked that $\{\hat{b}_1, \hat{b}_2, \hat{b}_3\}$ and $\{\hat{c}_1, \hat{c}_2\}$ are

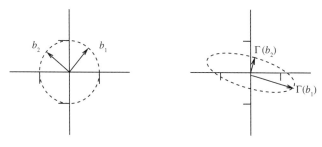

Figure 7.9. An example of the SVD (Example 7.5).

orthonormal sets, and that $M\hat{b}_1 = 4\vec{c}_1$, $M\hat{b}_2 = 0.5\vec{c}_2$, and $M\hat{b}_3 = \vec{0}$. Thus, $\hat{b}_1$, $\hat{b}_2$, and $\hat{b}_3$ are the right singular vectors; $\hat{c}_1$ and $\hat{c}_2$ are the left singular vectors; and $\sigma_1 = 4$, $\sigma_2 = 0.5$, $\sigma_3 = 0$ are the singular values.

7.7.1 Matrix Decomposition

Let M be an $m \times n$ matrix with rank r. Let $\sigma_1, \dots, \sigma_n$ be the singular values of M; let $\mathscr{B} = \{\hat{b}_1, \dots, \hat{b}_n\}$ be the corresponding right singular vectors; and let $\{\hat{c}_1, \dots, \hat{c}_r\}$ be the left singular vectors. If $r < m$, then let $\hat{c}_{r+1}, \dots, \hat{c}_m$ be an orthonormal basis for the orthogonal complement to Image(M) in $\mathbb{R}^m$; thus, $\mathscr{C} = \{\hat{c}_1, \dots, \hat{c}_m\}$ is an orthonormal basis for $\mathbb{R}^m$.

Now let $\vec{v}$ be any vector in $\mathbb{R}^n$. Let $p_1, \dots, p_n$ be the coordinates of $\vec{v}$ in $\mathscr{B}$. Thus, $\vec{v} = p_1 \cdot \hat{b}_1 + \dots + p_n \hat{b}_n$. So

$$\begin{aligned} M \cdot \vec{v} &= M \cdot (p_1 \cdot \hat{b}_1 + \dots + p_n \hat{b}_n) \\ &= p_1 M \cdot \hat{b}_1 + \dots + p_n M \cdot \hat{b}_n \\ &= p_1 \sigma_1 \hat{c}_1 + \dots + p_r \sigma_r \hat{c}_r. \end{aligned}$$

In the last step, the terms after r are all zero, and can be dropped. Thus, the coordinates of $M\vec{v}$ with respect to $\mathscr{C}$ are $\langle p_1\sigma_1, p_2\sigma_2, \dots, p_r\sigma_r, 0, 0, \dots \rangle$, filled out at the end with $(m - r)$ zeroes.

Therefore, we have Coords$(M\vec{v}, \mathscr{C}) = S \cdot$ Coords$(\vec{v}, \mathscr{B})$, where S is the $m \times n$ diagonal matrix with the singular values on the diagonal. Now let B be the $n \times n$ matrix with columns $\hat{b}_1, \dots, \hat{b}_n$ and let L be the $m \times m$ matrix with columns $\hat{c}_1, \dots, \hat{c}_m$. By Theorem 7.2 we have Coords$(\vec{v}, \mathscr{B}) = B^{-1}\vec{v}$ and $M \cdot \vec{v} = L \cdot$ Coords$(M \cdot \vec{v}, \mathscr{C})$. Let $R = B^{-1}$; since B is an orthonormal matrix $R = B^T$. Putting all of this together we have $M\vec{v} = (L \cdot S \cdot R)\vec{v}$. Since L, S, and R are independent of the choice of $\vec{v}$, this equation holds for all $\vec{v}$, so we must have $M = L \cdot S \cdot R$. We have thus proven Theorem 7.7.

Theorem 7.7. *Let M be any m × n matrix. Then there exist matrices L, S, R such that:*

- *S is an m × n diagonal matrix with elements on the diagonal in descending order,*

- *R is an n × n orthonormal matrix,*

- *L is an m × m orthonormal matrix,*

- *M = L · S · R.*

Specifically the diagonal elements of S are the singular values of M, the rows of R are the right singular vectors, and the columns of L are the left singular vectors.

The expression of M as the product $L \cdot S \cdot R$ is known as the *singular value decomposition* of M. Strictly speaking, this is a slight misnomer because the decomposition is not unique; in the case where the same singular value appears multiple times in S, Theorem 7.12 applies to all the singular value decompositions of M.

For instance, in Example 7.5, we have

$$\begin{bmatrix} .0752 & 1.236 \\ -0.636 & -0.048 \end{bmatrix} = \begin{bmatrix} 0.96 & 0.28 \\ -0.28 & 0.96 \end{bmatrix} \cdot \begin{bmatrix} 1.5 & 0 \\ 0 & 0.5 \end{bmatrix} \cdot \begin{bmatrix} 0.6 & 0.8 \\ -0.8 & 0.6 \end{bmatrix},$$

In Example 7.6, we have

$$\begin{bmatrix} 1.0447 & 1.4777 & 0.9553 \\ -2.4761 & -2.2261 & -0.9880 \end{bmatrix}$$

$$= \begin{bmatrix} 0.5 & 0.8660 \\ -0.8660 & 0.5 \end{bmatrix} \cdot \begin{bmatrix} 4.0 & 0 & 0 \\ 0 & 0.5 & 0 \end{bmatrix} \cdot \begin{bmatrix} 0.6667 & 0.6667 & 0.3333 \\ -0.6667 & 0.3333 & 0.6667 \\ 0.3333 & -0.6667 & 0.6667 \end{bmatrix}.$$

Now consider any $n \times n$ nonsingular matrix M. Then, Rank$(M) = n$, so all the singular values $\sigma_1, \ldots, \sigma_n$ are nonzero. Let $\vec{b}_1, \ldots, \vec{b}_n$ be the right singular vectors, and let $\vec{c}_1, \ldots, \vec{c}_n$ be the left singular vectors; thus, $M\vec{b}_i = \sigma_i \vec{c}_i$, and $M^{-1}\vec{c}_i = (1/\sigma_i)\vec{b}_i$. Therefore, the left singular vectors of M are the right singular vectors of M^{-1} and vice versa; the singular values of M^{-1} are the reciprocals of the singular values of M. (If we want to get the singular values in descending order, we have to reverse all of the numberings.)

Note that if we start with the SVD decomposition equation $M = LSR$ and we take the transpose of both sides, we get

$$M^T = (LSR)^T = R^T S^T L^T. \tag{7.1}$$

But since L and R are orthnormal matrices, their transposes are likewise orthonormal; and since S is a diagonal matrix, S^T is also a diagonal matrix. Therefore, Equation (7.1) is the SVD decomposition for M^T. In particular, the singular values of M^T are the same as those for M; as with most theorems about the transpose, this is not at all obvious from the fundamental definition.

7.7.2   Proof of Theorem 7.4 (Optional)

Lemma 7.8. *Let M be a matrix. Let $\hat{p}$ and $\hat{q}$ be orthogonal unit vectors. Let $\vec{u} = M \cdot \hat{p}$ and $\vec{v} = M \cdot \hat{q}$. If $\vec{u} \bullet \vec{v} > 0$, then there exists a unit vector $\hat{w}$ that is a linear sum of $\hat{p}$ and $\hat{q}$ such that $|M\hat{w}| > |M\hat{p}|$.*

Proof: We choose $\hat{w}$ to be the unit vector parallel to $\hat{p} + \epsilon\hat{q}$, where $\epsilon > 0$. (We specify the value of ϵ later.) Thus, $\vec{e} = \hat{p} + \epsilon\hat{q}$ and $\hat{w} = \vec{e}/|\vec{e}|$. Since $\hat{p}$ and $\hat{q}$ are orthogonal, $|\vec{e}| = \sqrt{1 + \epsilon^2}$, so $1/|\vec{e}|^2 = 1/(1 + \epsilon^2) > 1 - \epsilon^2$. Now,

$$M \cdot \hat{w} = M \cdot ((\hat{p} + \epsilon\hat{q})/|\vec{e}|) = (\vec{u} + \epsilon\vec{v})/|\vec{e}|,$$

so

$$(|M\hat{w}|)^2 = (M\hat{w}) \bullet (M\hat{w}) = (\vec{u} \bullet \vec{u} + 2\epsilon\vec{u} \bullet \vec{v} + \epsilon^2\vec{v} \bullet \vec{v})/|\vec{e}|^2 > (\vec{u} \bullet \vec{u} + 2\epsilon\vec{u} \bullet \vec{v} + \epsilon^2\vec{v} \bullet \vec{v})(1 - \epsilon^2)$$

$$= |\vec{u}|^2 + 2\epsilon\vec{u} \bullet \vec{v} \pm O(\epsilon^2).$$

So as long as $\epsilon \ll \vec{u} \bullet \vec{v}$, we have $|M \cdot \hat{w}|^2 > |\vec{u}|^2 = |M \cdot \hat{p}|^2$, so $|M \cdot \hat{w}| > |M \cdot \hat{p}|$. $\square$

Corollary 7.9. *Let M be an $m \times n$ matrix and let V be a subspace of $\mathbb{R}^n$. Let $\hat{p}$ be a unit vector in V such that $|M \cdot \hat{p}|$ is maximal, and let $\hat{z}$ be a unit vector in V orthogonal to $\hat{p}$. Then $M \cdot \hat{z}$ is orthogonal to $M \cdot \hat{p}$.*

Proof: Proof of the contrapositive: Suppose that $M\hat{z}$ is not orthogonal to $M\hat{p}$; thus, $(M\hat{z}) \bullet (M\hat{p}) \neq 0$. If $(M\hat{z}) \bullet (M\hat{p}) < 0$, then let $\hat{q} = -\hat{z}$; else, let $\hat{q} = \hat{z}$. Then $(M\hat{q}) \bullet (M\hat{p}) > 0$. By Lemma 7.8, there exists a unit vector $\hat{w}$ in V such that $|M\hat{w}| > |M\hat{p}|$. But then $\hat{p}$ is not the unit vector in V that maximizes $|M\hat{p}|$. $\square$

We now turn to the proof of Theorem 7.4. We prove that Theorem 7.4 is true by proving that the SVD algorithm (Algorithm 7.1) computes sets of right and left singular vectors and singular values that satisfy the theorem. This proof has three parts.

1. The algorithm can always be executed. The only part of the algorithm that is at all difficult is in regard to Step 3, the existence of a unit vector $\vec{b}$ in V that maximizes $|M \cdot \hat{b}|$. The existence of such a maximal value follows from a basic, general theorem of real analysis that a continuous function over a closed, bounded set has a maximal value (this theory is beyond the scope of this book); here, the function $|M \cdot \hat{b}|$ is the continuous function and the set of unit vectors in the vector space V is closed and bounded. Since the dimension of V is reduced by 1 at each iteration of the **for** loop, it works its way down from $\mathbb{R}^n$ at the start of the algorithm to the zero space at the end; therefore, the algorithm terminates.

2. The vectors $\hat{b}_1, \ldots, \hat{b}_n$ are orthogonal, since at the end of the ith iteration V is the orthogonal complement to $\hat{b}_1, \ldots, \hat{b}_i$, and all the remaining values of $\hat{b}$ are chosen from within this space.

3. All that remains to be shown is that the vectors $\hat{c}_1, \ldots, \hat{c}_r$ are orthogonal. Consider any two indices $i < j$. On the ith iteration of the loop, $\hat{b}_i$ is the unit vector in the current value of V that maximizes $|M \cdot \hat{b}_i|$. Since

$\hat{b}_j$ is in that value of V, and since $\hat{b}_j$ and $\hat{b}_i$ are orthogonal, it follows from Corollary 7.9 that $M\hat{b}_i$ and $M\hat{b}_j$ are orthogonal, so $\hat{c}_i$ and $\hat{c}_j$ are orthogonal.

7.8 Further Properties of the SVD

The SVD can be used to solve many other important minimization and maximization problems associated with a matrix or a linear transformation. We mention, without proof, some of the most important of these. What is remarkable is that all these different problems have such closely related solutions.

Definition 7.10. Let A and B be two $m \times n$ matrices. The *Frobenius distance* between A and B, denoted $d_2(A, B)$, is defined as

$$d_2(A, B) = \sqrt{\sum_{i,j} (A[i, j] - B[i, j])^2}.$$

The *absolute distance* between A and B, denoted $d_1(A, B)$, is defined as

$$d_1(A, B) = \sum_{i,j} |A[i, j] - B[i, j]|.$$

If we flatten A and B out to be two long, $m \cdot n$ dimensional vectors, then the Frobenius distance is just the Euclidean distance between these vectors. For example, if we let

$$A = \begin{bmatrix} 1 & 5 & 2 \\ 8 & 2 & 3 \end{bmatrix} \quad \text{and} \quad B = \begin{bmatrix} 2 & 8 & 2 \\ 6 & 4 & 2 \end{bmatrix},$$

then

$$d_2(A, B) = \sqrt{(1-2)^2 + (5-8)^2 + (2-2)^2 + (8-6)^2 + (2-4)^2 + (3-2)^2}$$
$$= \sqrt{19}$$
$$= 4.3589,$$
$$d_1(A, B) = |1-2| + |5-8| + |2-2| + |8-6| + |2-4| + |3-2|$$
$$= 9.$$

Definition 7.11. Let A be an $m \times n$ matrix; let $k \leq \min(m, n)$; and let L, S, R be the SVD decomposition of A. Thus, $A = L \cdot S \cdot R$. Let S^k be the diagonal matrix in which the first k diagonal elements are equal to the corresponding element of S and the remaining elements are 0. Let L^k be the $m \times m$ matrix in which the first k columns are equal to L and the rest are 0. Let R^k be the $n \times n$ matrix in which the first k rows are equal to R and the rest are 0. Then the *kth SVD reconstruction* of A is the product $LS^kR = L^kS^kR^k$.

Theorem 7.12. *Let A be an m × n matrix. The matrix of rank k that is closest to A is the kth SVD reconstruction of A when "closeness" is measured either by the Frobenius distance or by the absolute distance.*

Moreover, if the singular values that are dropped in S^k are small compared to those that are included, then these two distances are small compared to the inherent size of A. For details, see Trefethen and Bau (1997). Section 7.9.3, discusses how this theorem can be applied to data compression.

A MATLAB example follows.

```
>> A=[1,0,1,0;  2,4,6,7;  3,8,10,2]
A =
     1        0        1        0
     2        4        6        7
     3        8       10        2

>> [L,S,R]=svd(A)          % SVD Decomposition
L =
    -0.0590    0.0532   -0.9968
    -0.5878   -0.8090   -0.0084
    -0.8069    0.5854    0.0790
S =
    16.1103         0         0         0
          0    4.8394         0         0
          0         0    1.0185         0
R =
    -0.2269    0.0396   -0.7626   -0.6045
    -0.5466    0.2991    0.5875   -0.5163
    -0.7234    0.2178   -0.2527    0.6045
    -0.3556   -0.9282    0.0975   -0.0504

>> R=R';
>> L*S*R
ans =
     1.0000    0.0000    1.0000    0.0000
     2.0000    4.0000    6.0000    7.0000
     3.0000    8.0000   10.0000    2.0000

>> S1=zeros(3,4);
>> S1(1,1)=S(1,1)
S1 =
    16.1103         0         0         0
          0         0         0         0
          0         0         0         0



>> A1=L*S1*R                     % First SVD Approximation
A1 =
     0.2156    0.5194    0.6874    0.3379
     2.1485    5.1762    6.8504    3.3671
     2.9492    7.1052    9.4034    4.6219
```

```
>> rank(A1)
ans =
    1
>> norm(A1-A)                    % Frobenius distance
ans =
    4.8394
>> sum(sum(abs(A1-A)))          % Absolute distance
ans =
   11.9264

>> S2=S1;
>> S2(2,2)=S(2,2)
S2 =
   16.1103         0         0         0
         0    4.8394         0         0
         0         0         0         0
>> A2=L*S2*R                    % Second SVD Approximation
A2 =
    0.2258    0.5964    0.7435    0.0990
    1.9935    4.0050    5.9978    7.0008
    3.0613    7.9527   10.0203    1.9922
>> rank(A2)
ans =
    2
>> norm(A2-A)                    % Frobenius distance
ans =
    1.0185
>> sum(sum(abs(A2-A)))          % Absolute distance
ans =
    1.8774
```

We have already seen that the first right singular vector for matrix A is the unit vector $\hat{u}$ such that $|A \cdot \hat{u}|$ is maximal and that the last right singular vector is the vector $\hat{u}$ for which $|A \cdot \hat{u}|$ is minimal. Theorems 7.13 and 7.15 state that this generalizes to the first and last k singular vectors as a collection in two different ways.

Theorem 7.13. *Let A be an $m \times n$ matrix and let $k \le n$. For any orthonormal set $\{\hat{u}_1, \ldots, \hat{u}_k\}$ of k n-dimensional unit vectors, consider the sum $\sum_{i=1}^{k} |A \cdot \hat{u}_i|^2$. The set of k orthonormal vectors that maximizes this sum is the set of the first k right singular vectors of A. The set of k orthonormal vectors that minimizes this sum is the set of the last k right singular vectors of A.*

Section 14.6 discusses an application of Theorem 7.13 to statistical analysis.

Finally, Theorem 7.16 states that the space spanned by the k first right singular vectors of a linear transformation Γ is the k-dimensional subspace of $\mathbb{R}^n$ that is most expanded by Γ, and that the space spanned by the last k span is the space that is least expanded.

First, we need to define the amount that a linear transformation expands or contracts a subspace; this is done by using the determinant, as defined in Definition 7.14 and Theorem 7.15.

Definition 7.14. Let Γ be a linear transformation from $\mathbb{R}^n$ to $\mathbb{R}^m$. Let $k \leq$ Rank(Γ). Let V be any k-dimensional subspace of $\mathbb{R}^n$ such that $V \cap \text{Null}(\Gamma) = \{\vec{0}\}$. Let $\mathcal{B} = \langle \hat{v}_1, \ldots, \hat{v}_k \rangle$ be an orthonormal basis for V, and let $\mathcal{C} = \langle \hat{u}_1, \ldots, \hat{u}_k \rangle$ be an orthonormal basis for $\Gamma(V)$. Let G be the matrix such that, for any vector $\vec{v} \in V$, $\text{Coords}(\Gamma(\vec{v}), \mathcal{C}) = G \cdot \text{Coords}(\vec{v}, \mathcal{B})$. Then the *determinant of Γ restricted to V*, denoted $\text{Det}_V(\Gamma)$, is defined as $|\text{Det}(G)|$.

Here the value $\text{Det}_V(\Gamma)$ does not depend on the choice of the bases $\mathcal{B}$ and $\mathcal{C}$. The geometric significance of this quantity is given in Theorem 7.15.

Theorem 7.15. *Let Γ, k, and V be as in definition 7.14. Let* **R** *be a region in V. Then the k-dimensional measure (i.e., the k-dimensional analog of area or volume) of $\Gamma(\mathbf{R})$ is equal to $\text{Det}_V(\Gamma)$ times the k-dimensional measure of* **R**.

The determinant restricted to V is thus a measure of how much Γ expands or contracts regions in V.

We can now relate maximal and minimal expansion to the singular vectors.

Theorem 7.16. *Let A be an $m \times n$ matrix and let Γ be the corresponding linear transformation from $\mathbb{R}^n$ to $\mathbb{R}^m$. Let $r = \text{Rank}(\Gamma)$ and let $k < r$.*

- *Over all k-dimensional subspaces V such that $V \cap \text{Null}(\Gamma) = \{\vec{0}\}$, the maximum value of $\text{Det}_V(\Gamma)$ is achieved when V is the subspace spanned by the k first right singular vectors of A.*

- *Over all k-dimensional subspaces V such that V is orthogonal to $\text{Null}(\Gamma)$, the minimum value of $\text{Det}_V(\Gamma)$ is achieved when V is the subspace spanned by the k last singular vectors of A with nonzero singular value; that is, the vectors $\hat{u}_{r+1-k}, \ldots, \hat{u}_r$, where $\hat{u}_1, \hat{u}_2, \ldots, \hat{u}_n$ are the right singular vectors.*

7.8.1   Eigenvalues of a Symmetric Matrix

The theory of eigenvalues and eigenvectors is a very important aspect of matrix theory for applications in physics and engineering as well as purely mathematical analysis. We have largely omitted it in this textbook—first, because it is much less important in computer science applications; and second, because an adequate general discussion necessarily involves linear algebra over the field of complex numbers. However, for the particular case of a symmetric matrix, only real numbers are involved, and the theory is closely related (in fact, identical) to the theory of singular values, so we describe eigenvalues briefly here.

Definition 7.17. Let A be an $n \times n$ matrix. An n-dimensional vector $\vec{v}$ is an *eigenvector* of A with associated *eigenvalue* λ if $A \cdot \vec{v} = \lambda \cdot \vec{v}$.

Definition 7.18. A square matrix A is *symmetric* if $A^T = A$.

Theorem 7.19. *Let A be a symmetric matrix. Let $\hat{r}$ be a right singular vector with corresponding singular value σ and corresponding left singular vector $\hat{l}$. Then $\hat{r}$ is an eigenvector with eigenvalue $\pm\sigma$, and thus $\hat{l} = \pm\hat{r}$.*

The proof is omitted. The converse—if the right singular vectors are eigenvectors, then the matrix is symmetric—is uncomplicated, and left as Problem 7.4.

Corollary 7.20. *Let A be a symmetric $n \times n$ matrix. Then there exist n linearly independent real eigenvectors of A, with real eigenvalues. Moreover, the eigenvectors are all orthogonal.*

The significance of Corollary 7.20 within the general theory of eigenvalues is that for nonsymmetric matrices, eigenvalues and eigenvectors may be complex, eigenvectors may not be orthogonal, and there may not exist n linearly independent eigenvectors, even if complex vectors are considered. Thus, real symmetric matrices have eigenvectors that are strikingly well-behaved.

7.9 Applications of the SVD

This section discusses three applications of the SVD: condition number, approximate rank, and lossy compression. In Section 14.6, we discuss *principal component analysis*, which is an application of the SVD to statistics.

7.9.1 Condition Number

The *condition number* of a function is a measure of how errors in the input propagate to become errors in the output; specifically, it is the ratio between the relative error in the output and the relative error in the input. For instance, suppose that we want to calculate $f(x)$, and we have an estimate $\tilde{x}$ of x that is accurate to within 1% of $|x|$; that is, $|\tilde{x} - x| \leq .01x$. If the condition number of f is 4, then $f(\tilde{x})$ is within 4% of $f(x)$. If the condition number of f is 10, then $f(\tilde{x})$ is within 10% of $f(x)$. If the condition number of f is 1000, then $|f(\tilde{x}) - f(x)|$ may be 10 times as large as $|f(x)|$; in that case, $f(\tilde{x})$ is useless as an estimate for most purposes.

Symbolically,

$$\text{Condition}(f) = \sup_{x,\tilde{x}} \frac{|f(\tilde{x}) - f(x)|/|f(x)|}{|\tilde{x} - x|/|x|},$$

where x and $\tilde{x}$ range over all values for which none of the denominators involved are 0.

For a linear transformation, the condition number can be directly computed from the singular values.

Theorem 7.21. *If f is the linear transformation corresponding to matrix M, then the condition number of f is the ratio of the largest and smallest singular values: $\text{Condition}(f) = \sigma_1(M)/\sigma_n(M)$.*

It is easy to show that the condition number is at least as large as σ_1/σ_n. Let $\hat{b}_1$ and $\hat{b}_n$ be the right singular vectors corresponding to σ_1 and σ_n. Let $x = \hat{b}_n$ and let $\tilde{x} = \vec{b}_n + \epsilon \hat{b}_1$. Then $|x| = 1$; $|\tilde{x} - x| = \epsilon$, $|f(x)| = |f(\hat{b}_n)| = \sigma_n$ and $|f(\tilde{x}) - f(x)| = f(\tilde{x} - x) = f(\epsilon \hat{b}_1) = \epsilon \sigma_1$. So $\text{Condition}(f) \geq (\epsilon \sigma_1/\sigma_n)/(\epsilon/1) = \sigma_1/\sigma_n$.

Showing that the condition number is not larger than σ_1/σ_n takes a little more work, and we omit it here.

We can use this to gain more insight into the problems in Section 5.5 that gave MATLAB such grief. Recalling the first example in Section 5.5, let $d = 10^{-9}$ and let M be the matrix

$$M = \begin{bmatrix} 1+d & 1 \\ 1 & 1-d \end{bmatrix}.$$

It is easily shown that the singular vectors are approximately $\langle \sqrt{1/2}, \sqrt{1/2} \rangle$ with singular value 2 and $\langle \sqrt{1/2}, -\sqrt{1/2} \rangle$ with singular value d. Hence, the condition number is $2/d = 2 \cdot 10^9$, so this matrix is very ill-conditioned. In solving the linear equation $M\vec{x} = \langle 1, 1 \rangle$, note that the singular values of M^{-1} are 10^9, and $1/2$, and the condition number is again $2 \cdot 10^9$. Another example is discussed in Exercise 7.2.

7.9.2 Computing Rank in the Presence of Roundoff

As discussed in Section 5.2, the rank of matrix M can be computed by using Gaussian elimination to reduce M to row-echelon form and then counting the number of nonzero rows. This works very well when the coefficients of M are known exactly and the Gausssian elimination algorithm can be carried out precisely. However, if there is any noise in the coefficients of M, or any roundoff error in the execution of Gaussian elimination, then Gaussian elimination does not give any useful information about the rank. For example, it does not work to discard any rows that are "close to" $\vec{0}$; Gaussian elimination normalizes the first coefficient in each row to 1, so no nonzero row is close to $\vec{0}$.

In this situation, the following method works well; it is, in fact, the method that MATLAB uses to compute rank. We set a small threshhold, compute the singular values of M, and conjecture that singular values below the threshhold are actually zero and therefore the rank is equal to the number of singular values above the threshhold. This conjecture is reasonable in two senses. First, it often does give the true rank of M; it is indeed often the case that the singular values that are computed as small are actually zero, and they appear as nonzero only due to roundoff error. Second, even if the small singular values are actually nonzero, the rank r computed by this procedure is "nearly" right, in the sense that M is very close to a matrix M' of rank r; namely, the rth SVD reconstruction of M.

Choosing a suitable threshhold is a process of some delicacy; the right choice depends the size of M and on how M was constructed. The comments on Assignment 6.1 discuss a case in which MATLAB 's default threshhold turns out to be too small.

7.9.3   Lossy Compression

A *compression* algorithm Φ is one that takes as input a body of data D and constructs a representation $E = \Phi(D)$ with the following two constraints.

- The computer memory needed to record E is much less than the natural encoding of D.

- There is a decompression algorithm Ψ to reconstruct D from E. If $\Psi(E)$ is exactly equal to D then Φ is a *lossless* compression algorithm, If $\Psi(E)$ is approximately equal to D, then Φ is a *lossy* compression algorithm.

Whether a lossless compression algorithm is needed or a lossy compression algorithm will suffice depends on the application. For compressing text or computer programs, we need lossless compression; reconstituting a text or program with each character code off by 1 bit is not acceptable. For pictures, sound, or video, however, lossy compression may merely result in a loss of quality but still be usable; if every pixel in an image is within 1% of its true gray level, a human viewer will barely notice the difference.

The SVD can be used for lossy compression as follows. Suppose that your data are encoded in an $m \times n$ matrix M. Compute the SVD decomposition $M = LSR$. Choose a value k that is substantially less than m or n. Now let S' be the diagonal matrix with the first k singular values. Let R' be the matrix whose first k rows are the same as R and the rest 0. Let L' be the matrix whose first k columns are the same as L and the rest 0. Let $M' = L'S'R'$. By T heorem 7.12, M' closely approximates M, particularly if the singular values that have been dropped are small compared to those that have been retained.

The compressed encoding here is to record the nonzero elements of L', S' and R'. L' has km nonzero elements, S' has k, and R' has kn, for a total of $k(m+n+1)$; if k is small, then this is substantially smaller than the mn nonzero elements in M.

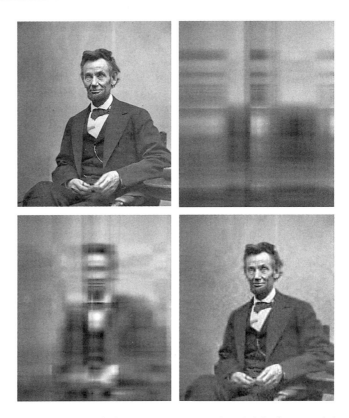

Figure 7.10. Using the SVD for lossy image compression. Original gray-scale image (upper left) and reconstructions with two singular values (upper right), five singular values (lower left), and 30 singular values (lower right).

Figures 7.10 and 7.11 show a gray-scale image and a line drawing, respectively, and a few of their approximations with different values of k.

Figure 7.11. Using the SVD for lossy image compression. Original line drawing of a bear (left) and reconstructions with 4 singular values (middle), and 10 singular values (right).

7.10   MATLAB

7.10.1   The SVD in MATLAB

The MATLAB function svd(M) can be used in two ways. If we request one value
to be returned, then the function returns the vector of singular values; if we
request three values to be returned, then it returns the three matrices L, S, R.

```
% Example 7.2 in the text
>> M=[1.0447, 1.4777, 0.9553; -2.4761, -2.2261, -0.9880]
M =
     1.0447    1.4777    0.9553
    -2.4761   -2.2261   -0.9880

>> svd(M)
ans =
     4.0000
     0.5000

>> [L,S,R]=svd(M)
L =
    -0.5000    0.8660
     0.8660    0.5000
S =
     4.0000         0         0
          0    0.5000         0
R =
    -0.6667   -0.6667    0.3333
    -0.6667    0.3333   -0.6667
    -0.3333    0.6667    0.6667

>> P=L*S*R'
P =
     1.0447    1.4777    0.9553
    -2.4761   -2.2261   -0.9880

>> norm(P-M)
ans =
     7.3723e-16
```

7.10.2   The DFT in MATLAB

MATLAB provides built-in functions fft and ifft for carrying out the fast Fourier
transform; however, these use the complex version of the FFT.

   If we want to use the real DFT, as described in Section 7.6, then we have
two choices. The conceptually easy way is to use the function trigBasis(K) to
construct the $2K \times 2K$ matrix M of Fourier component vectors. Then M\V will
convert a vector V in the real domain to the frequency domain, and M * V will

convert V from the frequency domain to the time domain. However, this loses the efficiency of the FFT.

To take advantage of the efficient FFT, two "black box" routines are provided here that compute the real DFT using MATLAB's built-in fft as a subroutine. The function dft(V) maps a column vector V from the time domain to the frequency domain; that is, dft(V) = trigBasis(K)\V. The function idft(V) maps a column vector V from the frequency domain to the time domain; that is, trigBasis(K) * V = idft(V).

In using the FFT, note that the algorithm works best if the dimension is a power of two or factorizable into small primes. It runs considerably slower if the dimension is a large prime or has a large prime factor. Of course, depending on the application, there may not be any choice about the dimension.

```
% File trigBasis.m
% Returns the matrix whose columns are the trigonometric Fourier components
% of size 2K
%
%       1  cos(pi/K)    sin(pi/K)    cos(2 pi/K) sin(2 pi/K) ... cos(pi)
%       1  cos(2 pi/K) sin(2 pi/K) cos(4 pi/K) sin(4 pi/K) ... cos(pi)
%       1  cos(3 pi/K) sin(3 pi/K) cos(6 pi/K) sin(6 pi/K) ... cos(pi)
%       ...

function b = trigBasis(K)
   N=2*K;
   b=zeros(N,N);
   b(:,1) = ones(N,1);
   for j=1:K
     for i=1:N
       b(i,2*j) = cos(i*j*pi/K);
       if (2*j ~= N) b(i,2*j+1) = sin(i*j*pi/K); end
     end
   end
end

% Example
>> trigBasis(2)
ans =
    1.0000    0.0000    1.0000   -1.0000
    1.0000   -1.0000    0.0000    1.0000
    1.0000   -0.0000   -1.0000   -1.0000
    1.0000    1.0000   -0.0000    1.0000
>> trigBasis(3)
ans =
    1.0000    0.5000    0.8660   -0.5000    0.8660   -1.0000
    1.0000   -0.5000    0.8660   -0.5000   -0.8660    1.0000
    1.0000   -1.0000    0.0000    1.0000   -0.0000   -1.0000
    1.0000   -0.5000   -0.8660   -0.5000    0.8660    1.0000
    1.0000    0.5000   -0.8660   -0.5000   -0.8660   -1.0000
    1.0000    1.0000   -0.0000    1.0000   -0.0000    1.0000
```

```
% File dft.m
% Converts column vector in time domain to vector in frequency domain
%    using the real Fourier components discussed in the text
% Let X = < x[1] ... x[n] >
% Let F = dft(X)
% Then X[i] = F[1] + F[2] cos(pi i/N) + F[3] sin(pi i/N) + ...
%        + F[N-2] cos(pi (N-1)/N) +F[N-1] sin(pi i (N-1)/N)) + F[N] cos(pi i).

% dft(x) is equal to trigBasis(N/2)\x but computes faster, since it uses
% the FFT.

function f = dft(x)
   n = size(x);
   n = n(1);
   y(1,1) = x(n,1);
   y(2:n,1) = x(1:n-1,1);% Rotate x to change from 1-based indexing to 0-based
   g = ifft(y);
   f(1,1)=g(1,1);
   k = floor(n/2);
   for j=1:k-1
     f(2*j,1) = g(j+1,1) + g(n+1-j);
     f(2*j+1,1) = (g(j+1,1) - g(n+1-j))/i;
   f(n,1)=g(k+1,1);
   end
end

% Examples
>> dft([1;2;2;1])'
ans =
    1.5000   -0.5000   -0.5000        0
>> (trigBasis(2)\[1;2;2;1])'
ans =
    1.5000   -0.5000   -0.5000        0

>> dft([1;4;2;8;5;7])'
ans =
    4.5000    0.6667   -2.3094        0        0    1.8333

>> (trigBasis(3)\[1;4;2;8;5;7])'
ans =
    4.5000    0.6667   -2.3094   -0.0000   -0.0000    1.8333


% Converts column vector in frequency domain to vector in time domain.
%    using the real Fourier components discussed in the text
% Let X = < x[1] ... x[n] >
% Let F = idft(X)
% Then F[i] = X[1] + X[2] cos(pi i/N) + X[3] sin(pi i/N) +
%        ... X[N-2] cos(pi (N-1)/N) +X[N-1] sin(pi i (N-1)/N)) + X[N] cos(pi i).

% idft(x) is equal to trigBasis(N/2)*x but computes faster, since it uses
% the FFT.
```

```
function f = idft(x)
   n = size(x);
   n = n(1);
   k = floor(n/2);
   y(1,1)=x(1,1);
   for j=1:k-1
     y(j+1,1) = (x(2*j,1) + i * x(2*j+1,1))/2;
     y(n+1-j,1) = (x(2*j,1) - i * x(2*j+1,1))/2;
   end;
   y(k+1,1) = x(n,1);
   g = fft(y);
   f(n,1) = g(1,1);
   f(1:n-1,1) = g(2:n,1); % Rotate g to change from 0 based indexing to 1 based
end

% Examples

>> idft([1;2;2;1])'
ans =
      2        0      -2       4
>> (trigBasis(2)*[1;2;2;1])'
ans =
   2.0000    0.0000   -2.0000    4.0000

>> idft([1;4;2;8;5;7])'
ans =
  -1.9378   -0.5981   -2.0000    4.5981  -14.0622   20.0000

>> (trigBasis(3)*[1;4;2;8;5;7])'
ans =
  -1.9378   -0.5981   -2.0000    4.5981  -14.0622   20.0000

>> idft(dft([1;4;2;8;5;7]))'
ans =
   1.0000    4.0000    2.0000    8.0000    5.0000    7.0000

>> dft(idft([1;4;2;8;5;7]))'
ans =
   1.0000    4.0000    2.0000    8.0000    5.0000    7.0000
```

Exercises

Exercise 7.1. Let $\mathscr{C}$ be the coordinate system with origin $\mathbf{o}$, unit length d, and unit coordinate arrows $\overrightarrow{x}$ and $\overrightarrow{y}$. Let $\mathscr{B}$ be the coordinate system with origin $\mathbf{p}$, unit length e, and unit coordinate arrows $\overrightarrow{i}$ and $\overrightarrow{j}$.

Suppose that $\mathrm{Coords}(\mathbf{p}, \mathscr{C}) = \langle 2,3 \rangle$, $\mathrm{Coords}(e, \mathscr{C}) = 5$, $\mathrm{Coords}(\overrightarrow{i}, \mathscr{C}) = \langle 4,3 \rangle$, and $\mathrm{Coords}(\overrightarrow{j}, \mathscr{C}) = \langle -3,4 \rangle$.

(a) Let **q** be the point with coordinates $\langle 1, 2 \rangle$ in $\mathcal{B}$. What are the coordinates of **q** in $\mathcal{C}$?

(b) Let **r** be the point with coordinates $\langle 1, 2 \rangle$ in $\mathcal{C}$. What are the coordinates of **r** in $\mathcal{B}$?

Note: You should be able to solve (a) by hand. For (b), try to solve it by hand; if you get stuck, use MATLAB.

Exercise 7.2. Consider the 8×8 ill-conditioned matrix discussed in Section 5.5:

```
1.001   1.002   1.003   1.004   1.005   1.006   1.007   1.009
1.002   1.003   1.004   1.005   1.006   1.007   1.008   1.010
1.003   1.004   1.005   1.006   1.007   1.008   1.010   1.011
1.004   1.005   1.006   1.007   1.008   1.010   1.011   1.012
1.005   1.006   1.007   1.008   1.010   1.011   1.012   1.013
1.006   1.007   1.008   1.010   1.011   1.012   1.013   1.014
1.007   1.008   1.010   1.011   1.012   1.013   1.014   1.015
1.008   1.010   1.011   1.012   1.013   1.014   1.015   1.016
```

(a) What is the condition number for this matrix? (*Note:* Copying this from the 8×8 matrix shown is error-prone; the matrix is not quite as regular as it appears at first glance. Copy the MATLAB code used to create it in Section 5.5.)

(b) In the example given on page 133, let x be the second vector of constant terms and let $\tilde{x}$ be the first vector. Then the ratio of relative error in the answer to relative error in the constant terms is about 84,000, which is large but much less than the condition number. Construct an example for which this ratio is close to the condition number.

Exercise 7.3. This exercise asks you to experiment with using singular value decomposition for image compression.

Step 1. Obtain a gray-scale image to use.

Step 2. Load the image into an $m \times n$ array Image, by using the supplied function[5] getimage.m.

```
Image = getimage('lincoln.png');
```

Step 3. Carry out the SVD decomposition on Image:

```
[L,S,R] = svd(Image);
R=R';
```

---

[5]This function is courtesy of Pascal Getreuer, in http://www.math.ucla.edu/~getreuer/ matlabimaging.html.

Step 4. For various small values of k, construct the kth order approximation of Image, as described in Section 7.9.3:

- Let $S1$ be the $k \times k$ matrix with the first k singular values in S.
- Let $L1$ be the $m \times k$ matrix with the first k columns of L.
- Let $R1$ be the $k \times n$ matrix with the first k rows of R.
- Let $Image1 = L1 \cdot S1 \cdot R1$.

Step 5. Print out the image by using the MATLAB function `imwrite`. It is recommended that images be output in .png format.

```
imwrite(Image1,'<filename>')
```

Step 6. Pick a couple of characteristics of the image, and find the smallest values of k for which these characteristics are recognizable in the reconstructed image. For instance, for Lincoln, you might choose to find the smallest values for which the image is recognizably a seated person, and the smallest value for which it is recognizably Lincoln.

You should submit the images you have found in Step 6, together with a statement of what features you were looking for.

If you want to be ambitious, use a color image instead. The function `getimage` will return an $m \times n$ array, one layer for each primary color. Do SVD decompositions on each of the layers separately, and proceed as above.

If you want to be really ambitious, do this with several images of different visual characteristics—high contrast versus low contrast, simple versus complex, gray scale versus line drawing—and try to get a feeling for when this works well, when badly, and so forth.

Problems

Problem 7.1. Let A be an $m \times n$ matrix and let $A = L \cdot S \cdot R$ be the singular value decomposition of A. The *Moore-Penrose pseudoinverse* of A is the $n \times m$ matrix computed as follows: Let S' be the $n \times m$ diagonal matrix in which each nonzero singular value σ in S is replaced by $1/\sigma$ and every zero singular value is left unchanged. Then the pseudoinverse A' is equal to $R^T \cdot S' \cdot L^T$.

For example, let

$$A = \begin{bmatrix} 13/9 & 11/9 & 13/9 & 11/9 \\ -23/18 & -25/18 & -23/18 & -25/18 \\ 5/9 & 7/9 & 5/9 & 7/9 \end{bmatrix}.$$

Then the SVD of A is $A = L \cdot S \cdot R$ where

$$
L = \begin{bmatrix} 2/3 & 2/3 & 1/3 \\ -2/3 & 1/3 & 2/3 \\ 1/3 & -2/3 & 2/3 \end{bmatrix}, \quad
S = \begin{bmatrix} 4 & 0 & 0 & 0 \\ 0 & 1/3 & 0 & 0 \\ 0 & 0 & 0 & 0 \end{bmatrix}, \quad
R = \begin{bmatrix} 1/2 & 1/2 & 1/2 & 1/2 \\ 1/2 & -1/2 & 1/2 & -1/2 \\ 1/2 & 1/2 & -1/2 & -1/2 \\ 1/2 & -1/2 & -1/2 & 1/2 \end{bmatrix}.
$$

Then

$$
S' = \begin{bmatrix} 1/4 & 0 & 0 \\ 0 & 3 & 0 \\ 0 & 0 & 0 \\ 0 & 0 & 0 \end{bmatrix},
$$

so

$$
A' = R^T \cdot S' \cdot L^T = \begin{bmatrix} 13/12 & 5/12 & -23/24 \\ -11/12 & -7/12 & 25/24 \\ 13/12 & 5/12 & -23/24 \\ -11/12 & -7/12 & 25/24 \end{bmatrix}.
$$

(Keep in mind that L and R are orthogonal matrices, so $L^T = L^{-1}$ and $R^T = R^{-1}$.)
Prove that, for *any* matrix A (*not* just the particular example above) if A' is the pseudoinverse, then the following properties hold:

(a) $A \cdot A' \cdot A = A$.

(b) $A' \cdot A \cdot A' = A'$.

(c) If A is a nonsingular square matrix, then $A' = A^{-1}$.

(d) $A \cdot A'$ and $A' \cdot A$ are both symmetric matrices. That is, if $B = A \cdot A'$ and $C = A' \cdot A$ then $B^T = B$ and $C^T = C$.

   The converse is also true: For any matrix A, the only matrix that satisfies the properties (a), (b), and (d) is the Moore-Penrose pseudoinverse of A. However, that is not easy to prove.

Problem 7.2. Section 6.4.7 stated that if A is a square matrix and Γ is the linear transformation corresponding to A, then, for a region R, Volume$(\Gamma(R))$ = $|\mathrm{Det}(A)| \cdot$Volume(R). Using this relation, show that, for any square matrix A, $|\mathrm{Det}(A)|$ is equal to the product of the singular values of A.

Problem 7.3. A transformation Γ is a *projection* if, for all vectors $\vec{v}$, $\Gamma(\Gamma(\vec{v}))$ = $\Gamma(\vec{v})$. (This is related to the geometric notion of projection, but a little different.) Characterize a projection in terms of its singular values and in terms of the relation between the right and left singular vectors.

Problem 7.4. Prove the following: If the right singular vectors of matrix A are all eigenvectors of A, then A is a symmetric matrix.

Problem 7.5. Any set $\mathscr{B}$ of n linearly independent n-dimensional vectors forms a basis for $\mathbb{R}^n$, and therefore any n-dimensional vector can be written as a linear sum over $\mathscr{B}$. Applying this logic to the three-dimensional vector space of colors would seem to suggest that if your palette is three somewhat different shades of red, you can mix them to form any desired color, such as blue. That seems implausible. Resolve this apparent paradox.

Programming Assignments

Assignment 7.1. A *low-pass filter* is a signal processor that removes, or lessens, the high-frequency Fourier components of a signal while leaving the low-frequency components unchanged. Conversely, a *high-pass filter* removes, or lessens, the low-frequency, leaving the high-frequency unchanged.

Write two functions LowPass(S,F) and HighPass(S,P), using input parameter S as a time signal over time interval [1,2N] and P as the cut-off period in one direction or the other. (The frequency is the reciprocal of the period.) The algorithms should proceeds as follows:

- Let T be the discrete Fourier transform of S.

- For the low-pass filter, set $T[k] \leftarrow 0$ for all $k > N/P$. For the high-pass filter, set $T[k] \leftarrow 0$ for all $k \le N/P$.

- Return the inverse transform of the modified value of T.

Test your program on the square wave $S[i] = 1$ for $i = 1, \ldots, 32$, $S[i] = -1$ for $i = 33, \ldots, 60$. Plot the original signal, the Fourier coefficients, and the results of low-pass and high-pass filters for various values of N.

II

Probability

Chapter 8

Probability

Most of mathematics deals with statements that are known with certainty, or as least as much certainty as can be attained by fallible human beings. The statement "2+2=4" is the archetype of a statement that is known to be true. In a similar way, the objective in many types of computer software is to be entirely reliable, or at least as reliable as a machine that is built of silicon and is running a commercial operating system can be. Indeed, we don't want compilers, or payroll databases, or control systems for nuclear reactors to give more or less correct answers most of the time—we want them to always work correctly. To a substantial degree, this is often an attainable goal.

The reality of life, however, is that we cannot work exclusively with information that is known with certainty. Often we must choose our actions as best we can on the basis of information that is partial and uncertain. Likewise, programs such as search engines, recommender systems, automatic translators, and the like, whose results are not reliably correct, occupy an increasingly large and important segment of the world of software.[1] Therefore, the mathematical theory of dealing with uncertain information has become increasingly important in many computer science applications. *Probability theory* is the fundamental theory of manipulating uncertainty.

8.1   The Interpretations of Probability Theory

Probability theory has to do with assigning numbers, called *probabilities*, to entities of a certain kind, called *events*. For the remainder of this book, we use the notation $P(E)$ to mean the probability of event E.

Unfortunately, there are multiple, conflicting interpretations of probability theory, each with a different idea about what kind of entity an "event" is, and

---

[1]Even with programs from which we expect and can get reliably correct answers, there may be other aspects from which a probabilistic analysis is useful. For example, a database may introduce a query optimization that, with high probability, speeds up the computation of the answer.

therefore a different idea of what probability theory is fundamentally *about*. (See Fine (1973) for a survey.) In this book, we look at two interpretations:

- The *sample space* or *frequentist* interpretation, which has a fixed universe of entities Ω, called a sample space. An event is a subset of the sample space. The probability of event E is the fraction of Ω that is in E.

- The *likelihood* or *subjectivist* interpretation, in which an event E is a sentence or a proposition. $P(E)$ is a judgment of how likely it is that E is true, as judged by a reasoner with some particular background knowledge.

We first discuss the sample space interpretation, which is conceptually the simpler of the two theories. We then turn to the likelihood interpretation, which is closer to the way that probability theory is used in the kinds of applications we later discuss. But we will see that the way that probability theory is used in applications does not fit perfectly with the likelihood interpretation either. There is, perhaps, yet another interpretation, the "application" interpretation, implicit here; if so, no one has yet clearly formulated what it is.

8.2   Finite Sample Spaces

In the sample space interpretation of probability theory, all discussions of probabilities are carried out relative to a *probability space*. Probability spaces can be either finite or infinite; both cases are important, and there are significant differences between them. In this chapter, we deal exclusively with finite spaces. (We deal with the infinite case in Chapter 9.)

A finite probability space consists of two parts: a finite set Ω, called the *sample space*, and a real-valued function $P(x)$ over Ω, called the *probability function* or *distribution*. The probability function has two properties:

F1. For all $x \in \Omega$, $0 \le P(x) \le 1$.

F2. $\sum_{x \in \Omega} P(x) = 1$.

An event is a subset of Ω. For any event E, the probability of E is the sum of the probabilities of its elements: $P(E) = \sum_{x \in E} P(x)$.

A set containing a single element of Ω is an *elementary event*.

A probability function that satisfies properly F2 is said to be *normalized*. It is often convenient to deal with a more general class of weight functions $w(x)$ that satisfy the following weaker conditions:

W1. For all $x \in \Omega$, $w(x) \ge 0$.

W2. For some $x \in \Omega$, $w(x) > 0$.

We define the weight of any event E as the sum of the weights of the elements of E: $w(E) = \sum_{x \in E} w(x)$.

Clearly, if $P(x)$ is a probability function satisfying F1 and F2, then the weight function $w(x) = P(x)$ satisfies W1 and W2. Conversely, if weight function $w(x)$ satisfies W1 and W2, then we can *normalize* it by defining the associated probability function $P(x) = w(x)/w(\Omega)$. It is immediate that $P(x)$ so defined satisfies F1 and F2 and that for any event E, $P(E) = w(E)/w(\Omega)$.

The following examples are standard starting points for discussions of probability.

Example 8.1. We flip a fair coin. The sample space $\Omega = \{$ H, T $\}$. The probability function is $P(H) = 1/2$; $P(T) = 1/2$. The weight function $w(H) = w(T) = 1$ gives the same probability function. There are only four events in this sample space: $P(\emptyset) = 0$; $P(\{H\}) = 1/2$; $P(\{T\}) = 1/2$; $P(\Omega) = 1$. This probability function is called the *uniform* or *equiprobable* distribution, because all the elements of Ω have the same probability.

Example 8.2. We flip a weighted coin that comes up heads $3/4$ of the time. The sample space $\Omega = \{$ H, T $\}$. The probability function is $P(H) = 3/4$; $P(T) = 1/4$.

Example 8.3. We roll a fair die. The sample space $\Omega = \{1, 2, 3, 4, 5, 6\}$. The probability function is the equiprobable distribution $P(x) = 1/6$ for each $x \in \Omega$. There are $64 = 2^6$ possible events here. Example probabilities include the following.

- To compute the probability of rolling an even number, we use the event $E = \{2, 4, 6\}$, and compute $P(E) = P(2) + P(4) + P(6) = (1/6) + (1/6) + (1/6) = 3/6$.

- To compute the probability of rolling at least a 3, we use the event $E = \{3, 4, 5, 6\}$, and compute $P(E) = P(3) + P(4) + P(5) + P(6) = (1/6) + (1/6) + (1/6) + (1/6) = 4/6$.

When the probability is uniformly distributed, as in Example 8.3, we can equally well use the weight function $w(x) = 1$ for each $x \in \Omega$. Then for any event E, $w(E) = \#E$, the number of elements in E, so $P(E) = \#E/\#\Omega$.

Example 8.4. We flip a fair coin three times. The sample space here is the set of all eight possible outcomes:

$$\Omega = \{\text{HHH, HHT, HTH, HTT, THH, THT, TTH, TTT}\}.$$

The probability function is the uniform distribution, so, as with the die in Example 8.3, above, for any event E, $P(E) = \#E/\#\Omega$. For instance, if we want to know the probability of getting at least two heads, then $E = \{$ HHT, HTH, THH $\}$, so $P(E) = \#E/\#\Omega = 3/8$.

8.3 Basic Combinatorial Formulas

Many probabilistic problems, such as computing the probability of getting a full house (three cards of one number and two of another) in a five-card poker hand, require the use of combinatorial reasoning. On the whole, combinatorics plays a smaller role in computer science applications of probability than it does in other kinds of applications; nonetheless, everyone who is dealing with probability should know some basic formulas.

8.3.1 Exponential

The number of sequences of k items out of a set S of size n, allowing repetition, is n^k. For example, let $S = \{a, b\}$, $n = |S| = 2$, $k = 3$. Then the set of choices $C = \{aaa, aab, aba, abb, baa, bab, bba, bbb\}$ so $|C| = 8 = 2^3$. The argument is that each of the choices can be made in n ways independently of all the others; hence, the total number of combinations is $n \cdot n \cdot \ldots \cdot n$ (k times) $= n^k$.

8.3.2 Permutations of n Items

The number of sequences of all the elements, without repetition, in a set S of size n elements is called "n factorial," denoted $n!$, and computed as $n \cdot n-1 \cdot \ldots \cdot 2 \cdot 1$. For instance, if $S = \{a, b, c, d\}$ and $|S| = n = 4$, then the set of permutations is

$$
\begin{array}{cccccc}
abcd & abdc & acbd & acdb & adbc & adcb \\
bacd & badc & bcad & bcda & bdac & bdca \\
cabd & cadb & cbad & cbda & cdab & cdba \\
dabc & dacb & dbac & dbca & dcab & dcba
\end{array}
$$

and the number of permutations is $24 = 4! = 4 \cdot 3 \cdot 2 \cdot 1$.

The argument is that the first item can be chosen to be any of the n items in S. Having chosen the first, the second item can be chosen to be any of the remaining $n - 1$. Having chosen the first two, the third item can be chosen to be any of the remaining $n - 2$, and so on. The total number of combinations is therefore $n \cdot (n - 1) \cdot (n - 2) \cdot \ldots \cdot 2 \cdot 1 = n!$.

The MATLAB function `factorial(n)` computes $n!$.

Stirling's formula gives a useful approximation for $n!$:

$$
n! \approx \sqrt{2\pi n} \left(\frac{n}{e}\right)^n .
$$

Taking the natural logarithm of both sides and keeping only the two largest terms gives $\ln(n!) \approx n \ln n - n$. For example, $10! = 3{,}628{,}800$; Stirling's formula gives $3{,}598{,}695.62$, a relative error of 0.8%.

8.3.3 Permutations of k Items out of n

The number of sequences of k items chosen from a set S of n items, with no repetitions, is known as the *permutations* of k out of n; it is sometimes notated $P(n, k)$. It is computed as $n \cdot (n - 1) \cdot (n - 2) \cdot \ldots \cdot (n - (k - 1)) = n!/(n - k)!$. The argument is that the first item can be chosen in n ways, the second item in $n - 1$ ways, all the way to the kth item in $n - (k - 1)$ ways. Note that the formula $P(n, k) = n!/(n - k)!$ works for the special case $k = n$ as well, if we posit that $0! = 1$; for this and other similar reason, it is standard to define $0! = 1$.

As an example, let $S = \{a, b, c, d\}$, so $n = |S| = 4$, and let $k = 2$. Then the collection of sequences of two items is equal to $\{ab, ac, ad, ba, bc, bd, ca, cb, cd, da, db, cd\}$. The size of this set is $P(4, 2) = 4!/2! = 4 \cdot 3 = 12$. (*Note:* In permutations, order matters; ab is not the same as ba.)

8.3.4 Combinations of k Items out of n

The number of subsets of size k out of a set of size n is known as the *combinations of k in n*. It is often notated

$$C(n, k) \quad \text{or} \quad C\left(\begin{array}{c} n \\ k \end{array} \right) \quad \text{or} \quad \left(\begin{array}{c} n \\ k \end{array} \right),$$

and is computed as $n!/k!(n - k)!$. The argument is first, considering the list of permutations of k out of n items, each subset of size k appears in that list in each of the permutations of its elements; it therefore appears $k!$ times. (*Note:* In combinations, ab is the same as ba; order doesn't matter.) Since the list of permutations has $n!/(n - k)!$ elements, the total number of subsets must be $(n!/(n - k)!)/k! = n!/(n - k)!k!$.

For example, let $S = \{a, b, c, d\}$ so $n = |S| = 4$, and let $k = 2$. Then the collection of subsets of size 2 is $\{a, b\}, \{a, c\}, \{a, d\}, \{b, c\}, \{b, d\}, \{c, d\}$. The number of such subsets is $4!/(2! \cdot 2!) = (4 \cdot 3)/(2 \cdot 1) = 6$.

The numbers $C(n, k)$ are also known as the *binomial coefficients* because $C(n, k)$ is the coefficient of the term $x^k y^{n-k}$ in the expansion of $(x + y)^k$. These are also the numbers in Pascal's triangle; they satisfy the double recurrence $C(n, k) = C(n - 1, k) + C(n - 1, k - 1)$. These numbers figure prominently in the discussion of the binomial distribution in Section 9.6.2.

The MATLAB function nchoosek(n,k) computes $C(n, k)$.

8.3.5 Partition into Sets

Let $n = k_1 + k_2 + \ldots + k_m$. The number of ways of dividing a set S of size n into disjoint subsets $S_1, \ldots, S_m$, where $|S_1| = k_1, |S_2| = k_2, \ldots, |S_m| = k_m$, is often denoted $C(n : k_1, \ldots, k_m)$ or as

$$\left(\begin{array}{c} n \\ k_1 \quad , \ldots, \quad k_m \end{array} \right).$$

It is computed as $n!/(k_1! \cdot k_2! \cdot \ldots \cdot k_m!)$. The argument is as follows. First choose subset S_1 of size k_1 out of the whole set S; there are $C(n, k_1) = n!/k_1!(n-k_1)!$ possible choices. Now there are $n - k_1$ elements remaining, so the number of ways of choosing a set S_2 of size k_2 out of those is $C(n - k_1, k_2)$. Continuing on in this way, when we come to choose S_p there are $n - k_1 - k_2 - \ldots - k_{p-1}$ elements remaining, so the number of ways of choosing S_p is $C(n - k_1 - k_2 - \ldots - k_{p-1}, k_p)$. Multiplying all these together we get

$$C(n, k_1) \cdot C(n - k_1, k_2) \cdot C(n - k_1 - k_2, k_3) \cdot \ldots$$

$$= \frac{n!}{k_1! \cdot (n-k_1)!} \cdot \frac{(n-k_1)!}{k_1! \cdot (n-k_1-k_2)!} \cdot \frac{(n-k_1-k_2)!}{k_3! \cdot (n-k_1-k_2-k_3)!} \cdot \ldots$$

$$= \frac{n!}{k_1! \cdot k_2! \cdot \ldots \cdot k_m!}.$$

For example, let $S = \{a, b, c, d\}$, so $n = |S| = 4$, and let $k_1 = 2$, $k_2 = 1$, $k_3 = 1$. Then the collection of partitions is

S_1	S_2	S_3		S_1	S_2	S_3
$\{a,b\}$	$\{c\}$	$\{d\}$		$\{a,b\}$	$\{d\}$	$\{c\}$
$\{a,c\}$	$\{b\}$	$\{d\}$		$\{a,c\}$	$\{d\}$	$\{b\}$
$\{a,d\}$	$\{b\}$	$\{c\}$		$\{a,d\}$	$\{c\}$	$\{b\}$
$\{b,c\}$	$\{a\}$	$\{d\}$		$\{b,c\}$	$\{d\}$	$\{a\}$
$\{b,d\}$	$\{a\}$	$\{c\}$		$\{b,d\}$	$\{c\}$	$\{a\}$
$\{c,d\}$	$\{a\}$	$\{b\}$		$\{c,d\}$	$\{b\}$	$\{a\}$

The number of partitions is $4!/(2! \cdot 1! \cdot 1!) = 12$.

8.3.6 Approximation of Central Binomial

As discussed further in Section 9.8.2, the central term of the combinatorial function $C(n, n/2)$ is approximately equal to $2^n \cdot \sqrt{2/\pi n}$. For example, C(20,10) = 184,756, and this approximation gives a value of 187,078.97, a relative error of 1.2%.

8.3.7 Examples of Combinatorics

Example 8.5. Say we deal a poker hand of five cards from a 52-card deck. Here the sample space Ω is the set of all five-card hands, and the probability function is the uniform distribution. A five-card hand is thus a combination of 5 items out of 52, and the size of Ω, #Ω, is $C(52, 5) = 52!/(47! \cdot 5!) = 2,598,960$.

If we want to know the probability that an unseen hand is a flush, then the event E is the set of all hands that are flushes. Since we are using the uniform distribution, $P(E) = \#E/\#\Omega$. To count the number of hands that are flushes, we reason that the flush may be in any of the four suits. To construct a flush in spades, for example, we must choose 5 cards out of the 13 spades; the number of ways of doing this is $C(13,5)$. The total number of flushes in any suit is therefore $\#E = 4 \cdot C(13,5) = 4 \cdot 13!/(8! \cdot 5!) = 5148$. Therefore $P(E) = 5148/2,598,960 = 0.002$.

Example 8.6. An urn contains four red balls and nine black balls. Three times we pick a ball out of the urn, note its color, and put it back in the urn. (This is called "sampling with replacement.")

To construct the sample space, we will give each ball a number; we will number the red balls 1 to 4 and we will number the black balls 5 to 13. The sample space Ω is then just all triples of balls, where a triple can include a repetition:

$$\Omega = \{\langle 1,1,1\rangle, \langle 1,1,2\rangle, \ldots, \langle 1,1,13\rangle, \langle 1,2,1\rangle, \ldots, \langle 13,13,12\rangle, \langle 13,13,13\rangle\}.$$

The probability function is again the uniform distribution. The event "exactly one ball is red" is the subset of all triples containing one red ball: $E = \{\langle 1,5,5\rangle, \langle 1,5,6\rangle, \ldots, \langle 13,13,4\rangle\}$. The total number of draws of three balls is $13^3 = 2197$. To count the number of draws that have exactly one red ball, we reason that the red ball may be either the first, the second, or the third drawn. If we consider the draws in which the first ball is red and other two are black, then there are four ways to choose the first ball, and nine ways to choose each of the other two black balls, for a total of $4 \cdot 9 \cdot 9$ possibilities. Since there are three possible positions of the red ball, the total number of draws with 1 red ball, $\#E = 3 \cdot 4 \cdot 9 \cdot 9 = 972$. Therefore, $P(E) = 972/2197 = 0.442$.

Example 8.7. The same urn contains four red balls and nine black balls. But now we pick three balls out of the urn without putting them back. (This is called "sampling without replacement.")

We number the balls as in Example 8.6. The sample space Ω is then just all triples of balls with no repetitions in the triples:

$$\Omega = \{\langle 1,2,3\rangle, \langle 1,2,4\rangle, \ldots, \langle 1,2,13\rangle, \langle 1,3,2\rangle, \ldots, \langle 13,12,10\rangle, \langle 13,12,11\rangle\}.$$

An element of Ω is a permutation of 3 balls out of the 13, so $\#\Omega = P(13,3) = 13 \cdot 12 \cdot 11 = 1716$.

Let E be the event that a draw has exactly one red ball. The calculation to compute $\#E$ is the same as in Example 8.6 except that, when we draw the second black ball, there are only eight possible choices, not nine. Therefore, the total number of such draws $\#E = 3 \cdot 4 \cdot 9 \cdot 8 = 864$ so $P(E) = \#E/\#\Omega = 0.503$.

The poker hand of Example 8.5 is also an example of sampling without replacement; the deck corresponds to the urn and the cards correspond to the balls. (Each card has a different color.) If we were to deal one card at a time, replace it in the deck, shuffle, and deal again, that would be sampling with replacement.

8.4   The Axioms of Probability Theory

The theory presented in Section 8.3 takes as its starting point the function $P(x)$ where $x \in \Omega$. However, this approach works only in the case of finite sample spaces. A more general approach to axiomatization of probability theory is based on events, which are the common coin of all probabilistic theories. In this approach, there are four axioms of probability theory.

P1. For any event E, $0 \le P(E) \le 1$.

P2. For any events E and F, if $E \cap F = \emptyset$ then $P(E \cup F) = P(E) + P(F)$.

P3. $P(\emptyset) = 0$.

P4. $P(\Omega) = 1$.

It is easy to prove that these four axioms are consequences of axioms F1 and F2, together with the definition $P(E) = \sum_{x \in E} P(x)$.

It is also possible to prove that P1 through P4 are *all* the axioms of probability theory that we need. Suppose we have a collection of facts about probabilities of different events and their combinations. For example, suppose we are told the following facts:

$$P(A) = 0.2,$$
$$P(B) = 0.7,$$
$$P(A \cap B) = 0.1,$$
$$P(C \cup B) = 0.9,$$
$$P(C \setminus (A \cap B)) = 0.4.$$

If this collection of facts is consistent with axioms P1–P4 then we can construct a sample space and a probability function and assign each event to a subset of the sample space in a way that makes all these statements true. In the terminology of mathematical logic, this set of axioms is *complete* for sample space models.

One model of the above system of constraints would be:

$$\Omega = \{v, w, x, y, z\}, \qquad P(v) = 0.1,$$
$$A = \{v, w\}, \qquad\qquad P(w) = 0.1,$$
$$B = \{v, x, y\}, \qquad\qquad P(x) = 0.4,$$
$$C = \{v, y, z\}; \qquad\qquad P(y) = 0.2,$$
$$P(z) = 0.2.$$

Then

$$
\begin{aligned}
P(A) &= P(\{v, w\}) &&= 0.2, \\
P(B) &= P(\{v, x, y\}) &&= 0.7, \\
P(A \cap B) &= P(\{v\}) &&= 0.1, \\
P(C \cup B) &= P(\{v, x, y.z\}) &&= 0.9, \\
P(C \setminus (A \cap B)) &= P(\{y.z\}) &&= 0.4.
\end{aligned}
$$

The following result is simple but important; it can easily be proved either from axioms P1–P4 or from axioms F1 and F2:

Definition 8.8. Two events E and F are *mutually exclusive* if $E \cap F = \emptyset$.

A collection of events $\{E_1, \ldots, E_k\}$ forms a *frame of discernment* if every element x is in exactly one E_i. Equivalently,

- for $i \neq j$, $E_i \cap E_j$ are mutually exclusive.

- $E_1 \cup E_2 \cup \ldots \cup E_k = \Omega$.

Theorem 8.9. *Let $\mathcal{E} = \{E_1 \ldots E_k\}$ be a collection of events. If every pair of events $E_i, E_j, i \neq j$, in $\mathcal{E}$ are mutually exclusive, then $P(E_1 \cup \ldots \cup E_k) = P(E_1) + \ldots + P(E_k)$. In particular, if $\{E_1, \ldots, E_k\}$ forms a frame of discernment, then $P(E_1) + P(E_2) + \ldots + P(E_k) = P(\Omega) = 1$.*

8.5 Conditional Probability

Let E and F be events, and assume that $P(F) > 0$. The *conditional probability of E given F*, denoted $P(E \mid F)$ is the fraction of F that is also E. In other words, we shrink the sample space from Ω to F while keeping the weight function unchanged, and we now consider the probability of E within this restricted space. Probabilities with no conditional, such as we have been considering previously, are called *absolute* probabilities.

Once we have restricted the space to be F, the part of E that is significant is just $E \cap F$. Therefore, the probability of E within the sample space F is $P(E \mid F) = w(E \cap F)/w(F)$. Now, if we divide both numerator and denominator by $w(\Omega)$, we have

$$
P(E \mid F) = \frac{w(E \cap F)/w(\Omega)}{w(F)/w(\Omega)} = \frac{P(E \cap F)}{P(F)}.
$$

This is the *conditional probability formula*. Multiplying through gives us the *conjunction formula*: $P(E \cap F) = P(F) \cdot P(E \mid F)$.

Example 8.10. (Example 8.3 revisited.) Let Ω be the sample space of all eight outcomes of flipping a coin three times, and let $P(x) = 1/8$ for each outcome x.

Let F be the event of getting exactly two heads, and let E be the event that the first flip is heads. Thus $E = \{$ HHH, HHT, HTH, HTT $\}$; $F = \{$ HHT, HTH, THH $\}$; and $E \cap F = \{$ HHT, HTH $\}$. Therefore $P(E|F) = P(E \cap F)/P(F) = (2/8)/(3/8) = 2/3$, and $P(F|E) = P(E \cap F)/P(E) = (2/8)/(4/8) = 1/2$.

Since conditioning probabilities on an event G is simply a change in the sample space, it follows that any formula that holds generally for absolute probabilities also holds if some event G is added as a condition across the board. For instance, we can add G as a condition to each of the axioms P1–P4; the results are the following axioms.

C1. For any event E, $0 \le P(E|G) \le 1$.

C2. For any events E and F, if $E \cap F \cap G = \emptyset$, then $P(E \cup F|G) = P(E|G) + P(F|G)$.

C3. If $E \cap G = \emptyset$, then $P(E|G) = 0$.

C4. $P(G|G) = 1$.

We can also add G as a condition in the conjunction formula:

$$P(E \cap F|G) = \frac{P(E \cap F \cap G)}{P(G)} = \frac{P(E \cap F \cap G)}{P(F \cap G)} \cdot \frac{P(F \cap G)}{P(G)} = P(E|F \cap G) \cdot P(F|G).$$

Note that this is the same as the formula $P(E \cap F) = P(E|F)P(F)$ with the additional condition G added across the board.

The conjunction formula can be extended to cover the conjunction of any number of events. For three events E, F, G, we have

$$P(E \cap F \cap G) = P(E|F \cap G)P(F \cap G) = P(E|F \cap G) \cdot P(F|G) \cdot P(G).$$

For k events, we have

$$P(E_1 \cup E_2 \cup \ldots \cup E_k) = P(E_1|E2 \cup \ldots \cup E_k) \cdot P(E_2|E3 \cup \ldots \cup E_k) \cdot \ldots \cdot P(E_{k-1}|E_k) \cdot P(E_k).$$

8.6   The Likelihood Interpretation

The sample space model of probability is clear and coherent and is well suited to many kinds of problems. However, many of the applications for which we need to reason with uncertain information do not fit this model at all well. Consider the following questions:

1. Given the text of an email message, what is the likelihood that it is spam?

2. Given a search query and the text of a web page, what is the likelihood that the page is relevant to the query?

3. Given the form of an image, what is the likelihood that it shows an airplane?

4. Given a patient's personal history and medical record, what is the likelihood that she has diabetes?

What happens if we try to analyze these kinds of examples in terms of a sample space? Consider question 4, about a medical diagnosis. The sample space Ω is presumably the set of all people; or, better, the set of all pairs of a person and a day, such as $\langle$Marilyn O'Reilly, 4/7/2008$\rangle$. We want to determine $P(E \mid F)$ where E is diabetes and F is the medical record. More precisely, in the sample space model, the event E is the set $\langle P, D \rangle$ such that P had diabetes on day D; and the event F is the set $\langle P, D \rangle$ of people who on day D had the specified medical record. The problem is that, if the medical record is detailed enough, with, say, a long history of precise test results, and a large set of answers to personally intrusive questions, then, in this history of the medical profession, only one person on one day has had exactly that medical record; namely, Marilyn O'Reilly herself, on April 7, 2008. So F is the singleton set $\{\langle$Marilyn O'Reilly, 4/7/2008$\rangle\}$ and $E \cap F$ is either equal to F if Marilyn had diabetes on that date, or equal to the empty set if she did not. So, if Marilyn had diabetes on that date, then $E = F$ and $P(E \mid F) = 1$; if she didn't, then $E = \emptyset$ and $P(E \mid F) = 0$. We don't know which until we determine whether she had diabetes. Clearly, this is not getting us where we want to go. The other three questions run into similar problems. Indeed, many standard probability textbooks assert specifically that it is meaningless to speak of "the probability that Marilyn O'Reilly has diabetes," or even, strictly speaking, "the probability that the next time I flip a coin it will come up heads"; probabilities, they say, are only meaningful as applied to general categories.

In order to apply probability to examples such as the diagnosis question, we use the likelihood interpretation of probability theory. In this theory, an *event* is a proposition, such as "Marilyn O'Reilly has diabetes," "Marilyn O'Reilly has a blood glucose level of 150 mg/dL," "Marilyn O'Reilly does vigorous exercise at the gym three times a week," and so on. There is an implicit (in practice, usually rather rich) background body of knowledge. For example, in the diagnosis domain, this background knowledge might include general information about the frequency of different conditions in the population and the general relation between diagnoses and medical test results and measurements, but no facts about individual patients. Then $P(E)$ is the likelihood that E is true given only the background knowledge. $P(E \mid F)$ is the likelihood that E is true if one learns F in addition to the background knowledge. In the diabetes example, E is the proposition "Marilyn O'Reilly has diabetes on 4/7/2008" and F is the conjunction of all the facts in her medical record. In the spam example (question 1), E is the proposition "Message 12472 is spam," and F is the proposition "The text of message 12472 is $\langle$ whatever it is $\rangle$."

So now it is a meaningful question to ask, "What is $P(E \mid F)$?" It means, "How likely is it that Marilyn has diabetes, if we are told her medical record in addition to our background knowledge?" This is certainly the right question to be asking, so we have at least made some progress over the sample space theory in which it is a meaningless question. However the analysis does not, in itself, provide any guidance whatever about how to calculate the answer.

It is not at all obvious that "likelihood" in this sense is a coherent enough notion to allow mathematical treatment. However, the theory of subjective probability posits that a mathematical analysis is indeed possible, and that, in fact, the measurement of likelihood observes the five axioms of probability S1–S5. A number of different kinds of arguments have been advanced to justify the use of these axioms; see Fine (1973). For our purposes we simply accept the probabilistic model of likelihood because (a) it is clearly a reasonable model in the case where the likelihood judgment is based on a sample space; and (b) it has been found to work well in practice. Before we state the axioms of probability, though, we must define some terms and symbols.

In this view, a probabilistic event is a *sentence* within a *propositional language*. A propositional language consists of three parts:

- The two *truth constants* $\top$ (true) and $\bot$ (false).

- A set of *propositional atoms*. We use boldface symbols, such as **Q**, **R**, for propositional atoms.

- The Boolean operators $E \wedge F$ (E and F), $E \vee F$ (E or F), $E \Rightarrow F$ (E implies F), $E \Leftrightarrow F$ (E if and only if F), and $\neg E$ (not E). These allow two simple sentences E and F to be combined into a compound sentence.

A *sentence* is a combination of propositional atoms and truth constants using the Boolean operators. The relations between sentences are defined in the *propositional calculus*, also called *Boolean logic*. We assume the readers are familiar with the propositional calculus.

The subjective theory of probabilities posits that the probability function satisfies the following axioms.

S1. For any sentence E, $0 \le P(E) \le 1$.

S2. If it is known that $\neg(E \wedge F)$, then $P(E \vee F) = P(E) + P(F)$.

S3. $P(\top) = 1$.

S4. $P(\bot) = 0$.

S5. If it is known that $E \Leftrightarrow F$, then $P(E) = P(F)$.

In particular, both S2 and S5 hold in the case where the sentence in question is known because it is a logical truth. For example, it is a theorem of propositional logic that $(A \vee B) \Leftrightarrow \neg(\neg A \wedge \neg B)$; therefore, one can apply axiom S5 and conclude that $P(A \vee B) = P(\neg(\neg A \wedge \neg B))$.

As we can see, these are largely the same as axioms P1–P4 in the frequentist theory, but there are two differences. First, the set operators $\cup$ and $\cap$ have been changed to the propositional operators $\vee$ and $\wedge$. Second, axiom S5 has been added. This reflects the fact that set theory and propositional logic use different conventions about equality; in set theory, two sets that contain the same elements are equal, whereas in propositional logic, two sentences that have the same truth conditions are considered unequal if they have different forms.

We define conditional probability in the same way as in the frequentist theory: $P(E \mid F) \equiv P(E \wedge F)/P(F)$.

8.7   Relation between Likelihood and Sample Space Probability

The likelihood theory and sample space theory are closely related; they end up in much the same place, although their approaches are from different directions.

Suppose that we have a finite set $\mathscr{S} = \mathbf{P}_1,\ldots,\mathbf{P}_k$ of propositional atoms in mind. Then we can define an *elementary event* to be the conjunction of either the positive or the negative of all the atoms; and we can define a sample space to be the set of all elementary events, with the associated probability.

For example, suppose the three propositional atoms are $\mathbf{P},\mathbf{Q},\mathbf{R}$. Then the associated sample space Ω is $\{E_1,\ldots,E_8\}$, where

$$E_1 = \mathbf{P} \wedge \mathbf{Q} \wedge \mathbf{R},$$
$$E_2 = \mathbf{P} \wedge \mathbf{Q} \wedge \neg\mathbf{R},$$
$$E_3 = \mathbf{P} \wedge \neg\mathbf{Q} \wedge \mathbf{R},$$
$$E_4 = \mathbf{P} \wedge \neg\mathbf{Q} \wedge \neg\mathbf{R},$$
$$E_5 = \neg\mathbf{P} \wedge \mathbf{Q} \wedge \mathbf{R},$$
$$E_6 = \neg\mathbf{P} \wedge \mathbf{Q} \wedge \neg\mathbf{R},$$
$$E_7 = \neg\mathbf{P} \wedge \neg\mathbf{Q} \wedge \mathbf{R},$$
$$E_8 = \neg\mathbf{P} \wedge \neg\mathbf{Q} \wedge \neg\mathbf{R}.$$

There are two important points about this. First, exactly one of these elementary events must be true. That is, it is a theorem of the propositional calculus that $E_1 \vee E_2 \vee E_3 \vee E_4 \vee E_5 \vee E_6 \vee E_7 \vee E_8$; and for each $i \neq j$, $\neg(E_1 \wedge E_j)$.

The analog of Theorem 8.9 can be proven from axioms S1–S5, so

$$P(E_1) + P(E_2) + P(E_3) + P(E_4) + P(E_5) + P(E_6) + P(E_7) + P(E_8) = 1.$$

Therefore, these probabilities are a valid probability function over the sample space Ω. This probability function is known as the *joint probability distribution* over the set of propositional atoms $\mathbf{P}, \mathbf{Q}, \mathbf{R}$.

Second, each of these elementary events corresponds to one line in a truth table. If we know that some particular elementary event E_i is true, then that determines the truth or falsehood of any sentence F over these propositional atoms. For example if the sentence $E_3 = \mathbf{P} \wedge \neg\mathbf{Q} \wedge \mathbf{R}$ is true, then $\mathbf{P}$ is true, $\mathbf{Q}$ is false, and $\mathbf{R}$ is true, so $\mathbf{P} \wedge \mathbf{Q}$ is true, $\neg\mathbf{P} \vee \mathbf{Q}$ is false $\neg(\mathbf{P} \wedge \mathbf{R}) \vee \mathbf{Q}$ is false, and so on. Therefore, any sentence F is logically equivalent to the disjunction of some of the elementary events; namely, all the sentences E_i that would make F true. As examples,

$$\mathbf{P} \wedge \mathbf{Q} \equiv E_1 \vee E_2, \tag{8.1}$$

$$\neg\mathbf{P} \vee \mathbf{Q} \equiv E_1 \vee E_2 \vee E_5 \vee E_6 \vee E_7 \vee E_8, \tag{8.2}$$

$$(\neg\mathbf{P} \wedge \mathbf{R}) \vee \mathbf{Q} \equiv E_1 \vee E_2 \vee E_5 \vee E_6 \vee E_7. \tag{8.3}$$

However, since the E_is are mutually exclusive, it follows that the probability of any disjunction of E_is is just the sum of their probabilities:

$$P(\mathbf{P} \wedge \mathbf{Q}) = P(E_1 \vee E_2) = P(E_1) + P(E_2),$$
$$P(\neg\mathbf{P} \vee \mathbf{Q}) = P(E_1 \vee E_2 \vee E_5 \vee E_6 \vee E_7 \vee E_8)$$
$$= P(E_1) + P(E_2) + P(E_5) + P(E_6) + P(E_7) + P(E_8),$$
$$P((\neg\mathbf{P} \wedge \mathbf{R}) \vee \mathbf{Q}) = P(E_1 \vee E_2 \vee E_5 \vee E_6 \vee E_7)$$
$$= P(E_1) + P(E_2) + P(E_5) + P(E_6) + P(E_7).$$

So we can identify the sentence $\mathbf{P} \wedge \mathbf{Q}$ with the set $\{E_1, E_2\}$; we can identify the sentence $\neg\mathbf{P} \vee \mathbf{Q}$ with the set $\{E_1, E_2, E_5, E_6, E_7, E_8\}$; and so on.

Thus, for any finite likelihood theory, there is a finite sample space, and vice versa. However, the two views develop the theory in opposite directions. With finite sample spaces, we start with the elements and combine them into sets to form events. With a finite propositional theory, we start with events and combine them with conjunction and negation to form elements. As shown in Chapter 9, in the more general setting, which includes infinite sample spaces, the theory starts and ends with events; it never reaches elements.

8.8   Bayes' Law

Bayes' law is a simple but powerful rule that governs reversing the direction of conditional probabilities.

By the rule of conjunction $P(E \wedge F) = P(E) \cdot P(F \mid E)$. Also $P(E \wedge F) = P(F) \cdot P(E \mid F)$. So $P(F) \cdot P(E \mid F) = P(E) \cdot P(F \mid E)$. Dividing through by $P(F)$ we get

$$P(E \mid F) = \frac{P(E) \cdot P(F \mid E)}{P(F)}.$$

That is Bayes' law.

This law is important in the cases where we want to know the number on the left, and we do know (or can guess) the numbers on the right. In many cases of that kind, you don't actually know $P(F)$ directly. Rather, you know that E is one of a frame of discernment $\{E = E_1, E_2, \ldots, E_k\}$ and you know $P(E_i)$ and $P(F \mid E_i)$ for each E_i. Then you can compute $P(F)$ as follows. Since $\{E_1, \ldots, E_k\}$ is a frame of discernment, it follows that

- for $i \neq j$, $F \wedge E_i$ and $F \wedge E_j$ are mutually exclusive;

- $F = (F \wedge E_1) \vee (F \wedge E_2) \vee \ldots \vee (F \wedge E_k)$.

Therefore, $P(F) = P(F \wedge E_1) + \ldots + P(F \wedge E_k)$. So we can rewrite Bayes' law as

$$P(E_i \mid F) = \frac{P(E_i) \cdot P(F \mid E_i)}{P(E_1) \cdot P(F \mid E_1) + \ldots + P(E_k) \cdot P(F \mid E_k)}.$$

As an example, a patient sees a doctor for a regular checkup, and the doctor decides that it would be a good idea to test the patient for disease D, a disease that affects 1 in every 10,000 people. A test T for D is 99% accurate; that is, if the patient has D, then with 99% probability the test will come up positive; if the patient does not have D, then with 99% probability the test will come up negative. Sadly, the test comes up positive. What is the probability that the patient has D?

By Bayes' law, we have

$$P(D \mid T) = \frac{P(D) \cdot P(T \mid D)}{P(D) \cdot P(T \mid D) + P(\neg D) \cdot P(T \mid \neg D)}.$$

We are given that $P(D) = 0.0001$, $P(T \mid D) = 0.99$, $P(T \mid \neg D) = 0.01$. Also $P(\neg D) = 1 - P(D) = 0.9999$. So $P(D \mid T) = (0.99 \cdot 0.0001)/((0.99 \cdot 0.0001) + (0.01 \cdot 0.9999)) = 0.0098$, slightly less than 1 in 100.

This seems surprising at first, but it actually is reasonable. Suppose the doctor tests 10,000 people. One of them has D, and the test will come up positive. The other 9,999 don't have D, but the test will come up positive for roughly 100 of those. So there will be 101 patients with positive results, only one of whom actually has the disease. In general, in this kind of reasoning, people tend to give too much weight to the accuracy of the test, $P(T|D)$, and not nearly enough to the base rate $P(D)$.

Note, however, that this applies only if the people whom the doctor is testing are a random sample of the population. If the doctor has some particular reason for testing the patient—for instance, the patient is complaining of symptoms or the patient is a member of some population who is at risk (e.g., age, ethnic group)—then this calculation does not apply because there is additional information that must be taken into account. In that circumstance, we face the problem of evidence combination, which we discuss in Section 8.9.1.

8.9   Independence

A critical component of probability theory is the observation that *most events have nothing whatever to do with one another.* Finding out whether it is raining in Poughkeepsie does not at all influence our estimate of the likelihood that the yen will gain against the dollar tomorrow. The usability of the theory of probability, and indeed the possibility of rational thought, rests on this disconnection. If we had to rethink the likelihood of everything each time we learned a new fact, then thinking would be pretty much impossible.

The mathematical notion here is *independence*:

Definition 8.11. Event E is *independent* of F if learning F does not affect the estimate of the likelihood of E. That is, $P(E \mid F) = P(E)$.

(Note that the natural definition is in terms of the likelihood interpretation of probability. The concept of independence seems much more arbitrary when viewed in terms of the sample space interpretation.)

The condition $P(E \mid F) = P(E)$ is equivalent to $P(E, F)/P(F) = P(E)$. This leads to two important consequences:

1. $P(E, F) = P(E) \cdot P(F)$. This is a simpler version of the general rule of conjunction. We emphasize that this applies *only* if E and F are independent.

2. $P(F \mid E) = P(E, F)/P(E) = P(F)$. Thus, if E is independent of F, then F is independent of E. So independence is a *symmetric* relation, which is not obvious from the original definition.

Note: Here we are using a standard notation of probability theory in which E, F within a probability operator means $E \wedge F$. Thus, $P(E, F)$ means $P(E \wedge F)$; $P(E \mid F, G)$ means $P(E \mid F \wedge G)$, and so on.

Like other aspects of probability, independence can be conditionalized:

Definition 8.12. Let E, F, G be events. Event E is *conditionally independent of F given G* if, after we have learned G, additionally learning F does not affect the likelihood of E. That is $P(E \mid F, G) = P(E \mid G)$.

By the same argument as Definition 8.12, if E is conditionally independent of F given G, then the following two statements hold:

1. $P(F \mid E, G) = P(F \mid G)$; that is, F is conditionally independent of E given G.

2. $P(E, F \mid G) = P(E \mid G) \cdot P(F \mid G)$.

Independence also applies to larger collections of events:

Definition 8.13. A collection of events $\{E_1, \ldots, E_k\}$ is independent if finding out the values of any subset does not affect the likelihood of the rest. That is, if S is a subset of the events and $E_i \notin S$, then $P(E_i \mid S) = P(E_i)$.

If the set $\{E_1, \ldots, E_k\}$ is independent, then $P(E_1, \ldots, E_k) = P(E_1) \cdot P(E_2) \cdots \cdot P(E_k)$.

It is possible to have a collection of events in which every pair is independent but the collection as a whole is not independent. An example where this is important is shown in Section 8.9.2.

Example 8.14. (Example 8.3 revisited.) We flip a coin three times. Let H_1, H_2, H_3 be the events that the first, second, and third flips come up heads. Then these three are independent, so $P(H_1, H_2, H_3) = 1/8$, agreeing with our earlier calculation from the sample space.

Example 8.15. (Example 8.6 revisited.) We have an urn with four red balls and nine black balls, and we sample with replacement three times. Let R_1, R_2, R_3 be the events that the three samples are red. Then these three are independent, since each time you are sampling from the same collection.

Example 8.16. (Example 8.7 revisited.) We have an urn with four red balls and nine black balls, and now we sample *without* replacement three times. Let R_1, R_2, R_3 be the events that the three samples are red. Then these three events are *not* independent. Since four out of the thirteen balls are red, and before you start, all the balls are equally likely to be picked on the second draw, the absolute probability $P(R_1) = 4/13$. If R_1 is true, then when you pick the second ball, there are twelve balls in the urn of which three are red, so $P(R_2 \mid R_1) = 3/12$. If R_1 is false, then when you pick the second ball, there are twelve balls in the urn of which four are red, so $P(R_2 \mid \neg R_1) = 4/12$.

8.9.1 Independent Evidence

Suppose that we want to determine the likelihood of some proposition X, and we have two pieces of evidence, E and F. Both E and F are each quite good evidence for the truth of X. Specifically, the absolute probability $P(X) = 1/3$, the conditional probabilities $P(X \mid E) = 0.8$, and $P(X \mid F) = 0.9$, so both E and F

very much raise the likelihood of X. What can we say about the probability of $P(X \mid E, F)$?

The answer is that we can't say anything at all. The probability $P(X \mid E, F)$ can have any value from 0 to 1 inclusive. The intuitive justification is that we can construct a model in which $E \cap F$ is only a very small fraction of E and of F, so it is consistent with the constraints on $P(X \mid E)$ and $P(X \mid F)$ either that X covers $E \cap F$ or that X is disjoint from $E \cap F$.

For example, we can satisfy the constraints $P(X) = 1/3$, $P(X|E) = 0.8$, $P(X|F) = 0.9$, $P(X \mid E, F) = 0.01$ by using the following probability space:

$$\Omega = \{a, b, c, d, e, f, g\},$$
$$E = \{a, b, c, d\},$$
$$F = \{a, b, e, f\},$$
$$X = \{a, c, e\},$$
$$w(a) = 1, \ w(b) = 99, \ w(c) = 799, \ w(d) = 101, w(e) = 899, \ w(f) = 1, \ w(g) = 3800.$$

This generalizes to any number of pieces of evidence. You can have evidence $E_1, \ldots, E_k$, in which each fact individually is good evidence for X, any pair of facts is strong evidence for X, any three facts is very strong evidence for X, all the way to any collection of $k - 1$ facts is overwhelming evidence for X, and yet $P(X \mid E_1, \ldots, E_k) = 0$. Theorem 8.17 states this fact.

Theorem 8.17. *Let X and $E_1, \ldots, E_k$ be variables ranging over events. Let $m = 2^k$, and let $S_1, \ldots, S_m$ be the m subsets of $\{E_1, \ldots, E_k\}$. Let $\langle c_1, \ldots, c_m \rangle$ be any 2^k-tuple of numbers strictly between 0 and 1. Then the system of equations $\{P(X \mid S_1) = c_1, \ldots, P(X \mid S_m) = c_m\}$ is consistent. That is, there exists a sample space Ω, a probability function P, and an assignment of X and of all the E_i to events in Ω that satisfies all these equations.*

The proof is left as a rather difficult exercise (Problem 8.2).

Let us return to our original problem: we have $P(X) = 1/3$, $P(X \mid E) = 0.8$, $P(X \mid F) = 0.9$, but we can't get any direct information about $P(X \mid E, F)$. What can we do?

This is a common—in fact, almost universal—situation in applied probabilistic reasoning: we have a situation involving a combination of many events, and we can't get direct information about the statistics of the entire combination. Standard operating procedure in this situation is that we use independence assumptions to split the large combination into smaller groups about which we do have information. In this case, we don't have information about the combination X, E, F, but we do have information about each of the pairs X, E and X, F, so we have to find a set of independence assumptions that will allow us to analyze X, E, F in terms of X, E and X, F.

In this particular case, the usual independence assumption is to posit that
E and F are conditionally independent, both given X and given $\neg X$. That is,
$P(E, F \mid X) = P(E \mid X) \cdot P(F \mid X)$ and $P(E, F \mid \neg X) = P(E \mid \neg X) \cdot P(F \mid \neg X)$. Section 13.6.2 presents an argument justifying the choice of this particular inde-
pendence assumption.

By Bayes' law,

$$P(X \mid E, F) = \frac{P(E, F \mid X) \cdot P(X)}{P(E, F)}. \tag{8.4}$$

Likewise,

$$P(\neg X \mid E, F) = \frac{P(E, F \mid \neg X) \cdot P(\neg X)}{P(E, F)}. \tag{8.5}$$

Dividing Equation (8.4) by Equation (8.5) gives

$$\frac{P(X \mid E, F)}{P(\neg X \mid E, F)} = \frac{P(E, F \mid X) \cdot P(X)}{P(E, F \mid \neg X) \cdot P(\neg X)}. \tag{8.6}$$

Applying the independence assumptions, we get

$$\frac{P(X \mid E, F)}{P(\neg X \mid E, F)} = \frac{P(E \mid X) \cdot P(F \mid X) \cdot P(X)}{P(E \mid \neg X) \cdot P(F \mid \neg X) \cdot P(\neg X)}. \tag{8.7}$$

For any events A, B, let us define the odds on A as $\text{Odds}(A) = P(A)/P(\neg A)$,
and the conditional odds on A given B, as $\text{Odds}(A \mid B) = P(A \mid B)/P(\neg A \mid B)$.
For example, if $P(A) = 3/4$, then $\text{Odds}(A) = (3/4)/(1/4) = 3$. (These are the "3-
to-1 odds" used at the racetrack.) Note that $\text{Odds}(A) = P(A)/P(\neg A) = P(A)/(1 -
P(A))$; solving for $P(A)$, we have $P(A) = \text{Odds}(A)/(1 + \text{Odds}(A))$. Likewise,
$\text{Odds}(A \mid B) = P(A \mid B)/(1 - P(A \mid B))$ and $P(A \mid B) = \text{Odds}(A \mid B)/(1 + \text{Odds}(A \mid B))$.
Moreover, by Bayes' law,

$$\text{Odds}(A \mid B) = \frac{P(A \mid B)}{P(\neg A \mid B)} = \frac{P(A) \cdot P(B \mid A)/P(B)}{P(\neg A) \cdot P(B \mid \neg A)/P(B)} = \text{Odds}(A) \cdot \frac{P(B \mid A)}{P(B \mid \neg A)}. \tag{8.8}$$

Let us define the "odds ratio" of A given B as $\text{OR}(A|B) = \text{Odds}(A|B)/\text{Odds}(A)$;
that is, the factor by which learning B affects the odds on A. For instance,
if $P(A) = 1/3$ and $P(A \mid B) = 3/4$, then $\text{Odds}(A) = 1/2$ and $\text{Odds}(A \mid B) = 3$, so
$\text{OR}(A \mid B) = 6$; learning B has increased the odds on A by a factor of 6. Then we
can rewrite Equation (8.8) in the form

$$\frac{P(B \mid A)}{P(B \mid \neg A)} = \frac{\text{Odds}(A \mid B)}{\text{Odds}(A)} = \text{OR}(A \mid B). \tag{8.9}$$

Now, using Equation (8.9), we can rewrite Equation (8.7) in the form

$$\text{Odds}(X \mid E, F) = \text{OR}(X \mid E) \cdot \text{OR}(X]F) \cdot \text{Odds}(X). \tag{8.10}$$

Dividing through by Odds(X) gives, finally,

$$OR(X \mid E, F) = OR(X \mid E) \cdot OR(X \mid F). \qquad (8.11)$$

We return again to our original problem: suppose that $P(X) = 1/3$, $P(X|E) = 0.8$, and $P(X|F) = 0.9$. Then Odds(X) $= 1/2$, Odds($X|E$) $= 4$, and Odds($X|F$) $= 9$, so OR($X \mid E$) $= 8$ and OR($X \mid F$) $= 18$.

Using the independence assumption, by Equation (8.11), OR($X \mid E, F$) $=$ OR($X|E$)·OR($X|F$) $= 144$. Therefore, Odds($X|E, F$) $=$ OR($X|E, F$)·Odds(X) $= 72$, and $P(X \mid E, F) =$ Odds($X \mid E, F$)$/(1 + $Odds($X \mid E, F$))$ = 72/73 = 0.986$.

As another example, in the trial of O.J. Simpson for murdering his ex-wife Nicole Brown and Ronald Goldman, there was evidence that Simpson had abused his wife on earlier occasions, and the question arose as to whether this evidence could be presented. The prosecution argued that the evidence was relevant because most spousal murders follow earlier abuse. The defense argued that it was irrelevant and merely prejudicial because most men who abuse their wives do not end up murdering them. From our analysis, we can see that both arguments are incomplete. The real question is, how much more frequent, in terms of odds update, is spousal murder among the population of spousal abusers than among the general population of married men? (It makes sense to separate by gender here because the incidence of both abuse and murder of men against women is hugely higher than the reverse.) X here is murder, E is abuse, and F is all the rest of the evidence.

The details of the derivation presented are not important. What is important is the combination of Bayes' law and independence assumptions for combining evidence; Bayes' law is used to munge the formulas into a state where the independence assumptions can be applied. We see this situation again, with wide range of less lurid applications, in Section 8.11.

8.9.2   Application: Secret Sharing in Cryptography

Suppose that we have a secret text, and we want to split it among n people in such a way that if all n get together, they can reconstruct the secret, but no subset of $n - 1$ people can reconstruct the secret or any part of it. For example, suppose the secret is the code to launch nuclear missiles, and we want to make sure that they are not launched unless everyone who is supposed to agree to do so does indeed agree.

Let us first consider the case for which the secret is a single bit B. Then we carry out the following algorithm: for ($i = 1$ to $n-1$) $Q_i \leftarrow$ a random bit; $Q_n = (B + Q_1 + Q_2 + \ldots + Q_{n-1})$ mod 2.

Now tell person i the value of the bit Q_i. Note that

$$Q_1 + \ldots + Q_n \bmod 2 = Q_1 + \ldots + Q_{n-1} + (Q_1 + \ldots + Q_{n-1} + B) \bmod 2$$
$$= (Q_1 + Q_1) + (Q_2 + Q_2) + \ldots + (Q_{n-1} + Q_{n-1}) + B \bmod 2$$
$$= B.$$

So the n participants can reconstruct B by adding up their individual pieces mod 2.

Let's show that B is independent of any subset S of size $n-1$ of the Q_i. This is clearly true if S does not contain Q_n, as $Q_1, \ldots, Q_{n-1}$ were just randomly chosen bits. Suppose that S does contain Q_n but does not contain Q_i. To simplify the notation, let us choose $i = 1$; the other cases are obviously the same. Now,

$$Q_n = ((Q_2 + \ldots + Q_{n-1} + B) + Q_1) \bmod 2. \tag{8.12}$$

Since $Q_1, \ldots, Q_{n-1}$ are chosen randomly, if you know $Q_2, \ldots, Q_{n-1}$, then for either value of B, Q_1 has a 50/50 chance of being 1 or 0. But by Equation (8.12), if $Q_2 + \ldots + Q_{n-1} + B = 1 \bmod 2$, then $Q_n = 1 - Q_1$, and if $Q_2 + \ldots + Q_{n-1} + B = 0 \bmod 2$, then $Q_n = Q_1$. Therefore if you have been told the values of $B, Q_2, \ldots, Q_{n-1}$, then Q_n has a 50/50 chance of being 0 or 1. But before we were told anything, Q_n likewise had a 50/50 chance of being 0 or 1. Thus, Q_n is independent of $B, Q_2, \ldots, Q_{n-1}$. By the symmetry of independence, B is independent of $Q_2, \ldots, Q_n$.

In short, if we have all n pieces, we can construct B, but having any set of $n-1$ pieces gives no information at all about B.

If the secret is a string of k bits, then we can do this for each bit, and hand a bitstring of k random bits to each of the n participants. To reconstruct the secret, add up each place in the bit string mod 2. This is known as "bitwise XOR (exclusive or)"; exclusive OR is the same as addition modulo 2.

Note that the argument does not require any assumptions about the distribution of B. It may be that we start with some external information about the distribution of B; what the argument shows is that we don't learn anything more by finding out $n-1$ of the keys. For the same reason, $Q_1, \ldots, Q_n$ are not necessarily collectively independent absolutely; if we have external information about B, then that will create a dependency.

In older probability books, the fact that pairwise independence does not imply overall independence is often described as a somewhat annoying anomaly of no practical importance. The moral is that practical importance changes over time.

8.10   Random Variables

A *random variable* is a variable that can have a value within some associated domain of values. If the domain of values is finite, then each possible value for the random variable constitutes an event and has a probability. The different values for a single random variable constitute a frame of discernment; that is, exactly one of them must be true, so the sum of their probabilities is equal to 1.

A good way to think about random variables is that a random variable is a dartboard.[2] The different values that the variable can take are marked on the dartboard. Throwing a dart and seeing which area it lands in is the event.

Example 8.18. (Example 8.4 revisited.) Let's flip a coin three times. This can be described by using three random variables F_1, F_2, F_3, one for each flip. Each of these takes values within the set {H, T}. So the pair of events F_1=H and F_1=T form a frame of discernment; so does the pair F_2=H and F_2=T, and likewise for the pair F_3=H and F_3=T.

If we have a finite collection of random variables, then an elementary event is an assignment of one value to each variable. For example, $F_1 = \text{H} \wedge F_2 = \text{H} \wedge F_3 = \text{T}$ is one elementary event, $F_1 = \text{T} \wedge F_2 = \text{T} \wedge F_3 = \text{T}$ is another elementary event, and so on. The set of elementary events again forms a frame of discernment. An assignment of a probability to each elementary event is called a *joint distribution* over the random variables. For example, one joint distribution over $\{F_1, F_2, F_3\}$ would be

$$P(F_1 = \text{H}, F_2 = \text{H}, F_3 = \text{H}) = 0.4,$$
$$P(F_1 = \text{H}, F_2 = \text{H}, F_3 = \text{T}) = 0.1,$$
$$P(F_1 = \text{H}, F_2 = \text{T}, F_3 = \text{H}) = 0.07,$$
$$P(F_1 = \text{H}, F_2 = \text{T}, F_3 = \text{T}) = 0.02,$$
$$P(F_1 = \text{T}, F_2 = \text{H}, F_3 = \text{H}) = 0.1,$$
$$P(F_1 = \text{T}, F_2 = \text{H}, F_3 = \text{T}) = 0.03,$$
$$P(F_1 = \text{T}, F_2 = \text{T}, F_3 = \text{H}) = 0.08,$$
$$P(F_1 = \text{T}, F_2 = \text{T}, F_3 = \text{T}) = 0.2.$$

In these formulas, the commas between events mean conjunction; this is standard notation in probability theory. Thus, $P(F_1 = \text{H}, F_2 = \text{H}, F_3 = \text{H})$ means $P(F_1 = \text{H} \wedge F_2 = \text{H} \wedge F_3 = \text{H})$, $P(E \mid F, G)$ means $P(E \mid F \wedge G)$, and so on.

As in Section 8.7, any event E is the disjunction of elementary events, and $P(E)$ is the sum of the probabilities of these elementary events. For example,

$$(F_1 = \text{H} \wedge F_3 = \text{T}) \equiv (F_1 = \text{H} \wedge F_2 = \text{H} \wedge F_3 = \text{T}) \vee (F_1 = \text{H} \wedge F_2 = \text{T} \wedge F_3 = \text{T}),$$

So

$$P(F_1 = \text{H}, F_3 = \text{T}) = P(F_1 = \text{H}, F_2 = \text{H}, F_3 = \text{T}) + P(F_1 = \text{H}, F_2 = \text{T}, F_3 = \text{T}).$$

---

[2]Thanks to Alan Siegel for this metaphor.

Thus, absolute probability of any event can be calculated from the joint distribution. Since $P(E \mid F) = P(E \wedge F)/P(F)$, any conditional probability can be calculated from absolute probabilities. Therefore, *the joint probability distribution determines all absolute and conditional probabilities associated with these random variables.*

Two random variables E and F are *independent* if, for every value u in the domain of E and v in the domain of F, the event $E = u$ is independent of the event $F = v$. That is,

$$P(E = u, F = v) = P(E = u) \cdot P(F = v). \qquad (8.13)$$

Note: When a relation such as this holds for all values in the domains of the random variable involved, it is common in probabilistic writings to write the relation purely in terms of the random variable, omitting the values. For instance, Equation (8.13) would be written as $P(E, F) = P(E) \cdot P(F)$. This abbreviated form is certainly convenient, but it can also be quite confusing, and takes some getting used to. It is not used in this book, but is mentioned here because the reader may well see it elsewhere.

In general, a probabilistic model consists of

- a choice of random variables,

- a set of independence assumptions,

- numerical values for the probabilities involved in the dependency relationships, or methods for obtaining these.

Regarding independence, unless the number of random variables is small, then the dependencies in the model had better be sparse (that is, almost everything must be conditionally or absolutely independent of almost everything else).

8.11   Application: Naive Bayes Classification

This section discusses how Bayes' law and independence assumptions are used in the naive Bayes method of constructing classifiers from labeled data.

A *classification* problem has the following form: we have a collection of *instances*, and for each instance, we are given some information, which can be viewed as values of a set of *predictive attributes*. Our task is to assign the instance to one or another categories; each category is a possible value of a *classification attribute*. This application has several examples:

Character recognition.   An instance is an image of a printed symbol. The predictive attributes are various geometric features of the image. The classification attribute has values that are the various characters in the character sets; that is, the space of values is { 'A', 'B', ...}.

Computer vision. Many other computer vision problems likewise can be formulated as classification problems. In *face recognition,* the values of the
classification attribute are different people. In *image labeling* the set of
values of the classification attribute is some large vocabulary of categories; for example, {building, human, car, tiger, flower …}. In *pornog-
raphy filtering* (don't snicker—this is a critical task for web browser support), the classification attribute has values {pornography, decent}.

Medical diagnosis. An instance is a patient. The predictive attributes are test
results and features of the medical history. For each disease, there is a
classification attribute whose values are true or false.

Finance. An instance is an applicant for a loan. The predictive attributes are
income, assets, collateral, purpose of the loan, previous credit history,
and so on. The classification attribute has values approve and disap-
prove.

Text classification. An instance is a piece of text. The predictive attributes are
the words of the text (or other information about it). The classification
attribute is some important characteristic of the text. For instance, in
email filtering, the attribute might have values {spam, ordinary, urgent}.
In information retrieval or web search, the classification attribute is set
by the query, and has values {relevant to query Q, irrelevant to query Q}.

A *classifier* is a program or algorithm that carries out a classification task.
Most automated classifiers use some form of machine learning from a *labeled
corpus* (this is known as *supervised* learning). A labeled corpus is a table of instances for which the correct value of the classification attribute is given; for
instance, a set of images of characters labeled with the actual character, or a
set of email messages labeled "spam" or "not spam," and so on. The machine
learning task is to use the corpus to construct a classifier that works well. The
use of corpus-based machine learning techniques to construct classifiers probably constitutes the majority of work in machine learning, both applied and
theoretical.

One way to think about the classification task is to view it probabilistically:
we want to find the most *likely* value of the classification attribute, given the
predictive attributes. That is, we take the predictive attributes to be random
variables $A_1, \ldots, A_k$ and we take the classification attribute to be a random variable C. For a new instance, **x**, we are given the predictive attributes $a_1, \ldots, a_k$,
and we wish to find the value of c that maximizes $P(C = c \mid A_1 = a_1, A_2 = a_2 \cdot \ldots \cdot$
$A_k = a_k)$.

We estimate the probability of events by their frequency in T, a table of
labeled instances. For any event E, let us write $\#_T(E)$ to mean the number of
instances of E in T, and $\text{Freq}_T(E)$ to be the frequency of E in T; thus, $\text{Freq}_T(E) =$

$\#_T(E)/\#T$. Likewise, we write $\mathrm{Freq}_T(E|F) = \#_T(E \wedge F)/\#_T(F)$. Thus, we estimate $P(E) \approx \mathrm{Freq}_T(E)$ and $P(E|F) \approx \mathrm{Freq}_T(E|F)$. However, this works only if $\#_T(E)$ and $\#_T(E \cap F)$ are not too small. In particular, if these are zero, this estimate is fairly useless.

Suppose, for example, there are only two predictive attributes and table T is large, so every combination $C = c, A_1 = a_1, A_2 = a_2$ has numerous instances in T. Then we can estimate

$$P(C = c \mid A_1 = a_1, A_2 = a_2) \approx \underset{T}{\mathrm{Freq}}(C = c \mid A_1 = a_1, A_2 = a_2)$$
$$= \#_T(C = c, A_1 = a_1, A_2 = a_2)/\#_T(A_1 = a_1, A_2 = a_2),$$

and we are done. However, for the example problems we listed, there may be hundreds or thousands or more attributes, and each instance is unique; this particular collection of predictive attributes has *never* before appeared in the table. In that case, we cannot use the approximation

$$P(C = c \mid A_1 = a_1, \ldots, A_k = a_k) \approx \underset{T}{\mathrm{Freq}}(C = c \mid A_1 = a_1, \ldots, A_k = a_k)$$
$$= \#_T(C = c, A_1 = a_1, \ldots, A_k = a_k)/\#_T(A_1 = a_1, \ldots, A_k = a_k)$$

because both numerator and denominator are zero.

Instead, in the naive Bayes method, we make an independence assumption; as it happens, this is very similar to the one we made in Section 8.9.1. We assume that the predictive attributes are all conditionally independent, given the classification attribute. Having done that, we can now proceed as follows. By Bayes' law,

$$P(C = c \mid A_1 = a_1, \ldots, A_k = a_k) = P(C = c) \cdot P(A_1 = a_1, \ldots, A_k = a_k \mid C = c)/P(A_1 = a_1, \ldots, A_k = a_k),$$

which, by the independence assumption, equals

$$P(C = c) \cdot P(A_1 = a_1 \mid C = c) \cdot \ldots \cdot P(A_k = a_k \mid C = c)/P(A_1 = a_1, \ldots, A_k = a_k).$$

However, since what we are looking for is the value of c that maximizes this expression, and since c does not appear in the denominator, we can ignore the denominator. The denominator is just a normalizing expression; without it, we have an equivalent unnormalized set of weights. So our problem now is to find the value of c that maximizes $P(C = c) \cdot P(A_1 = a_1 \mid C = c) \cdot \ldots \cdot P(A_k = a_k \mid C = c)$. But now (hopefully) all of these probabilities can be estimated from frequencies in table T, as long as T is large enough that the pair $A_i = a_i, C = c$

---

function LearnClassifier(**in** T:table; C: attribute): **return** classifier;
{ **for each** (value c of C)
    compute $\text{Freq}_T(C = c)$
    for each (attribute $A \neq C$ in T)
      for each (value a of A)
        compute $\text{Freq}_T(A = a, C = c)$
    endfor endfor endfor
return a table of all these frequencies (this table is the classifier) }

function ApplyClassifier(**in** Q:classifier, X:unlabelled instance; C: attribute) **return** value of C;
return $\text{argmax}_{c:\text{valueof}C}\ \text{Freq}_T(C = c) \cdot \Pi_{A:\text{attribute}}\ \text{Freq}_T(A = X.A \mid C = c)$

/* In the above expression, $X.A$ denotes the A attribute of instance X.
Π_A means the product over all the different attributes.
$\text{argmax}_c E(c)$ means return the value of c that maximizes the expression $E(c)$ */

Algorithm 8.1. Naive Bayes algorithm.

is represented reasonably often for each value of a_i and c. So we estimate

$$P(C = c) \cdot P(A_1 = a_1 \mid C = c) \cdot \ldots \cdot P(A_k = a_k \mid C = c)$$
$$\approx \text{Freq}_T(C = c) \cdot \text{Freq}_T(A_1 = a_1 \mid C = c) \cdot \ldots \cdot \text{Freq}_T(A_k = a_k \mid C = c).$$

Thus, we have two simple algorithms: one to learn the classifier Q for classification attribute C from the table T of labeled data, and the other to apply the classifier to a new instance X (see Algorithm 8.1).

For example, let us take the case of email classification. One way to apply naive Bayes[3] is to take every word that appears in the table to be a predictive attribute of a message: "Hi" is one attribute, "conference" is another attribute, "OMG" is another attribute, and so on. The attributes are Boolean; attribute W of message M is true if W appears in M and false otherwise. So the three numbers Freq_T("conference" = true$|C$ = spam), Freq_T("conference" = true$|C$ = ordinary), and Freq_T("conference" = true$|C$ = urgent) are the fractions of spam, ordinary, and urgent messages that contain the word "conference." Since (in my email) the word "conference" appears much more frequently in ordinary messages than in either spam or urgent messages, if it appears in a new message, it will give a much stronger vote for considering this new message to be ordinary than to be either spam or urgent.

---

[3]There are a number of different ways to apply naive Bayes to text classification; the one described here, called the Bernoulli model, fits most neatly into our schema.

There are two problems here, one technical and one fundamental. The technical problem is that this application does not support the hopeful assumption mentioned that every pair of a value of a predictive attribute with a value of the classification attribute appears some reasonable number of times in the table. We have too many predictive attributes. If an instance contains an attribute value $A_i = a_i$ that has never appeared together with some classification value $C = c$ in the table, then the factor $\text{Freq}_T(A_i = a_i \mid C = c)$ is equal to zero, and therefore the product will be equal to zero, whatever other evidence is present. For example, I have never received an urgent message containing the word "Newfoundland"; therefore, $\text{Freq}_T(\text{"Newfoundland"} = \text{true} \mid C = \text{urgent}) = 0$. But if I now receive the email

> URGENT: You must get me the information you promised me a month ago ASAP. This is vitally important. I need it today, because I'm going to a funeral in Newfoundland tomorrow.

I would want the classifier to notice the words "URGENT," "ASAP," "vitally" and "important" and classify this as urgent, rather than saying that, because it contains the word "Newfoundland" it can't possibly be urgent.

This is known as the *sparse data problem;* it is a common problem in deriving probabilistic models from data. The solution here is to replace zero values by values that are small but nonzero; this is called smoothing. But, of course, we can't treat a zero value as better than a very small nonzero value, so we have to move up all small values. The usual solution, known as the Laplacian correction, is to use an estimate of the form $P(A = a \mid C = c) = (\text{Freq}_T(A = a \mid C = c) + \epsilon)/(1 + N\epsilon)$, where N is the number of possible values of A and $\epsilon > 0$ is a small parameter.

The fundamental problem is that the independence assumption bears no relation to reality. It is not remotely true that the predictive attributes are conditionally independent given the classification attribute. "Conference" is associated with "proceedings," "paper," "deadline," and "hotel"; "meeting" is associated with "agenda," "minutes,' "report"; and "late," "husband," "million," "unable," "collect," "account," "number," and names of various foreign countries go together; as do "Viagra" and—well, fill that one in for yourself.

The use of independence assumptions that have no basis in reality is characteristic of applied probabilistic models generally and of naive Bayes in particular; that is why it is called "naive." One clear consequence in naive Bayes applications is that the computed probabilities tend to be much too one-sided; that is to say, the classifier assigns a much higher confidence to its predictions than its actual accuracy warrants. Essentially, the naive Bayes takes as independent evidence what is actually just the same piece of evidence repeated a number of times. Nonetheless, in many applications, naive Bayes gives answers of a very good quality, computed quite cheaply.

We now look at a couple of further technical observations about the naive Bayes algorithm. In practice, rather than multiply the frequencies, it is common to add the logs of the frequencies. Thus,

$$\log(P(C = c \mid A_1 = a_1, \ldots, A_k = a_k))$$
$$= \log(\text{Freq}_T(C = c) \cdot \text{Freq}_T(A_1 = a_1 \mid C = c) \cdot \ldots \cdot \text{Freq}_T(A_k = a_k \mid C = c))$$
$$= \log(\text{Freq}_T(C = c)) + \log(\text{Freq}_T(A_1 = a_1 \mid C = c)) + \ldots + \log(\text{Freq}_T(A_k = a_k \mid C = c)).$$

Using the logarithm has two advantages. First, multiplying a long list of numbers between 0 and 1 (often mostly rather small numbers) results in underflow rather quickly; the use of the logarithm avoids this.

Second, this formulation demonstrates that the naive Bayes classifier has a simple, and familiar, form; it is just a linear discriminator. Define a vector space with one dimension for each value of each predictive attribute; that is, for each attribute A and for each possible value v, there is a separate dimension $d(A, v)$. Let n be the number of such dimensions. For each value c of the classification attribute, define $\vec{w}_c$ in $\mathbb{R}^n$ such that $\vec{w}[d(A, v)] = \log(\text{Freq}_T(A = v \mid C = c))$, and define a real quantity $t_c = \log(\text{Freq}_T(C = c))$. With any instance x, associate a Boolean vector $\vec{x}$ in $\mathbb{R}^n$ such that $\vec{x}[d(A, v)] = 1$ if $x.A = v$, and $\vec{x}[d(A, v)] = 0$ if $x.A \neq v$. Then, when the classifier classifies an instance $\vec{x}$, it simply computes the value of the expression $t_c + \vec{w}_c \bullet \vec{x}$ for each category c, and it picks the value of c for which this expression is largest.

Finally, let's look at running time. Generally, in machine learning, the learning part is executed separately from the actual task execution. The learning part is said to be "offline"; it is allowed to run quite slowly. The task executor that it creates is usually what the end user is interacting with; it is required to run quickly, "online." The two running times are considered separately. In the case of naive Bayes, the learning algorithm goes through the table T and generates the vectors $\vec{w}_c$ and the quantities t_c; this can be implemented to run in time $O(|T|)$, where $|T|$ is the size of the table. The classifier computes the value of $t_c + \vec{w}_c \cdot \vec{x}$ for each value of c; this runs in time $O(kc)$ where k is the number of attributes and c is the number of possible values for the classification attribute. In the case where $\vec{x}$ is sparse, as in text classification, it runs in time $c \cdot NZ(\vec{x})$, where $NZ(\vec{x})$ is the number of nonzero elements. So the learning algorithm runs reasonably quickly, and the classifier runs extremely quickly.

Exercises

Exercise 8.1. Let Ω be the sample space with six elements, $\Omega = \{a, b, c, d, e, f\}$. Let P be the probability distribution

$$P(a) = 0.05, P(b) = 0.1, P(c) = 0.05, P(d) = 0.1, P(e) = 0.3, P(f) = 0.4.$$

Consider the events $X = \{a, b, c\}$; $Y = \{b, d, e\}$, $Z = \{b, e, f\}$. Evaluate the following:

(a) $P(X)$.

(b) $P(Y)$.

(c) $P(Z)$.

(d) $P(X, Y)$.

(e) $P(X \mid Y)$.

(f) $P(X \mid Z)$.

(g) $P(Y \mid Z)$.

(h) $P(Y \mid X, Z)$.

(i) $P(X, Y \mid Z)$.

(j) Are X and Y absolutely independent?

(k) Are X and Z absolutely independent?

(l) Are X and Y conditionally independent given Z?

Exercise 8.2. Let A, B, C, D, E be five Boolean random variables. Assume that you are given the independence assumptions:

- A and B are independent absolutely,

- D is conditionally independent of both A and C, given B,

- E is conditionally independent of A, B, and D, given C.

Assume you also are given the following probabilities:

$$P(A = T) = 0.9, \qquad P(A = F) = 0.1,$$
$$P(B = T) = 0.6, \qquad P(B = F) = 0.4,$$
$$P(C = T \mid A = T, B = T) = 0.9, \qquad P(C = F \mid A = T, B = T) = 0.1,$$
$$P(C = T \mid A = T, B = F) = 0.8, \qquad P(C = F \mid A = T, B = F) = 0.2,$$
$$P(C = T \mid A = F, B = T) = 0.4, \qquad P(C = F \mid A = F, B = T) = 0.6,$$
$$P(C = T \mid A = F, B = F) = 0.1, \qquad P(C = F \mid A = F, B = F) = 0.9,$$
$$P(D = T \mid B = T) = 0.8, \qquad P(D = F \mid B = T) = 0.2,$$
$$P(D = T \mid B = F) = 0.2, \qquad P(D = F \mid B = F) = 0.8,$$
$$P(E = T \mid C = T) = 0.1, \qquad P(E = F \mid C = T) = 0.9,$$
$$P(E = T \mid C = F) = 0.8, \qquad P(E = F \mid C = F) = 0.2.$$

Compute the following probabilities (you may use MATLAB):

(a) $P(A = T, B = T)$.

(b) $P(A = T \lor B = T)$ (either A or B or both are T).

(c) $P(C = T)$.

(d) $P(C = T \mid A = T)$.

(e) $P(C = T, D = T)$.

(f) $P(A = T \mid C = T)$.

(g) $P(A = T \mid B = T, C = T)$.

(h) $P(E = T \mid A = T)$.

Exercise 8.3.

(a) Given that $P(E) = 1/2$, $P(E \mid F) = 3/4$, $P(E \mid G) = 4/5$, and assuming that F and G are independent evidence for E (that is, they are conditionally independent, given E and given $\neg E$), what is $P(E \mid F, G)$?

(b) Construct a joint probability distribution over E, F, G such that $P(E) = 1/2$, $P(E \mid F) = 3/4$, $P(E \mid G) = 4/5$, but $P(E \mid F, G) = 1/10$.

Exercise 8.4. Suppose that you have a coin that comes up heads with probability 3/4, and you flip it three times. For $i = 1, \ldots, 3$, let F_i be the Boolean random variable which is true if the ith flip comes up heads. Let C be the random variable with values 0,1,2,3, which indicate the total number of heads in the three flips.

(a) What is the associated sample space Ω? What is the probability distribution P over Ω?

(b) What is the probability distribution of C?

(c) What is the conditional probability distribution of F_1, given that $C = 1$?

(d) Are F_1 and F_2 conditionally independent, given that $C = 1$?

Exercise 8.5. Figure 8.1 shows a network with four nodes and five edges. Suppose that each connection fails with probability 1/10 and that failures of connections are all independent events.

(a) What is the probability that there is an active path from B to C? *Hint:* There are no active paths from B to C if all three paths B-A-C, B-C, and B-D-C have failed; these are independent events. This is known as a *series-parallel decomposition*.

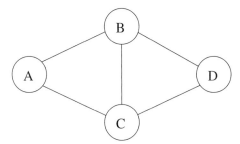

Figure 8.1. Network.

(b) What is the probability that there is a path from node A to D? *Hint:* In this case, it is not possible to do a series-parallel decomposition, and it is very easy, unless one is systematic, to miss a case or double-count. Rather, there are 32 elementary events (i.e., assignments of "working" or "failing" to each of the five edges). You should begin by enumerating all 32, implicitly or explicitly.

(c) What is the probability that there is an active path between all pairs of nodes?

Exercise 8.6. (You may use MATLAB for this exercise.) In Example 8.5 (p. 229), we computed the number of five-card poker hands, and the probability that a random hand is a flush.

(a) How many different hands are straights (numbers in sequence; suits arbitrary)? What is the probability that a random hand is a straight?

(b) How many hands are straight flushes (both a straight and a flush)? What is the probability that a random hand is a straight flush? Are the two events "straight" and "flush" independent?

(c) (This part is more difficult.) Suppose that two hands are dealt. Given that the first is a flush, what is the probability that the second is also a flush? Are the two events, "the first hand is a flush" and "the second hand is a flush," independent?

Problems

Problem 8.1. Prove axioms P1–P4 from axioms F1, F2.

Problem 8.2. (This is difficult.) Prove Theorem 8.17.

Programming Assignments

Assignment 8.1. Write two MATLAB functions SplitSecret(S,N) and Recover-Secret(Q) that implement the secret-sharing technique described in Section 8.9.2. Specifically, SplitSecret(S,N) takes as argument a character string S and a number N. It returns an N × 8|S| array Q of 1s and 0s, where each row is the part of the secret given to one of the sharers. RecoverSecret(Q) takes this matrix Q as input and computes the string S. *Note:* In MATLAB a character string can be converted to a vector of the corresponding ASCII codes (number between 0 and 255) by applying an arithmetic operation to it. The MATLAB function char(N) does the reverse conversion. For example,

```
>> format compact
>> 'cat'-0
ans =
    99    97    116
>> [char(99),char(97),char(116)]
ans =
cat
```

Assignment 8.2.   Generalize Exercise 8.5(c) as follows.   Write a function ProbConnect(P) that computes the probability that all nodes are connected in a network, where the probability that the node from I to J is active is given by P[I,J]. That is, P is an N × N matrix, where P[I,J] is the probability that the arc from I to J is working (connections may not be symmetric). If P[I,J]=0, then there is no connection from I to J. Assume that each arc is independent of all the others.

For instance, in the particular example considered in Exercise 8.5, P would be the matrix

$$P = \begin{bmatrix} 0 & 0.1 & 0.1 & 0 \\ 0.1 & 0 & 0.1 & 0.1 \\ 0.1 & 0.1 & 0 & 0.1 \\ 0 & 0.1 & 0.1 & 0 \end{bmatrix}.$$

You should just implement this in the naive way; enumerate all 2^E ways to label the arcs working and nonworking (E is the number of nonzero edges), and for each one, calculate its probability. (There are certainly more efficient algorithms than this, but all known algorithms are exponential in the worst case.)

Assignment 8.3. Consider a simplified variant of the card game blackjack. Two players are alternately dealt cards, which are random integers between 1 and 10, one at a time. (Assume that each deal is independent; the numbers are dealt by rolling a 10-sided die, not by dealing from a finite deck.) At each stage, a player may either request a card, or may pass and not get assigned a card. If a player's total is exactly 21, he immediately wins. If a player's total exceeds 21, he immediately loses. If a player has passed on one round, he is permitted to continue drawing cards on later rounds. If both players have passed, then the player with the higher total wins. In the case of a tie, the first player to pass wins.

For instance, consider the following sequences with two players:

Sequence 1.

> Player A draws 8.
> Player B draws 3.
> Player A draws 7.
> Player B draws 5.
> Player A draws 8, and loses (exceeds 21).

Sequence 2.

> Player A draws 8.
> Player B draws 3.
> Player A draws 9.
> Player B draws 7.
> Player A passes.
> Player B draws 9.
> Player A draws 3.
> Player B draws 7 and loses (exceeds 21).

We now generalize the above in two ways. First, the number of card values NCards may be different from 10. Second, instead of a single target value 21, we have a target range, from LTarget to UTarget. The player who reaches a total between LTarget and UTarget, inclusive, immediately wins. The player whose total exceeds UTarget immediately loses.

The optimal strategy can be computed by using a dynamic programming implementation of a probabilistic calculation. First, we note the following:

(a) If it is player X's turn to play, then his optimal move is determined by a game state consisting of three parts: whether or not player Y has just passed, X's total points, and Y's total points.

(b) If Y did not pass on the previous turn, and X's total is less than Y's, then X should definitely draw because if X passes, Y can immediately win.

(c) It will never happen that the game ends with both players passing because whichever player would lose will do better to take a chance on drawing.

In view of these observations, the optimal strategy and the player's chances of winning in any given situation can be expressed in two arrays, of size

LTarget × LTarget, indexed from 0 to LTarget − 1. The Boolean array Play[XT,
YT] gives the optimal move for player X in the case where player Y did not
pass on the previous move, where XT and YT are the current totals for X and Y:
Play[XT,YT]=1 if X should draw, 0 if X should pass. The array Prob[XT,YT] is
the probability that X will win if he makes the recommended move.

The assignment, then, is to write a function Blackjack(NCards, LTarget,
UTarget) that returns the two arrays Play and Prob.

The two arrays are filled in together, working backward from the game's
end to its beginning. For instance, if LTarget = 21, the algorithm computes
first Prob[20,20] and Play[20,20], then [19,20] and [20,19], then [18,20],
[19,19], and [20,18], and so on.

The value of the two arrays is filled in as follows:

```
if (rules B or C determine the winning move)
     then compute the probability for that move
     else compute the probabilities for both moves
               and use the move with higher probability;
save the move in Play[XT,YT] and the probability in Prob[XT,YT].
endif
```

The probability that X will win if he draws in state XT, YT can be computed
by considering each possible deal. If X draws a card with value $CARD$ and neither wins nor loses, then it will be Y's turn, and Y will be in the state $\langle YT, XT +$
$CARD \rangle$. The probability that Y wins in that state is Prob[YT,XT+CARD]; hence,
the probability that X wins if Y is in that state is $1 -$ Prob[YT,XT+CARD].

```
% Computing the probability that X will win if he draws in
% state XT,YT:
{ ProbWinning = 0.0
  for (CARD=1:NCards) {
     if (XT+CARD > UTarget) then ProbYWins = 1;
        elseif (XT+CARD >= LTarget) then ProbYWins = 0;
        else ProbYWins = Prob[YT,XT+CARD];
     ProbWinning = ProbWinning + (1-ProbYWins)/NCards;
  endfor }
  return ProbWinning
}
```

The probability that X will win if he passes after Y drew on the previous
turn is

```
if (YT > XT) then 0
else 1-Prob[YT,XT]
```

Chapter 9

Numerical Random Variables

Random variables whose domain is a set of numbers are important in a wide range of applications. Most of mathematical probability theory deals the properties of numerical random variables.

Example 9.1. (Example 8.2 revisited.) Say we roll a fair die. The random variable D has domain $\{1,2,3,4,5,6\}$. It is uniformly distributed, so $P(D = x) = 1/6$ for each value x.

Example 9.2. On May 5, 2010, we poll 1,000 randomly chosen Americans on their opinion of President Obama. Let us say that the question is Boolean: favorable or unfavorable. We can define a numerical random variable X whose value is the number of people in the poll who answer "favorable." The domain of the random variable is $\{0, 1, \ldots, 1000\}$. (We discuss the binomial probability function in Section 9.6.2.) Assuming that we are careful not to ask the same person twice, then this is an instance of *sampling without replacement*, discussed in Example 8.7.

Most numerical random variables have a domain in one of three categories.

(a) The domain is a finite set of numbers.

(b) The domain is an infinite set of integers.

(c) The domain is the real line, or an interval in the real line.

This chapter begins with category (a); Section 9.5 introduces category (b), which is not very different from (a); and Section 9.7 discusses category (c), which is substantially different from (a) or (b).

The fundamental concepts and properties of numerical random variables are largely the same for all three categories, so we introduce them here in the simplest context of random variables with finite domains. The first concept is the idea of a function of a random variable. If X is a random variable and f is a function over the domain of X, then $f(X)$ is also a random variable; the value of $f(X)$ is always f applied to the value of X. Likewise, if $X_1,\ldots,X_k$ are random variables, and $f(x_1,\ldots,x_k)$ is a function of k arguments, then $f(X_1,\ldots,X_k)$ is a random variable. Therefore, the event $f(X_1,\ldots,X_k) = c$ is the union of the events $f(X_1 = d_1,\ldots,X_k = d_k)$ over all tuples $\langle d_1,\ldots,d_k \rangle$ for which $f(d_1,\ldots,d_k) = c$.

Example 9.3. (Example 8.2 revisited.) As in Example 9.2, let D be the result of rolling a fair die. Then $E = D + 1$ is a random variable; the domain of E is $\{2,3,4,5,6,7\}$, and it is uniformly distributed. Moreover, the value of E is tied to the value of D. For example,

$$P(E = 3 \mid D = 2) = 1,$$
$$P(E = 3 \mid D = 3) = 0.$$

Example 9.4. We roll two fair dice. Now there are two random variables: D_1 for the first roll, and D_2 for the second. Each is uniformly distributed, and the two random variables are independent. Therefore, for each pair of values u, v, $P(D_1 = u, D_2 = v) = P(D_1 = u) \cdot P(D_2 = v) = (1/6) \cdot (1/6) = 1/36$.

Let $S = D_1 + D_2$. S is a random variable with domain $\{2,3,\ldots 12\}$. The probability that $S = w$ is equal to the sum of $P(D_1 = u, D_2 = v)$ over all u, v such that $u + v = w$. For instance,

$$P(S = 3) = P(D_1 = 1, D_2 = 2) + P(D_1 = 2, D_2 = 1) = \frac{2}{36} = \frac{1}{18},$$
$$P(S = 4) = P(D_1 = 1, D_2 = 3) + P(D_1 = 2, D_2 = 2) + P(D_1 = 3, D_2 = 1) = \frac{3}{36} = \frac{1}{12}.$$

Thus, S is *not* uniformly distributed.

Note that S is not the same as the random variable $F = D_1 + D_1$. F has the domain $\{2,4,6,8,10,12\}$ and is uniformly distributed over that range. $P(F = 4 \mid D_1 = 2) = 1$, whereas $P(S = 4 \mid D_1 = 2) = 1/6$. This illustrates that D_1 and D_2 are two *different* random variables, even though they have identical domains and probability distributions.

Theorem 9.5 is trivial but useful.

Theorem 9.5. *If X and Y are independent random variables, f is a function over the domain of X, and g is a function over the domain of Y, then $f(X)$ and $g(Y)$ are independent.*

Proof: Let u be a value of $f(X)$ and let v be a value of $g(Y)$. Let $A = f^{-1}(u) = \{a \mid u = f(a)\}$, the set of all values a that f maps into u; and let $B = g^{-1}(v) =$

$\{b \mid v = g(b)\}$, the set of all values b that g maps into v. Then

$$
\begin{aligned}
P(f(X) = u, g(Y) = v) &= P(X \in A, Y \in B) \\
&= \sum_{a \in A, b \in B} P(X = a, Y = b) \\
&= \sum_{a \in A, b \in B} P(X = a) \cdot P(Y = b) \\
&= \left[\sum_{a \in A} P(X = a) \right] \cdot \left[\sum_{b \in B} P(Y = b) \right] \\
&= P(X \in A) \cdot P(Y \in B) \\
&= P(f(X) = u) \cdot P(g(Y) = v). \qquad \square
\end{aligned}
$$

9.1   Marginal Distribution

Suppose that we have two random variables (not necessarily numeric): X with domain $\{u_1, \ldots, u_m\}$ and Y with domain $\{v_1, \ldots, v_n\}$. We can then construct an $m \times n$ matrix M, where $M[i, j] = P(X = u_i, Y = v_j)$. If we sum up each row, we get a column vector of length m; this is the overall probability distribution over X. If we sum up each row, we get a row vector of length n; this is the overall probability distribution over Y. These are known as the marginal distributions over the table, because we write them in the right and bottom margins. Table 9.1 shows an example of such a table, where X has domain $\{a, b, c, d\}$ and Y has domain $\{e, f, g, h\}$. We can see that $P(X = a, Y = g) = 0.10$, $P(X = b) = 0.30$, and $P(Y = f) = 0.22$.

The proof that this table works is simple. We take as an example event $Y = f$, which is equivalent to the disjunction $(Y = f \wedge X = a) \vee (Y = f \wedge X = b) \vee (Y = f \wedge X = c) \vee (Y = f \wedge X = d)$. Therefore,

$$P(Y = f) = P(Y = f, X = a) + P(Y = f, X = b) + P(Y = f, X = c) + P(Y = f, X = d).$$

The general case is exactly analogous, so we now have Theorem 9.6.

Theorem 9.6. *Let X and Y be random variables and let u be a value of X. Then $P(X = u) = \sum_v P(X = u, Y = v)$.*

$X \backslash Y$	e	f	g	h	$P(X)$
a	0.01	0.05	0.10	0.03	0.19
b	0.20	0.02	0.06	0.02	0.30
c	0.08	0.08	0.04	0.02	0.22
d	0.01	0.07	0.11	0.10	0.29
$P(Y)$	0.30	0.22	0.31	0.17	

Table 9.1. Marginal distribution.

9.2 Expected Value

The *expected value* of a random variable X, denoted $\mathrm{Exp}(X)$, also called the *mean* of X, is the weighted average of possible outcomes, where each outcome is weighted by the probability that it will occur:

$$\mathrm{Exp}(X) = \sum_v v \cdot P(X = v).$$

Note that $\mathrm{Exp}(X)$ is not a random variable; it is just a number.

For instance, if D is the random variable for the roll of a fair die, then

$$\mathrm{Exp}(D) = 1 \cdot P(D=1) + 2 \cdot P(D=2) + 3 \cdot P(D=3) + 4 \cdot P(D=4) + 5 \cdot P(D=5) + 6 \cdot P(D=6)$$

$$= (1 \cdot 1/6) + (2 \cdot 1/6) + (3 \cdot 1/6) + (4 \cdot 1/6) + (5 \cdot 1/6) + (6 \cdot 1/6)$$

$$= 21/6$$

$$= 7/2.$$

Example 9.7. (Example 9.4 revisited.) Let X be the roll of an unfair coin that comes up heads 3/4 of the time, and let us associate the value 1 with heads and the value 0 with tails. Then $\mathrm{Exp}(X) = 1 \cdot P(X = 1) + 0 \cdot P(X = 0) = 3/4$.

Suppose that the possible values of X are $v_1, \ldots, v_k$ with associated probabilities $p_1, \ldots, p_k$. Define the vectors $\vec{v} = \langle v_1, \ldots, v_k \rangle$ and $\vec{p} = \langle p_1, \ldots, p_k \rangle$. Then $\mathrm{Exp}(X) = \vec{v} \bullet \vec{p}$. Clearly, for any constant c, $\mathrm{Exp}(X + c) = \mathrm{Exp}(X) + c$, and $\mathrm{Exp}(c \cdot X) = c \cdot \mathrm{Exp}(X)$.

Expected value satisfies the following simple and important theorem.

Theorem 9.8. *Let X and Y be random variables. Then* $\mathrm{Exp}(X + Y) = \mathrm{Exp}(X) + \mathrm{Exp}(Y)$.

What is remarkable about this theorem is that it does *not* require that X and Y be independent. That means that it can be applied in cases where determining the distribution of $X + Y$ and carrying out the summation is difficult.

For example, consider the problem of sampling without replacement. Suppose that we have an urn with r red balls and b black balls, and we choose a sample of s balls without replacement. Let Q be the random variable, which is the number of red balls in the sample. What is $\mathrm{Exp}(Q)$? To do this problem directly, we would have to find the distribution of Q (which we will do in Section 9.6.2) and then we would have to perform the summation $\sum_{i=0}^{r+b} i \cdot P(Q = i)$. It can be done, but it takes some work and some combinatorics. But we can avoid all that just by using Theorem 9.8. Define the numerical random variables $Q_1, \ldots, Q_s$ as follows:

$$Q_i = 1 \text{ if the } i\text{th ball in the sample is red,}$$
$$Q_i = 0 \text{ if the } i\text{th ball in the sample is black.}$$

A priori, each ball in the urn has an equal chance of being chosen as the ith ball. Therefore, $P(Q_i = 1) = r/(r + b)$ and $P(Q_i = 0) = b/(r + b)$, so $\text{Exp}(Q_i) = 1 \cdot r/(r+b) + 0 \cdot b/(r+b) = r/(r+b)$. Clearly, $Q = Q_1 + \ldots + Q_s$, so, by Theorem 9.8,

$$\text{Exp}(Q) = \text{Exp}(Q_1) + \ldots + \text{Exp}(Q_s) = sr/(r + b).$$

This makes sense: the fraction of balls that are red is $r/(r + b)$, so in a sample of s balls, we would expect that about $(r/(r + b)) \cdot s$ would be red. As discussed in Section 8.9, the variables Q_i are not independent, but Theorem 9.8 can nonetheless be applied.

For example, in the case of the Obama poll (Example 9.4), suppose the total population is z and the fraction of the population supporting Obama is f. We represent the pro-Obama faction as red balls and the anti-Obama faction as black balls, so $r = fz$ and $b = z - fz$. So if N is the random variable representing the number of people who approve Obama in a poll of 1000, then $\text{Exp}(N) = 1000fz/z = 1000f$.

We can get the same result in a different way. Number the red balls $1, \ldots, r$, and define the random variables $X_1, \ldots, X_r$ as follows:

$$X_i = 1 \text{ if ball } i \text{ is in the sample,}$$
$$X_i = 0 \text{ if ball } i \text{ is not in the sample.}$$

Since the sample is a random selection of s balls out of $r + b$ total balls, each ball i has probability $P(X_i = 1) = s/(r + b)$ of being chosen for the sample. Therefore, $\text{Exp}(X_i) = s/(r + b)$. Clearly, $Q = X_1 + \ldots + X_r$, so $\text{Exp}(Q) = \text{Exp}(X_1) + \ldots + \text{Exp}(X_r) = sr/(r + b)$.

Having demonstrated the usefulness of Theorem 9.8, let us now prove it. The proof just involves rearranging the order of summation and using marginal probabilities. In the formula in this proof, the summation over u ranges over the domain of X and the summation over v ranges over the domain of Y.

Proof: Proof of Theorem 9.8:

$$\begin{aligned}
\text{Exp}(X + Y) &= \sum_{u,v} (u + v) \cdot P(X = u, Y = v) \\
&= \sum_{u,v} u \cdot P(X = u, Y = v) + \sum_{u,v} v \cdot P(X = u, Y = v) \\
&= \sum_{u} \sum_{v} u \cdot P(X = u, Y = v) + \sum_{v} \sum_{u} v \cdot P(X = u, Y = v) \\
&= \sum_{u} u \cdot \left(\sum_{v} P(X = u, Y = v) \right) + \sum_{v} v \cdot \left(\sum_{u} \cdot P(X = u, Y = v) \right) \\
&= \sum_{u} u \cdot P(X = u) + \sum_{v} v \cdot P(Y = v) = \text{Exp}(X) + \text{Exp}(Y),
\end{aligned}$$

by using marginal summation.   □

The conditional expected value of random variable X, given event E, is the expected value computed by using probabilities conditioned on E: $\text{Exp}(X|E) = \sum_v v \cdot P(X = v | E)$.

Similarly, Theorem 9.9 holds for the product of X and Y, *but only if X and Y are independent.*

Theorem 9.9. *If X and Y are independent, then* $\text{Exp}(X \cdot Y) = \text{Exp}(X) \cdot \text{Exp}(Y)$.

Proof: This is just the independence assumption combined with the distributive law:

$$\text{Exp}(X \cdot Y) = \sum_{u,v} (u \cdot v) \cdot P(X = u, Y = v)$$

$$= \sum_{u,v} (u \cdot v) \cdot P(X = u) \cdot P(Y = v)$$

$$= \left(\sum_u u \cdot P(X = u) \right) \cdot \left(\sum_v v \cdot P(Y = v) \right) = \text{Exp}(X) \cdot \text{Exp}(Y). \qquad \square$$

Suppose that random variable X represents the outcome of a random process that we can run repeatedly in independent trials, and suppose that we run the process N times, where N is a large number. Let $S = \langle s_1, \ldots, s_N \rangle$ be the sample of output values. Then with very high probability, each value v appears approximately $N \cdot P(X = v)$ times in the sample. Therefore, with high probability, the total value of S is $\sum_v v \cdot (N \cdot P(X = v))$, so the average value of elements in S is $\text{Exp}(X)$.

9.3   Decision Theory

Decision theory is the application of probability theory to the choice of action. The fundamental premise is that an agent should choose the action, or the strategy, that maximizes the expected value of the *utility* of the outcome. (Utility in this sense is a measure of the goodness of an outcome on some kind of numeric scale.) This is known as the *maximum expected utility* principle.

Let us start with a simple example. We have the opportunity to buy a lottery ticket. The ticket costs \$1. If it is the winning ticket, it will pay \$100; otherwise, it will pay nothing. The probability that the ticket will win is $1/1000$.

We can analyze this as follows. We have two possible actions:

A1. Buy a ticket.

A2. Hold onto our money. (In decision theory, doing nothing is a form of action.)

If we carry out action A1, then there is a probability of 999/1000 that you will end up \$1 poorer and a probability of 1/1000 that you will end up \$99 richer. The *expected gain* is therefore $(999/1000) \cdot -1 + (1/1000) \cdot \$99 = -0.9$. Another way to look at this is that if we buy N tickets, where N is a large number, all of which have these same terms and all of which are independent, then with high probability we will end up poorer by $0.9N$.

In contrast, if we carry out action A2, then with certainty our gain is zero, so our expected gain is 0. Therefore, A2 is the preferred action.

Let's consider another example. Suppose that a student has class in an hour, and has two choices for spending the next hour: (a) studying for the pass/fail quiz at the start of class; or (b) playing a video game. Let's make the following assumptions:

- The video game is worth 10 utils (the unit of utility) of enjoyment.

- The activity of studying is worth −5 utils of enjoyment.

- Passing the exam is worth 4 utils.

- Failing the exam is worth −20 utils.

- The net utility is the sum of the utility associated with the next hour's activity plus the utility associated with the result of the exam. (We choose the action on the basis of the net utility. This is generally taken to be the sum of the utility of the component parts, but there is no logical necessity for this to hold.)

- If the student studies, the probability is 0.75 that he or she will pass.

- If the student does not study, the probability is 0.5 that he or she will pass.

Putting this together, if the student studies, then there is a 0.75 probability of a net utility of $4 + (-5) = -1$ and a 0.25 probability of a net utility of $4 + (-20) = -16$; the expected net utility is therefore $0.75 \cdot -1 + 0.25 \cdot -16 = -4.75$. If the student doesn't study, then there is a 0.5 probability of a net utility of $10 + 4 = 14$ and a 0.5 probability of a net utility of $10 + (-20) = -10$; the expected net utility is therefore $0.5 \cdot 14 + 0.5 \cdot -10 = 2$. So the rational choice is to play the video game.

9.3.1 Sequence of Actions: Decision Trees

In a more complex situation, an agent may have to carry out a sequence of actions, and the choice of later actions may depend on the earlier actions.

For example, Joe wants to eat lunch out and get back to the office as quickly as possible. An Indian restaurant is a 5 minute walk north of him and a Chinese restaurant is a 10 minute walk south of him. Either restaurant is in one of two states, quiet or busy, throughout the lunch hour. The Chinese restaurant is busy with probability 0.5, and the Indian restaurant is busy with probability 0.8; these two states are independent events and do not change throughout the lunch hour. At both restaurants, if the restaurant is quiet, then lunch takes 10 minutes; if the restaurant is busy, then with probability 0.25, lunch will take 30 minutes, and with probability 0.75, lunch will take 60 minutes. Assume that Joe's measure of utility is the total elapsed time before he gets back to the office.

Clearly, four plans are worth considering:

1. John walks to the Indian restaurant, and stays there whether or not it is busy. The probability is 0.2 that lunch will take 10 minutes, for a total elapsed time of 20 minutes; the probability is $0.8 \cdot 0.25 = 0.2$ that lunch will take 30 minutes, for a total elapsed time of 40 minutes; and the probability is $0.8 \cdot 0.75 = 0.6$ that lunch will take 60 minutes, for a total elapsed time of 70 minutes. Thus, the expected time is $0.2 \cdot 20 + 0.2 \cdot 40 + 0.6 \cdot 70 = 54$ minutes.

2. John walks to the Indian restaurant. If it is busy, he then walks to the Chinese restaurant. He stays there, whether or not it is busy. The expected time is $0.2 \cdot 20 + 0.8 \cdot 0.5 \cdot 40 + 0.8 \cdot 0.5 \cdot 0.25 \cdot 60 + 0.8 \cdot 0.5 \cdot 0.75 \cdot 90 = 53$ minutes. (Note that the total walking time to go from the office to one restaurant to the other to the office is 30 minutes.)

3. John walks to the Chinese restaurant, and stays there whether or it is busy. The expected time is $0.5 \cdot 30 + 0.5 \cdot 0.25 \cdot 50 + 0.5 \cdot 0.75 \cdot 80 = 51.25$ minutes.

4. John walks to the Chinese restaurant. If it is busy, he walks to the Indian restaurant. He stays there, whether or not it is busy. The expected time is $0.5 \cdot 30 + 0.5 \cdot 0.2 \cdot 40 + 0.5 \cdot 0.8 \cdot 0.25 \cdot 60 + 0.5 \cdot 0.8 \cdot 0.75 \cdot 90 = 52$ minutes.

Therefore, the optimal plan is the third choice, to walk to the Chinese restaurant and stay there.

The situation for this scenario can be illustrated in a *decision tree*. A decision tree is a tree with a root, branches (outarcs emanating from nodes), and leaves (outcomes).

*   The root is the starting state.

*   Some internal nodes are *decision nodes*, represented with squares. The outarcs from a decision node correspond to actions.

*   The remaining internal nodes are *chance nodes*, represented with circles. The outarcs from a chance node correspond to events; they are labeled with the probability of the event.

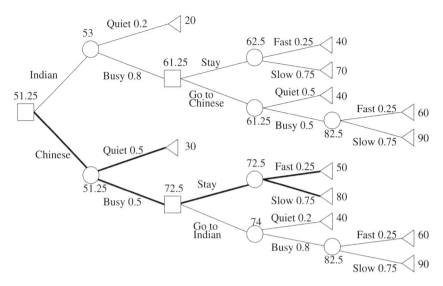

Figure 9.1. Decision tree. Squares denote decision nodes, circles denote chance nodes, and the bold outarcs from each decision node signify the optimal decision at that point.

- The leaves, commonly represented as triangles, are the outcomes and are labeled with the utility.

The value of a decision node is the value of the best of its children, and the prescribed action to take is the outarc that leads to the best child. The value of a chance node is the expected value of its children. For a large tree, this is a more effective way of structuring the calculation than the enumeration of strategies above.

Figure 9.1 shows the decision tree for the restaurant problem. Decision trees are commonly drawn left to right because a path from the root to a leaf moves forward in time. Figure 9.1 shows the *optimal strategy* highlighted with thicker arcs. The optimal strategy is obtained by deleting every suboptimal action and all its descendent nodes.

It should be noted that this is not the *complete* decision tree for this problem; that would also show actions that make no sense, such as going from one restaurant to another even if the first is quiet. In fact, the complete decision tree is actually infinite, since the agent can go back and forth between restaurants arbitrarily many times. In this case, all but the actions shown in Figure 9.1 are obviously suboptimal, so they need not be shown. (To be precise, for this probability model, these actions are easily to be seen to be outside the optimal strategy, no matter what their probabilities and waiting times are. Therefore, they can be excluded without doing any numerical calculations.)

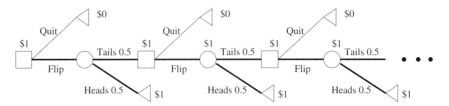

Figure 9.2. Infinite decision tree.

In other problems, the tree of reasonable actions actually is infinite. For a simple example, suppose that someone makes the following offer: We can flip a coin as many times as we want, and he will pay $1 the first time the coin comes up heads. The expected value of this offer is $1, since the probability is 1 that eventually some flip will come up heads. The decision tree for this case is shown in Figure 9.2. Several kinds of data structures and algorithms have been developed that are more effective than decision trees for these kinds of problems, but they are beyond the scope of the discussion here.

To represent the probabilistic aspects of reasoning about complex strategies, we use a random variable, Outcome(A), whose value is the utility of strategy A.[1] For instance, let A be the plan "Go to the Chinese restaurant and stay there even if it is busy," let B be the event, "The Chinese restaurant is busy," and let F be the event, "Service is reasonably fast." Then we are given the following information:

$$P(B) = 0.5,$$
$$P(F \mid B) = 0.25,$$
$$\text{Outcome}(A) = \{20 \text{ if } \neg B; \; 50 \text{ if } B \wedge F; \; 80 \text{ if } B \wedge \neg F\}.$$

Therefore,

$$P(\text{Outcome}(A) = 20) = P(\neg B) = 0.5,$$
$$P(\text{Outcome}(A) = 50) = P(B, F) = P(F \mid B) \cdot P(B) = 0.125,$$
$$P(\text{Outcome}(A) = 80) = P(B, \neg F) = P(\neg F \mid B) \cdot P(B) = 0.375.$$

So $\text{Exp}(\text{Outcome}(A)) = 0.5 \cdot 30 + 0.125 \cdot 50 + 0.375 \cdot 80 = 51.25$.

9.3.2 Decision Theory and the Value of Information

Decision theory also allows a value to be assigned to gaining information, in terms of the information's usefulness in helping to make decisions with better

---

[1]The problem of systematically representing strategies and calculating their various outcomes is the problem of *plan representation and reasoning;* it is beyond the scope of this book. For more information, see Ghallab, Nau, and Traverso (2004) and Russell and Norvig (2009).

outcomes. For example, suppose that a book manuscript has been submitted to a publisher, with the following simplifying assumptions.

1. A book is either a success or a failure. If the book is a success, the publisher will gain $50,000. If the book is a failure, the publisher will lose $10,000.

2. The probability that a manuscript will succeed is 0.2, and the probability that it will fail is 0.8.

The publisher has a choice of two actions: publish the manuscript, in which case the expected gain is $0.2 \cdot \$50,000 - 0.8 \cdot \$10,000 = \$2,000$; or reject the manuscript, in which case the expected gain is $0. So the publisher's preferred plan is to publish the manuscript, for an expected gain of $2,000.

Now suppose the publisher has another option; namely, to consult with a reviewer. Consider, first, the case in which the publisher knows an *infallible* reviewer, who can always successfully judge whether a book will succeed or fail. In that case, the publisher has the option of carrying out the following strategy:

Consult with the reviewer;
if (the reviewer recommends the book)
   then publish it;
   else reject it;

The expected value of this strategy can be calculated as follows. With probability 0.2 the book will be a success, the reviewer will approve it, and the publisher will publish it and will gain $50,000. With probability 0.8, the book will be a failure, the reviewer will reject it, and the publisher will reject it and will gain $0. Therefore, the expected value of the outcome of this strategy, at the outset, is $0.2 \cdot \$50,000 + 0.8 \cdot 0 = \$10,000$. Therefore, the reviewer's opinion is worth $\$10,000 - \$2,000 = \$8,000$ to the publisher; if the reviewer's fee is less than $8,000, then it is worthwhile to the publisher to pay for a review.

Unfortunately, reviewers are not actually infallible. The most that a publisher can realistically expect is that the reviewer's opinion bears some relation to the actual outcome. Let R be a Boolean random variable representing the reviewer's opinion; and let S be a random variable representing whether the book would be a success if the publisher publishes it. Suppose we are given the following additional conditional probabilities: $P(R = T \mid S = T) = 0.7$; and $P(R = T \mid S = F) = 0.4$.

Let $A1$ be the strategy already discussed: Publish if $R = T$, reject if $R = F$. Then we can evaluate Exp(Outcome(S)) as follows:

$$\text{Exp(Outcome}(A1)) = \text{Exp(Outcome}(A1)|R = T) \cdot P(R = T) + \text{Exp(Outcome}(A1)|R = F) \cdot P(R = F).$$

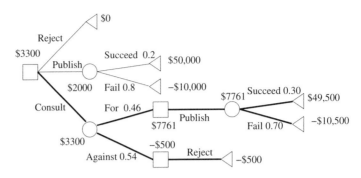

Figure 9.3. Decision tree for a publisher consulting with one reviewer.

If $R = F$ then the publisher does not publish; so $\text{Exp}(\text{Outcome}(A1) | R = F) = 0$. To evaluate the first term above we proceed as follows

$$\text{Exp}(\text{Outcome}(A1) | R = T) \cdot P(R = T)$$
$$= \$50,000 \cdot P(S = T | R = T) \cdot P(R = T) - \$10,000 \cdot P(S = F | R = T) \cdot P(R = T)$$
$$= \$50,000 \cdot P(R = T | S = T) \cdot P(S = T) - \$10,000 \cdot P(R = T | S = F) \cdot P(S = F)$$
$$= \$50,000 \cdot 0.7 \cdot 0.2 - \$10,000 \cdot 0.4 \cdot 0.8 = \$3,800.$$

So $\text{Exp}(\text{Outcome}(A1)) = \$3,800$. Thus, the value of the reviewer's opinion is $\$3,800 - \$2,000 = \$1,800$; as long as the reviewer charges less than $\$1,800$, it is worth consulting him.

The decision tree for this problem is shown in Figure 9.3. The reviewer's fee is taken to be \$500. Again, actions that are obviously senseless are omitted. For instance, there is no point in consulting unless the publisher is going to follow the reviewer's advice, so "reject" is not an option if a favorable review has been received, and "publish" is not an option if an unfavorable review has been received.

More complex versions of this problem, in which the publisher has the option of hiring several reviewers, are considered in Problems 9.2–9.3 and Programming Assignment 9.2.

As this kind of example gets more complex, the space of strategies becomes more difficult to characterize. (A small example of this is seen in Exercise 9.3). Likewise, the problem of determining whether a strategy is executable becomes more complex. Consider, for example, "strategy" $A2$: if $S = T$ then publish; else, reject. Clearly, this is not actually a strategy. (The archetype of this kind of strategy is the advice, "Buy low, sell high.") But why is strategy $A1$ executable and $A2$ not executable? Or, more concretely, how does one develop a problem representation that will allow all and only executable strategies to be considered? Partial solutions to this question are known, and the area is one of active research.

9.4   Variance and Standard Deviation

The expected value of X characterizes the center of distribution X. The variance $\mathrm{Var}(X)$ and the standard deviation $\mathrm{Std}(X)$ characterize how broadly the distribution is spread around the center.[2]

Let X be a random variable and let v be a value. We define the *spread* of X around v, denoted $\mathrm{Spread}(X, v)$, as the expected value of $(X - v)^2$. Note that this number $(X - v)^2$ is always positive and gets larger the farther X is from v.

Clearly, the most reasonable value of v to use to measure the inherent spread of X is the one that minimizes $\mathrm{Spread}(X, v)$. As luck would have it, that value of v is exactly $\mathrm{Exp}(X)$; and the associated value of the spread is called the *variance* of X, denoted $\mathrm{Var}(X)$.

As a measure of spread, however, the variance has the problem that it is in the wrong units; if X is measured in feet, for example, $\mathrm{Var}(X)$ is in square feet. To get a measure of the spread of X that is comparable to the values of X, we take the square root of the variance. This is the *standard deviation* of X, denoted $\mathrm{Std}(X)$.

Definition 9.10. Let X be a numeric random variable. Let $\mu = \mathrm{Exp}(X)$. Then

$$\mathrm{Var}(X) = \mathrm{Exp}((X - \mu)^2) = \sum_u P(X = u)(u - \mu)^2,$$

$$\mathrm{Std}(X) = \sqrt{\mathrm{Var}(X)}.$$

For example, let D be the roll of a single die. We calculated in Section 9.2 that $\mathrm{Exp}(D) = 7/2$. Therefore,

$$\begin{aligned}
\mathrm{Var}(D) &= (1/6)(1 - 7/2)^2 + (1/6)(2 - 7/2)^2 + (1/6)(3 - 7/2)^2 \\
&\quad + (1/6)(4 - 7/2)^2 + (1/6)(5 - 7/2)^2 + (1/6)(6 - 7/2)^2 \\
&= 1/6 \cdot (25/4 + 9/4 + 1/4 + 1/4 + 9/4 + 25/4) \\
&= 70/24 \\
&= 2.9167,
\end{aligned}$$

$$\mathrm{Std}(D) = \sqrt{\mathrm{Var}(D)} = 1.7079.$$

(You may ask, why use $\mathrm{Exp}((X - v)^2)$ rather than just $\mathrm{Exp}(|X - v|)$? That is also a useful number; the value of v that minimizes it is the *median* of X (see Problem 9.5). But $(X - v)^2$ has a number of advantages. The basic one is that it is differentiable, which simplifies many symbolic manipulations such as minimization. Also, the variance satisfies Theorem 9.12, which is an important theorem; there is no comparable theorem that holds for $\mathrm{Exp}(|X - v|)$.)

---

[2]The expected value, variance, and standard deviation of a *random variable,* discussed in this chapter, should not be confused with the related concepts of the mean, variance, and standard deviation of a *dataset,* discussed in Chapter 14.

In probability theory, the symbol μ is often used for the expected value and the symbol σ is used for the standard deviation, so σ^2 is the variance.

Clearly, for any constant c, $\text{Var}(X + c) = \text{Var}(X)$, $\text{Std}(X + c) = \text{Std}(X)$, $\text{Var}(c \cdot X) = c^2 \cdot \text{Var}(X)$, and $\text{Std}(c \cdot X) = c \cdot \text{Std}(X)$. That is, the variance and standard deviation are invariant under translation and the standard deviation is linear under scalar multiplication.

The significance of the standard deviation is illustrated in Theorem 9.11, known as Tschebyscheff's inequality (the name is also spelled Chebyshev, Čebisev, and other variations).

Theorem 9.11. *Let X be a random variable with mean μ and standard deviation σ. Then for any $w > \sigma$, $P(|X - \mu| \geq w) \leq \sigma^2 / w^2$. That is, if w is substantially greater than σ, it is very unlikely that X is more than w from μ, where "very unlikely" means the probability is not more than σ^2 / w^2.*

Proof: Let S be the set of all values of X, and let U be the subset of S of all values v of X such that $|v - \mu| \geq w$. Then for any value $u \in U$, $(u - \mu)^2 \geq w^2$, so $P(u) \cdot (u - \mu)^2 \geq P(u) \cdot w^2$. Therefore,

$$\sigma^2 = \sum_{v \in S} P(v) \cdot (v - \mu)^2 \geq \sum_{v \in U} P(v) \cdot (v - \mu)^2 \geq \sum_{v \in U} P(v) \cdot w^2 = w^2 \cdot \sum_{v \in U} P(v) = w^2 P(U).$$

So $P(|X - \mu| \geq w) = P(U) \leq \sigma^2 / w^2$.   □

The variance of the sum of two random variables, $\text{Var}(X + Y)$, satisfies a theorem similar to Theorem 9.8, *but it applies only if X and Y are independent:*

Theorem 9.12. *Let X and Y be independent random variables. Then $\text{Var}(X + Y) = \text{Var}(X) + \text{Var}(Y)$.*

Proof: Let $\mu_X = \text{Exp}(X)$ and $\mu_Y = \text{Exp}(Y)$; these are constants. Let $\bar{X}$ and $\bar{Y}$ be the random variables $\bar{X} = X - \mu_X$, $\bar{Y} = Y - \mu_Y$. Then $\text{Exp}(\bar{X}) = \text{Exp}(X - \mu_X) = \text{Exp}(X) - \mu_X = 0$, and likewise, $\text{Exp}(\bar{Y}) = 0$. Also $\bar{X} + \bar{Y} = X + Y - \mu_X - \mu_Y$, so $\text{Var}(\bar{X} + \bar{Y}) = \text{Var}(X + Y)$. By Theorem 9.5, $\bar{X}$ and $\bar{Y}$ are independent.

So we have

$$\text{Var}(X + Y) = \text{Var}(\bar{X} + \bar{Y})$$
$$= \text{Exp}((\bar{X} + \bar{Y})^2)$$
$$= \text{Exp}(\bar{X}^2 + 2\bar{X}\bar{Y} + \bar{Y}^2)$$
$$= \text{Exp}(\bar{X}^2) + \text{Exp}(2\bar{X}\bar{Y}) + \text{Exp}(\bar{Y}^2)$$
$$= \text{Exp}(\bar{X}^2) + 2\,\text{Exp}(\bar{X}) \cdot \text{Exp}(\bar{Y}) + \text{Exp}(\bar{Y}^2).$$

By Theorem 9.9, since $\bar{X}$ and $\bar{Y}$ are independent, $\text{Exp}(\bar{X} \cdot \bar{Y}) = \text{Exp}(\bar{X}) \cdot \text{Exp}(\bar{Y}) = 0$ so we have

$$\text{Exp}(\bar{X}^2) + \text{Exp}(\bar{Y}^2) = \text{Var}(\bar{X}) + \text{Var}(\bar{Y}) = \text{Var}(X) + \text{Var}(Y).$$ □

Using Theorems 9.11 and 9.12, we can derive an important consequence, Theorem 9.13.

Theorem 9.13. *Let* $X_1, \ldots, X_N$ *be independent random variables, each with mean* μ *and standard deviation* σ; *thus,* $\mathrm{Var}(X_i) = \sigma^2$. *Let* V *be the average of these:* $V = (X_1 + \ldots + X_N)/N$. *Then*

(a) $\mathrm{Exp}(V) = (\mathrm{Exp}(X_1) + \ldots + \mathrm{Exp}(X_N))/N = \mu$,

(b) $\mathrm{Var}(V) = \mathrm{Var}(X_1 + \ldots + X_N/N) = (\mathrm{Var}(X_1) + \ldots + \mathrm{Var}(X_N))/N^2 = \sigma^2/N$,

(c) $\mathrm{Std}(V) = \sqrt{\mathrm{Var}(V)} = \sigma/\sqrt{N}$.

The key point is statement (c); the spread, as measured by the standard deviation, goes down proportional to the square root of the number of repetitions.

For example, suppose that we flip a coin 10,000 times. Let X_i be the random variable for the ith flip, with heads=1 and tails=0. Let V be the average of the X_i; thus, V is the fraction of heads in the flips. We have $\mathrm{Exp}(X_i) = 1/2$ and $\mathrm{Std}(X_i) = 1/2$, so $\mathrm{Exp}(V) = 1/2$ and $\mathrm{Std}(V) = 1/200$. Using Theorem 9.11, we can conclude that $P(|V - 1/2| \geq 1/100) \leq ((1/200)/(1/100))^2 = 1/4$. Thus there is at least a 0.75 chance that V is between 0.49 and 0.51. As we show in Section 9.8.2, this is a substantial underestimate—the true probability is 0.9545—but we have been able to derive it with surprising ease.

9.5   Random Variables over Infinite Sets of Integers

The second category of numerical random variables are those whose domain is an infinite set of integers. (*Note:* This section requires an understanding of infinite sequences and infinite sums.)

Example 9.14. Suppose we repeatedly flip a coin until it turns up heads; then we stop. Let C be the random variable whose value is the number of flips we make. As in Example 9.4, let $F_1, F_2, \ldots$ be random variables corresponding to the successive hits; the difference is that now we need an infinite sequence of random variables. Then

$$P(C = 1) = P(F_1 = \mathrm{H}) = 1/2,$$
$$P(C = 2) = P(F_1 = \mathrm{T}, F_2 = \mathrm{H}) = 1/4,$$
$$P(C = 3) = P(F_1 = \mathrm{T}, F_2 = \mathrm{T}, F_3 = \mathrm{H}) = 1/8,$$

$$\vdots$$

$$P(C = k) = 1/2^k, \ldots$$

In general, we can let the sample space Ω be the set of natural numbers. A probability function over Ω is a function $P(i)$ such that $\sum_{i=1}^{\infty} P(i) = 1$. An event is a subset of Ω. The probability of event E is $P(E) = \sum_{x \in E} P(x)$. An unnormalized weight function $w(i)$ is any function such that the $\sum_{i=1}^{\infty} w(i)$ converges to a finite value.

Almost everything that we have done with finite random variables transfers over to this case, requiring only the change of finite sums to infinite sums. Otherwise, the definitions are all the same, the theorems are all the same, and the proofs are all the same, so we will not go through it all again. There are only three issues to watch out for.

First, axiom P2 in the event-based axiomatization of probability (Section 8.4) has to be extended to infinite collections of events, as follows:

P2′ Let $\{E_1, E_2, \ldots\}$ be a sequence of events, indexed by integers, such that, for all $i \neq j$, $E_i \cap E_j = \emptyset$. Then $P(E_1 \cup E_2 \cup \ldots) = P(E_1) + P(E_2) + \ldots$.

The property described in this axiom is called *countable additivity*.

Second, there is no such thing as a uniform distribution over Ω. If all values in Ω have the same weight w, if $w > 0$ then the total weight would be infinite; if $w = 0$, the total weight is zero. Neither of these is an acceptable alternative.

Third, a probability distribution may have an infinite mean; or a finite mean and an infinite variance. For example, the function distribution

$$P(1) = 1/(1 \cdot 2) = 1/2,$$
$$P(2) = 1/(2 \cdot 3) = 1/6,$$
$$P(3) = 1/(3 \cdot 4) = 1/12,$$
$$\vdots$$
$$P(k) = 1/k \cdot (k+1), \ldots$$

is a legitimate probability distribution, since

$$P(1) + P(2) + P(3) + \ldots + P(k) + \ldots = 1/(1 \cdot 2) + 1/(2 \cdot 3) + 1/(3 \cdot 4) + \ldots + 1/k(k+1) + \ldots$$
$$= (1 - 1/2) + (1/2 - 1/3) + (1/3 - 1/4) + \ldots + (1/k - 1/(k+1)) + \ldots$$
$$= 1.$$

However, the expected value is

$$1 \cdot P(1) + 2 \cdot P(2) + 3 \cdot P(3) + \ldots + k \cdot P(k) + \ldots = 1/(1 \cdot 2) + 2/(2 \cdot 3) + 3/(3 \cdot 4) + \ldots + k/k(k+1) + \ldots$$
$$= 1/2 + 1/3 + 1/4 + \ldots + 1/(k+1) + \ldots,$$

which is a divergent series.

Similarly, we can define a random variable X such that $P(X = k) = 1/k^2 - 1/(k+1)^2$ for $k = 1, 2, \ldots$. This is a legitimate probability distribution and has a finite mean, but the variance is infinite.

Therefore, to extend any of the theorems about $\mathrm{Exp}(X)$ or $\mathrm{Var}(X)$ here to the case of random variables with infinite domains, we have to add the condition that $\mathrm{Exp}(X)$ or $\mathrm{Var}(X)$ exists and is finite.

9.6   Three Important Discrete Distributions

In the theory of probability, mathematicians have studied many different specific distributions with different kinds of applications and mathematical properties. These are generally families of distributions, each with a set of real- or integer-valued parameters; each assignment of values to the parameters gives a different distribution. In this book, we briefly discuss a few distributions that are particularly important in computer science applications. In this section, we discuss three discrete distributions: the Bernoulli distribution, the binomial distribution, and the Zipf distribution. Continuous distributions are discussed in Section 9.8.

9.6.1   The Bernoulli Distribution

The Bernoulli distribution is associated with a single flip of a weighted coin. It has one parameter p, the probability that the flip comes up heads. The domain is the set $\{0, 1\}$. The distribution is defined as

$$P(X = 1) = p,$$
$$P(X = 0) = (1 - p).$$

The expectation $\mathrm{Exp}(X) = p$. The variance $\mathrm{Var}(X) = p(1 - p)$. The Bernoulli distribution is a very simple one in itself, but it is a basic component of more complex distributions.

9.6.2   The Binomial Distribution

Suppose that we have a weighted coin that comes up heads with probability p, and we flip it n times. Let X be the random variable whose value is the number times the coin comes up heads in the n flips. Then the probability distribution for X is the *binomial distribution*.

The binomial distribution has two parameters: n and p. It takes on values in $\{0, 1, \ldots, n\}$. We can calculate $P(X = k)$ as follows: Any particular sequence with k heads and $n - k$ tails has probability $p^k \cdot (1 - p)^{n-k}$. For example, if $p = 0.75$, $k = 2$, and $n = 5$, then the probability of the particular sequence HTTHT is

$$p \cdot (1 - p) \cdot (1 - p) \cdot p \cdot (1 - p) = p^2 \cdot (1 - p)^3 = 0.0088.$$

the	6,187,267	in	1,812,609	was	923,948	you	695,498	on	647,344
of	2,941,444	to	1,620,850	to	917,579	he	681,255	that	628,999
and	2,682,863	it	1,089,186	I	884,599	be	662,516	by	507,317
a	2,126,369	is	998,389	for	833,360	with	652,027	at	478,162

Table 9.2. Frequencies of the 20 most common words in the BNC database.

Any sequence of k heads and $n - k$ tails corresponds to one way of choosing the k positions for the heads out of the n positions in the sequence; the number of such sequences is $C(n, k) = n!/(k! \cdot (n - k)!)$ For example, the number of sequences with two heads and three tails is $5!/(2! \cdot 3!) = 10$. The total probability of getting k heads in a sequence of n flips is therefore

$$P(X = k) = B_{n,p}(k) = C(n, k) \cdot p^k \cdot (1 - p)^{n-k} = \frac{n!}{k!(n - k)!} \cdot p^k \cdot (1 - p)^{n-k}.$$

For example, for $n = 5, p = 3/4$, the distribution is

$$P(X = 0) = 5!/(5! \cdot 0!) \cdot (3/4)^0 (1/4)^5 = 1 \cdot 1/4^5 \quad = 0.0098,$$
$$P(X = 1) = 5!/(4! \cdot 1!) \cdot (3/4)^1 (1/4)^4 = 5 \cdot 3^1/4^5 \quad = 0.0146,$$
$$P(X = 2) = 5!/(3! \cdot 2!) \cdot (3/4)^2 (1/4)^3 = 10 \cdot 3^2/4^5 = 0.0878,$$
$$P(X = 3) = 5!/(2! \cdot 3!) \cdot (3/4)^3 (1/4)^2 = 10 \cdot 3^3/4^5 = 0.2636,$$
$$P(X = 4) = 5!/(1! \cdot 4!) \cdot (3/4)^4 (1/4)^1 = 5 \cdot 3^4/4^5 \quad = 0.3955,$$
$$P(X = 5) = 5!/(0! \cdot 5!) \cdot (3/4)^5 (1/4)^0 = 1 \cdot 3^5/4^5 \quad = 0.2373.$$

The random variable X is the sum of n independent random variables B_i, each with the Bernoulli distribution. Therefore, by Theorems 9.8 and 9.12,

$$\text{Exp}(X) = \text{Exp}(B_1) + \ldots + \text{Exp}(B_n) = n \cdot \text{Exp}(B_1) = np,$$
$$\text{Var}(X) = \text{Var}(B_1) + \ldots + \text{Var}(B_n) = n \cdot \text{Var}(B_1) = np(1 - p).$$

9.6.3   The Zipf Distribution

Take a large corpus of English text, count the frequency of every word, and list them in decreasing order. For example, the British National Corpus (BNC; Kilgarriff, 2010) is a collection of representative English texts. It is 100,106,029 words long and contains 938,972 different words. (An occurrence of a word is called a "token"; different words are called "types.") Table 9.2 shows the number of occurrences of the 20 most common words in the corpus.[3]

Let us consider the number of occurrences as a function of the rank of the word; that is, $f(1) = 6, 187, 267$, $f(2) = 2, 941, 444$, and so on. It turns out that

---

[3]The BNC database separates occurrences of words by part of speech; thus, the first "to" in the table is as a particle and the second is as a preposition. The figure for "that" is as a conjunction.

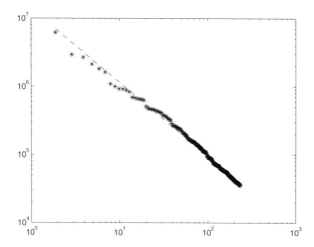

Figure 9.4. Occurrences of 200 common words by rank on a log-log plot. The points are shown in asterisks, and the solid line is the best fit.

the function $y(n) = 13,965,000 \cdot (n+0.8948)^{-1.0854}$ fits these data very well. This can be seen by using a log-log plot, where $x = n + 0.8948$; the curve becomes the straight line $\log(y) = -1.0854 \log(x) + \log(13,965,000)$. Figure 9.4 shows the first 200 data points and the line; as can be seen, the agreement is very close except for the value $n = 2$.

These data can be turned into a probability distribution by normalization; the probability of word w is the number of its occurrences in the corpus divided by the length of the corpus.

At the other end of frequency data, let us consider the most uncommon words in the corpus; we thus count the number of words that occur a very small number of times. In this corpus, there are 486,507 words that occur once; 123,633 words that occur twice, and so on. (Not all of these are words in the usual sense; this count includes numerals and some other nonword textual elements.) Table 9.3 shows the data up to words that occur 20 times.

These data fit very well to the curve $f(n) = 399,000 \cdot n^{-1.669}$. If we plot the data on a log-log scale (Figure 9.5), the fit to a straight line appears so perfect that it is not necessary to draw the line.

1	486,507	5	25,065	9	9,765	13	5,534	17	3,588
2	123,633	6	19,109	10	8,551	14	5,079	18	3,389
3	58,821	7	14,813	11	7,419	15	4,506	19	3,089
4	36,289	8	11,905	12	5,817	16	4,016	20	2,856

Table 9.3. Number of words that occur rarely in the BNC database.

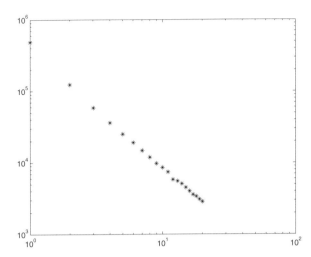

Figure 9.5. Number of words with very low frequency on a log-log plot. The data points are marked with asterisks. An asterisk at coordinates $\langle x, y \rangle$ means that there are y different words that occur exactly x times in the corpus.

Distributions of the form $\alpha(n + \beta)^{-\gamma}$ are called *Zipf* distributions; they are also known as the *inverse power law*, or *long-tail* or *fat-tail* distributions.[4] Many diverse phenomena follow a Zipf distribution, including word frequency, city population, company size, number of pages published by scientists, magazine circulation, movie popularity, number of accesses to web pages, number of in-links to web pages, number of outlinks from web pages, size of strongly connected components in the web, and personal wealth and personal income (see Sinha and Pan, 2006).[5]

A number of important features of the Zipf distribution should be noted. We describe these and illustrate them by again using the BNC word distribution, but they apply generally.

First, the top elements account for a substantial proportion of the total. In the word frequency example, the 10 most common words comprise 21% of all tokens; the 20 most common words comprise 28% and the top 175 words comprise 50%.

Second, infrequent elements, collectively, also account for a substantial proportion of the total. In the word frequency example, words that appear only once comprise 0.5% of the total, words that appear at most 10 times comprise 1.7%, and words that appear at most 20 times comprise 2.3%. This is the *long-*

---

[4]This is also sometimes called Zipf's law, but the only "law" is that many things follow the distribution.

[5]It is debatable whether some of these may actually be very similar distributions, such as log-normal distributions, but for our purposes here, it hardly matters.

tail or *fat-tail* effect: the probability that a random token in a corpus is a very low-ranked word is much greater than with other common distributions.

To underline the force of the Zipf distribution, let us compare what would happen if the words were evenly distributed; that is, if each token had equal probability of being any of the 938,000 types. Let $n \approx 10^8$ be the total length of the corpus; and let $p \approx 10^{-6}$ be the frequency of each word. For each word w, let X_w be a random variable whose value is the number of occurrences of w in the corpus; then X follows the binomial distribution $P(X = k) = B_{n,p}(k)$. By using an asymptotic analysis, we can show (these are order of magnitude calculations; that is, the exponent is within a factor of 2 or so):

- $P(X = 1) \approx 10^2 e^{-100} \approx 10^{-41}$. Therefore, the probability that the corpus would contain any word that occurs only once is about 10^{-35}. The probability that the corpus would contain 500,000 such words is about $10^{-1.7 \cdot 10^7}$.

- $P(X \geq 6 \cdot 10^6) \approx 10^{-2.4 \cdot 10^7}$. Therefore, the probability that the corpus contains any word occurring six million or more times in the corpus is likewise about $10^{-2.4 \cdot 10^7}$.

Small values of the exponent γ favor the second effect—the tail is long; large values of γ favor the first effect—the large elements are very large.

Third, power law distributions tend to give rise to computational problems in which it is easy to make great strides at the start and then progressively more difficult to make improvements. For example, consider the probability p_q that, after reading the first q words of the corpus, the next word we read is one that we have seen already. If the words follow a power law with an exponent of approximately -1, we can show that, for a wide range of C, p_q increases *logarithmically* as a function of q, and thus the C required to achieve a given value of p_C increases exponentially as a function of p.

Table 9.4 illustrates how this probability increases with the number of words read, using the case of a total vocabulary of 100,000 different words and a distribution proportional to $1/k$. Note that at each stage, doubling the number of words read increases the probability by about 5.5%. In this table q is the number of words that have been read. The value p is the probability that the next word read will be one that you have seen, as given by a Monte Carlo simulation; these numbers are accurate to within the two digits given here. (See Assignment 12.2 for the programming of this simulation.) The value $\tilde{p}$ is the value estimated by the approximate theoretical computation at the end of this section.

The consequence of this is a phenomenon often encountered in artificial intelligence research; you can easily get results that seem promising, and with a reasonable amount of work you can get results that are fairly good, but getting

q	p	$\tilde{p}$	q	p	$\tilde{p}$	q	p	$\tilde{p}$	q	p	$\tilde{p}$
100	0.28	0.21	200	0.33	0.27	400	0.39	0.32	800	0.44	0.38
1600	0.50	0.44	3200	0.56	0.49	6,400	0.61	0.55	12,800	0.66	0.61
25,600	0.73	0.67	51,200	0.78	0.72	102,400	0.84	0.78	204,800	0.89	0.84

Table 9.4. Probability that we have seen the next word.

really good results requires a lot more work, and excellent results seem to be entirely out of reach.

For actual individual words, as we discuss here, these results don't actually matter so much, for two reasons. First, a corpus of 1,000,000 words or even 100,000,000 words is actually reasonably easy to collect and to do basic forms of analysis, so it's not that daunting. Second, individual words are largely arbitrary; broadly speaking, words have to be learned one at a time, so there is no way to learn the properties of a word until we have seen it. If we want to collect a lexicon with 500,000 different words, we will have to work through a corpus that contains 500,000 different words; there is no choice. By contrast, if you have an application that requires knowing the properties of one billion different 3-grams (triples of consecutive words; see Section 10.3), it may not be necessary to examine a text corpus that contains one billion different 3-grams, because 3-grams are not arbitrary; there is a logic to the possible or probable sequences of words.

Power law distributions with small exponents have anomalous behavior with regard to the mean and the variance. A distribution $p_n = \alpha n^{-\gamma}$ has an infinite variance if $\gamma \le 3$; it has an infinite mean if $\gamma \le 2$; and it is not a probability distribution at all if $\gamma \le 1$ because the sum $\sum 1/n^\gamma$ diverges for $\gamma \le 1$. In this last case, we must assume that there are only a finite number of values, as we have done in computing Table 9.4. If γ is just above the critical values 3 or 2, then the variance and the mean are anomalously large. Therefore, if we are computing mean and variance from statistical values, and one or both seem strangely large and very unstable from one sample to another, then we should suspect that we may be dealing with a power law distribution and that computations based on mean and variance may not be meaningful.

Deriving the probability having previously seen the next word. Let H_k be the harmonic sum $H_k = \sum_{i=1}^{k}(1/i)$. Then it is known that $H_k = \ln(k)+\gamma+O(1/k)$, where $\gamma \approx 0.5772$ is known as Euler's constant.

Let W be the size of the vocabulary, and let w_k be the kth ranked word, for $k = 1,\ldots,W$. Assume that the frequency of w_k in the text is proportional to $1/k$. Then the normalization factor is H_W, so the probability that a random token is w_k is $p_k = 1/(k \cdot H_W)$.

Suppose that we have read a random text T of q tokens, where $H_W \ll q \ll W H_W$. Suppose that the last token we read was a word we have seen before in the text; then it is one of the words that occurs at least twice in the text. Therefore, the probability that the last word we read was one we had already seen is equal to the fraction of tokens in T whose word occurs at least twice in T. We can roughly assume that the text contains at least two occurrences of word w_k if $qp_k > 3/2$; this holds if $k \le 2q/3H_W$. Let $r = 2q/3H_W$. Then the fraction of tokens in T whose word occurs at least twice in T is $H_r/H_W = \ln(3q/2H_W)/H_W$.

Table 9.4 shows the estimate $\bar{p}$ derived from this argument and the much more accurate estimate p derived from Monte Carlo simulation, for various values of n. Over the entire range of n, $\bar{p} \approx p - 0.05$; thus, when plotting p against $\log(n)$, the argument gets the y-intercept wrong by about 0.05 but gets the slope almost exactly correct.

9.7 Continuous Random Variables

The third category of numerical random variables are random variables that take values over the real line. The theory here is different and more difficult than with finite domains or integer domains. (*Note:* This section requires calculus, and at one critical point it requires multivariable calculus.)

The sample space Ω is the real line $\mathbb{R}$. However, unlike the sample spaces we have looked at earlier, the probability of an event is not derived from the probabilities of individual elements; typically, each individual element has probability zero. Rather, the probability of events is a characteristic of the interval as a whole, just as the length of an interval is not the sum of the lengths of the points in the interval.

Specifically, we posit that there is a large[6] collection of subsets of $\mathbb{R}$ called the *measurable* sets. If X is a random variable with domain $\mathbb{R}$ and E is a measurable set, then $X \in E$ is a probabilistic event, and $P(X \in E)$ is a probabilistic event satisfying axioms P1, P2′, P3, and P4 (see Sections 8.4 and 9.5).

Notice that this is something like what we did in the likelihood interpretation of probability (Section 8.6), in the sense that we took the probability of events as a starting point. However, in that earlier analysis, we were able to form a finite sample space of elements by combining events; here that is not possible.

From the description so far, it sounds as though, in order to specify a probability distribution, we would have to specify $P(X \in E)$ separately for all measurable E, which would be a large undertaking. Fortunately, because of the axioms constraining probability distributions, there is a simpler approach: the probability of all events can be specified by specifying the *cumulative distribution*.

---

[6]The question of *what* subsets of $\mathbb{R}$ can be considered measurable is a very deep one. However, certainly any finite union of intervals is measurable, which is all that we need.

Definition 9.15. Let $c(t)$ be a continuous function from $\mathbb{R}$ into the closed interval [0,1] such that

- function $c(t)$ is monotonically nondecreasing; that is, if $t_1 \leq t_2$, then $c(t_1) \leq c(t_2)$;

- for any v such that $0 < v < 1$, there exists t such that $c(t) = v$.

Let X be a numeric random variable. Then c is the *cumulative distribution function (cdf)* for X if, for every t, $P(X \leq t) = P(X < t) = c(t)$.

(A more general definition allows c to be discontinuous from the left and also weakens some of the other conditions; however, we do not need that level of generality here.)

If we have specified a cumulative distribution $c(t)$ for a random variable X, then that determines the probability that X lies within any specified interval $[u, v]$:

$$P(X \in [u, v]) = P(X \leq v \wedge \neg(X < u)) = P(X \leq v) - P(X < u) = c(v) - c(u).$$

We can then use axiom P2$'$ to calculate $P(X \in E)$, where E is the union of a finite set of intervals or of an infinite set of intervals.

Example 9.16. Let $c(t)$ be the following function (Figure 9.6)

$$c(t) = \begin{cases} 0 & \text{for} \quad t < 0, \\ t & \text{for} \quad 0 \leq t \leq 1, \\ 1 & \text{for} \quad t > 1. \end{cases}$$

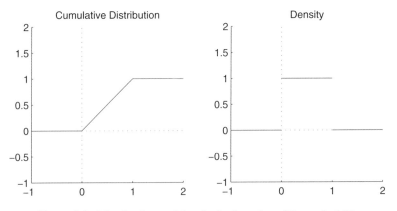

Figure 9.6. Distribution and density for function of Example 9.16.

Clearly, $c(t)$ satisfies the conditions of Definition 9.15. Let X be a random variable X with distribution $c(t)$. For that distribution, we have

$$P(X \in [u,v]) = c(v) - c(u) = \begin{cases} 0 & \text{if} \quad u \le v \le 0, \\ v & \text{if} \quad u \le 0 \quad \text{and} \quad 0 \le v \le 1, \\ v - u & \text{if} \quad 0 \le u \le v \le 1, \\ 1 - u & \text{if} \quad 0 \le u \le 1 \quad \text{and} \quad v \ge 1, \\ 0 & \text{if} \quad 1 \le u \le v. \end{cases}$$

If E is the union of disjoint intervals, then $P(E)$ is the sum of their probabilities. For example $P(X \in [-1, 1/6] \cup [1/3, 2/3] \cup [5/6, 2]) = 1/6 + 1/3 + 1/6 = 2/3$.

A more useful, although slightly less general, way to represent continuous distribution is in terms of the *probability density*.

Definition 9.17. Let X be a random variable with cumulative distribution $c(t)$. If $c(t)$ is piecewise differentiable (i.e., differentiable at all but isolated points), then the derivative of c, $f(t) = c'(t)$ is the *probability density function* (pdf) associated with X. (The value of f at the points where c is not differentiable does not matter.)

Another, more direct definition of f is as follows: For any point t,

$$f(t) = \lim_{u \to t, \epsilon \to 0^+} \frac{P(u < X < u + \epsilon)}{\epsilon}.$$

So $f(t)$ is the probability that X lies in a interval of length ϵ near t, divided by the length ϵ. (If this does not approach a unique limit, then the value of $f(t)$ is arbitrary and may be assigned arbitrarily. For reasonable probability distributions, this happens only at isolated points.)

A density function is essentially the limit of a series of histograms, with the subdivision on the x-axis getting smaller and smaller. At each stage, the y-axis is rescaled to be inversely proportional to the division, so that the height of each rectangle remains roughly constant.

For example, consider the density function $\tilde{P}(X = t) = (3/4)(1 - t^2), -1 \le t \le 1$. The corresponding cumulative function is the integral

$$c(t) = \int_{-1}^{t} \frac{3}{4}(1 - x^2)\,dx = -\frac{1}{4}t^3 + \frac{3}{4}t + \frac{1}{2}.$$

Thus,

$$P(t - \Delta \le X \le t + \Delta) = c(t + \Delta) - c(t - \Delta) = -\frac{3t^2\Delta}{2} - \frac{\Delta^3}{2} + \frac{3\Delta}{2}.$$

Rescaling by the width 2Δ, we have

$$\frac{P(t - \Delta \le X \le t + \Delta)}{2\Delta} = \frac{3 - 3t^2 - \Delta^2}{4}.$$

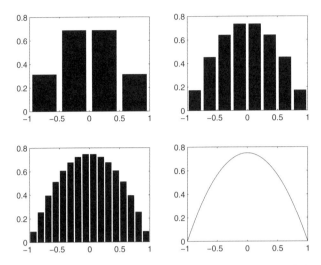

Figure 9.7. Probability density as a limit of histograms: 4 bars (top left), 8 bars (top right), 16 bars (bottom left), and density function (bottom right).

Note that as Δ goes to zero, this converges to the density $(3/4)(1-t^2)$. Figure 9.7 shows a series of bar charts for $P(t-\Delta \le X \le t+\Delta)/2\Delta$ for $\Delta = 1/4, 1/8, 1/16$, and the density function with $-1 \le t \le 1$.

We write $\tilde{P}(X = t)$ for the value at t of the probability density of X. (This is a slight abuse of notation, but should not cause confusion.) Thus, we have

$$\tilde{P}(X = t) = \frac{d}{dt}P(X \le t).$$

The probability $P(X \in E)$ is the integral of the pdf of X over E.

Theorem 9.18. *Let X be a random variable over $\mathbb{R}$ and let E be a measurable subset of $\mathbb{R}$. Then*

$$P(X \in E) = \int_{t \in E} \tilde{P}(X = t)dt.$$

Example 9.19. The random variable defined in Example 9.16 has the probability density $f(t)$, where

$$f(t) = \begin{cases} 0 & \text{for} \quad t < 0, \\ 1 & \text{for} \quad 0 \le t \le 1, \\ 0 & \text{for} \quad t > 1. \end{cases}$$

Since the cumulative function $c(t)$ is not differentiable at $t = 0$ and $t = 1$, the values of f at those points is not defined and may be chosen arbitrarily.

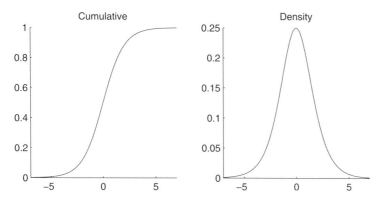

Figure 9.8. Curves for Example 9.20. The cdf (left) is $e^t/(1+e^t)$; pdf (right) is $e^t/(1+e^t)^2$.

Example 9.20. Let X be a random variable with cumulative distribution $c(t) = e^t/(1+e^t)$. Since e^t goes from 0 to ∞ as t goes from $-\infty$ to ∞, and e^t is a continuous increasing function, it is easily checked that $c(t)$ satisfies the conditions of Definition 9.15. Then the pdf of X is $\tilde{P}(X = t) = dc/dt = e^t/(1+e^t)^2$ (Figure 9.8).

In many ways, the probability density function $f(t)$ over elements $t \in \mathbb{R}$ looks a lot like a probability function over a finite sample space Ω, where summation in the finite case is replaced by integration in the continuous case. Before we discuss those parallels, however, let us emphasize that they are *not* the same thing, and that there are important differences. One difference is that a probability density function may be greater than 1, as long as the integral is 1. Example 9.21 illustrates this difference.

Example 9.21. Function $h(t)$ is a legitimate probability density function:

$$h(t) = \begin{cases} 0 & \text{for} \quad t < 0, \\ 2 & \text{for} \quad 0 \le t \le 1/2, \\ 0 & \text{for} \quad t > 1/2. \end{cases}$$

Second, probability densities behave differently from probabilities on elements with regard to functions of a random variable. If X is a random variable and g is an injection (invertible function), then for any value v, $P(g(X) = g(v)) = P(X = v)$. This is not the case for probability densities; $\tilde{P}(g(X) = g(v)) \ne \tilde{P}(X = v)$ because g changes not only the value of v but also the scale in the neighborhood of t. For instance, let X be the random variable of Example 9.16. Let $g(v) = v/2$. Then the random variable $g(X)$ has the density $h(t)$ discussed in Example 9.21; thus, $\tilde{P}(g(X) = g(t)) = 2 \cdot \tilde{P}(X = t)$. The explanation is as

Probability functions	Density functions
For all $v, P(v) \geq 0$	For all $v, \tilde{P}(v) \geq 0$
$1 = \sum_v P(v)$	$1 = \int_{-\infty}^{\infty} \tilde{P}(v)\,dv$
$\text{Exp}(X) = \sum_v v \cdot P(v)$	$\text{Exp}(Y) = \int_{-\infty}^{\infty} v \cdot \tilde{P}(v)\,dv$
$\text{Var}(X) = \sum_v (v - \text{Exp}(X))^2 \cdot P(v)$	$\text{Var}(Y) = \int_{-\infty}^{\infty} (v - \text{Exp}(Y))^2 \cdot \tilde{P}(v)\,dv$

Table 9.5. Similarities between probability functions and probability density functions.

follows. Let t be any value such that $0 < t < 1/2$, and let ϵ be small. Then

$$\tilde{P}(g(X) = g(t)) \approx \frac{1}{\epsilon} \cdot P(g(t) \leq g(X) \leq g(t) + \epsilon)$$

$$= \frac{1}{\epsilon} \cdot P\left(\frac{t}{2} \leq \frac{X}{2} \leq \frac{t}{2} + \epsilon\right)$$

$$= \frac{1}{\epsilon} \cdot P(t \leq X \leq t + 2\epsilon)$$

$$= 2 \cdot \frac{1}{2\epsilon} \cdot P(t \leq X \leq t + 2\epsilon) \approx 2 \cdot \tilde{P}(X = t).$$

An alternative derivation (essentially equivalent, but perhaps easier) involves the cumulative distribution function $c(t)$ defined in Example 9.16:

$$c(t) = \begin{cases} 0 & \text{for} \quad t < 0, \\ t & \text{for} \quad 0 \leq t \leq 1, \\ 1 & \text{for} \quad t > 1. \end{cases}$$

Let X be a random variable whose cdf is $c(t)$; that is, $P(X \leq t) = c(t)$. Then

$$\tilde{P}(X = t) = \frac{d}{dt} P(X \leq t) = \frac{d}{dt} c(t) = h(t).$$

Let $Y = g(X)$. Then

$$\tilde{P}(Y = u) = \frac{d}{du} P(Y \leq u) = \frac{d}{du} P(X \leq 2u) = \frac{d}{du} c(2u) = 2h(2u) = 2\tilde{P}(X = 2u).$$

So if $t = 2u$, then $\tilde{P}(Y = u) = \tilde{P}(g(X) = g(t)) = 2\tilde{P}(X = t)$.

Generalizing the above calculation, we can show that, for any differentiable invertible function $g(v)$, if X has density $f(t)$ then $g(X)$ has density $f(g(t))/g'(t)$.

In many ways, however, probability density functions do work in the same way as probability functions, replacing summation by integration. Let X be a discrete numeric random variable with probability function $P(x)$ and let Y be a continuous random variable with pdf $\tilde{P}(t)$. Table 9.5 shows the parallels between probability functions and pdfs.

Moreover, as with discrete probabilities, it is often convenient to use an unnormalized weight function $w(v)$. A function $w(v)$ can serve as a weight function if it satisfies the conditions that $w(v) \geq 0$ for all v, and that $\int_{-\infty}^{\infty} w(v) dv$ is finite but greater than 0. The pdf associated with weight function $w(v)$ is $f(v) = w(v) / \int_{-\infty}^{\infty} w(v) dv$.

The theorems that combine several random variables also carry over to the continuous case; however, to make them meaningful, we need to define the meanings of joint probability distribution and joint density function. (It is possible to define a joint cumulative distribution function, but it is awkward and not very useful.)

We posit that there exists a collection of measurable subsets of $\mathbb{R}^n$. A probability function over the measurable subsets of $\mathbb{R}^n$ is a function $P(E)$ mapping a measurable set E to a value in $[0,1]$ satisfying axioms P1, P2', P3, and P4.

Definition 9.22. Let $\langle X_1, X_2, \ldots, X_n \rangle$ be a finite sequence of continuous random variables. Then the *joint distribution of* $\langle X_1, \ldots, X_n \rangle$, denoted $P(\langle X_1, \ldots, X_n \rangle \in E)$, is a probability function over the measurable sets E.

Definition 9.23. Let $\langle X_1, X_2, \ldots, X_n \rangle$ be a finite sequence of continuous random variables, and let $P(\langle X_1, \ldots, X_n \rangle \in E)$ be its probability function. Let $\vec{t} = \langle t_1, \ldots, t_n \rangle$ be a vector in $\mathbb{R}^n$. The probability density function of $\langle X_1, X_2, \ldots, X_n \rangle$ satisfies the following: For any $\vec{t}$

$$\tilde{P}(X_1 = t_1, \ldots, X_n = t_n) = \lim_{\vec{u} \to \vec{t}, \epsilon \to 0^+} \frac{P(u_1 \leq X_1 \leq u_1 + \epsilon \wedge \ldots \wedge u_n \leq X_n \leq u_n + \epsilon)}{\epsilon^n},$$

if that limit exists. That is, the probability density at $\vec{t}$ is the probability that $\langle X_1, \ldots, X_n \rangle$ lies in an n-dimensional box of side ϵ near t, divided by the volume of the box, ϵ^n.

If there is no unique limit at a point $\vec{t}$, then the value of $f(\vec{t})$ does not matter. If P is a "well-behaved" function, then the set of points where the limit does not exist has measure zero.

As in the one-dimensional case, the probability can be obtained from the density by integration, as shown in Theorem 9.24.

Theorem 9.24. *Let* $X_1, X_2, \ldots, X_n$ *be continuous random variables and let E be a measurable subset of* $\mathbb{R}^n$. *Then*

$$P(\langle X_1, \ldots, X_n \rangle \in E) = \int, \ldots, \int_E \tilde{P}(X_1 = t_1 \ldots X_n = t_n) dt_1 \ldots dt_n.$$

Virtually all of the theorems we have proven for discrete probabilities now carry over to continuous probabilities, replacing summation by integration. For simplicity, we state them here in terms of two random variables, but they all generalize in an obvious way to n random variables. In all of the following theorems, X and Y are continuous random variables.

Theorem 9.25. *Variable X is independent of Y if and only if for all values t and u, $\tilde{P}(X = t, Y = u) = \tilde{P}(X = t) \cdot \tilde{P}(Y = u)$.*

The conditional density distribution is defined as $\tilde{P}(X = t \mid Y = v) = \tilde{P}(X = u, Y = v)/\tilde{P}(Y = v)$.

Theorem 9.26 is the basic theorem about marginal probabilities, analogous to Theorem 9.6.

Theorem 9.26. *For any value t of X,*

$$\tilde{P}(X = t) = \int_{-\infty}^{\infty} \tilde{P}(X = t, Y = u) du.$$

Theorem 9.27. *Let v be a particular value of Y such that $\tilde{P}(Y = v) \neq 0$. Then the conditional density $\tilde{P}(X = u \mid Y = v) = \tilde{P}(X = u, Y = v)/\tilde{P}(Y = v)$ is a probability density as a function of u.*

The wording of Theorems 9.8–9.13 is exactly the same in the continuous case as in the discrete infinite case. The proofs are exactly the same as in the finite case, replacing sums by integrals:

Theorem 9.8′. *If $\text{Exp}(X)$ and $\text{Exp}(Y)$ are finite then $\text{Exp}(X + Y) = \text{Exp}(X) + \text{Exp}(Y)$.*

Theorem 9.9′. *If $\text{Exp}(X)$ and $\text{Exp}(Y)$ are finite and X and Y are independent, then $\text{Exp}(X \cdot Y) = \text{Exp}(X) \cdot \text{Exp}(Y)$.*

Theorem 9.11′. *Let X be a random variable with mean μ and standard deviation σ. Then for any $w > 0$, $P(|X - \mu| \geq w) \leq \sigma^2/w^2$.*

Theorem 9.12′. *If $\text{Var}(X)$ and $\text{Var}(Y)$ are finite and X and Y are independent, then $\text{Var}(X + Y) = \text{Var}(X) + \text{Var}(Y)$.*

Theorem 9.13′. *Let $X_1 \ldots X_N$ be independent random variables, each with mean μ and standard deviation σ; thus $\text{Var}(X_i) = \sigma^2$. Let V be the average of these: $V = (X_1 + \ldots + X_N)/N$. Then*

(a) $\text{Exp}(V) = (\text{Exp}(X_1) + \ldots + \text{Exp}(X_N))/N = \mu$,

(b) $\text{Var}(V) = \text{Var}(X_1 + \ldots + X_N/N) = (\text{Var}(X_1) + \ldots + \text{Var}(X_N))/N^2 = \sigma^2/N$,

(c) $\text{Std}(V) = \sqrt{\text{Var}(V)} = \sigma/\sqrt{N}$.

9.8   Two Important Continuous Distributions

9.8.1   The Continuous Uniform Distribution

The uniform distribution is a generalization of Examples 9.16 and 9.21. The uniform distribution has two parameters: a lower bound L and upper bound U, where $L < U$; it has constant density $1/(U-L)$ at points in the interval $[L,U]$ and density zero outside $[L,U]$.

Let X be a random variable with a uniform distribution between L and U. Then X has the following features:

$$\tilde{P}(X = t) = \begin{cases} 1/(U-L) & \text{if } L \le t \le U, \\ 0 & \text{else}; \end{cases}$$

$$P(X \le t) = \begin{cases} 0 & \text{if } t \le L, \\ (t-L)/(U-L) & \text{if } L < t < U, \\ 1 & \text{if } U \le t; \end{cases}$$

$$\text{Exp}(X) = \frac{U+L}{2};$$

$$\text{Var}(X) = \frac{(U-L)^2}{12}.$$

The uniform distribution has a natural generalization to regions in $\mathbb{R}^n$. Let R be a bounded region in $\mathbb{R}^n$ whose volume is greater than zero. Then the uniform distribution over R has density $1/\text{volume}(R)$, where volume is the k-dimensional volume. If X is uniformly distributed over R, then, for any measurable subset $S \subset R$, $P(X \in S) = \text{measure}(S)/\text{measure}(R)$.

The uniform distribution is the natural distribution to assume if an event is known to lie within a given region. For instance, if it is known that a meteor fell somewhere in the state of Michigan, but no information is available about where, then it is reasonable to assign the event a uniform distribution over the extent of Michigan.

The uniform distribution over $[0,1]$ is also important computationally because it is easy to generate. To generate a 32-bit approximation of the uniform distribution between 0 and 1, just construct the number $0.b_1 b_2 \ldots b_{32}$ where each bit b_i has $1/2$ probability of being 0 or 1. If we want to generate random numbers corresponding to some other distribution, the usual procedure is to generate a uniform distribution and then apply a function to turn it into a different distribution.

9.8.2   The Gaussian Distribution

The *Gaussian* or *normal* distribution is the most commonly used continuous distribution. It arises in all kinds of contexts and applications.

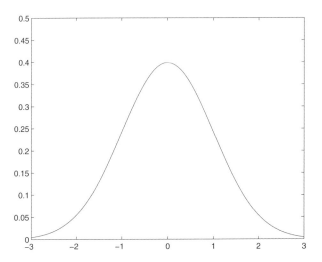

Figure 9.9. Gaussian density function.

The Gaussian distribution has two parameters: the mean μ and the standard deviation σ. The density function denoted $N_{\mu,\sigma}(t)$ is shown in Figure 9.9. It is defined as

$$N_{\mu,\sigma}(t) = \frac{1}{\sqrt{2\pi}\sigma} e^{-(t-\mu)^2/2\sigma^2}.$$

The denominator $\sqrt{2\pi} \cdot \sigma$ is just a normalizing factor. The density function is known as the "bell curve" because of its shape. It is symmetric around the mean μ; that is, for any x, $N_{\mu,\sigma}(\mu + x) = N_{\mu,\sigma}(\mu - x)$. It stays fairly flat within one standard deviation of μ; then it drops off rapidly going from one to two standard deviations; then it approaches 0 asymptotically very rapidly. Specifically,

$$N_{\mu,\sigma}(\mu + 0.5\sigma) = 0.88 \cdot N_{\mu,\sigma}(\mu),$$
$$N_{\mu,\sigma}(\mu + \sigma) = 0.60 \cdot N_{\mu,\sigma}(\mu),$$
$$N_{\mu,\sigma}(\mu + 1.5\sigma) = 0.32 \cdot N_{\mu,\sigma}(\mu),$$
$$N_{\mu,\sigma}(\mu + 2\sigma) = 0.135 \cdot N_{\mu,\sigma}(\mu),$$
$$N_{\mu,\sigma}(\mu + 2.5\sigma) = 0.0439 \cdot N_{\mu,\sigma}(\mu),$$
$$N_{\mu,\sigma}(\mu + 3\sigma) = 0.01 \cdot N_{\mu,\sigma}(\mu),$$
$$N_{\mu,\sigma}(\mu + 4\sigma) = 0.0003 \cdot N_{\mu,\sigma}(\mu),$$
$$N_{\mu,\sigma}(\mu + 5\sigma) = 0.000004 \cdot N_{\mu,\sigma}(\mu).$$

Likewise, most of the distribution lies within two standard deviations of the mean, and virtually all of it lies within three standard deviations. Specifically, if

X follows the distribution $N_{(\mu,\sigma)}$, then

$$P(\mu - 0.5\sigma \le X \le \mu + 0.5\sigma) = 0.38,$$
$$P(\mu - \sigma \le X \le \mu + \sigma) = 0.68,$$
$$P(\mu - 1.5\sigma \le X \le \mu + 1.5\sigma) = 0.86,$$
$$P(\mu - 2\sigma \le X \le \mu + 2\sigma) = 0.95,$$
$$P(\mu - 2.5\sigma \le X \le \mu + 2.5\sigma) = 0.987,$$
$$P(\mu - 3\sigma \le X \le \mu + 3\sigma) = 0.997,$$
$$P(\mu - 4\sigma \le X \le \mu + 4\sigma) = 1 - 6.3 \cdot 10^{-5},$$
$$P(\mu - 5\sigma \le X \le \mu + 5\sigma) = 1 - 5.7 \cdot 10^{-7}.$$

Thus, an event three standard deviations from the mean is rare (3 in 1,000); four standard deviations is very rare (less than one in 10,000); and five standard deviations is almost unheard of (less than one in 10,000,000).

If X follows the Gaussian distribution with mean μ and standard deviation σ, then variable $aX + b$ follows the Gaussian distribution with mean $a\mu + b$ and standard deviation $a\sigma$. In particular, the variable $Y = (X - \mu)/\sigma$ follows the distribution $N_{0,1}$. Therefore, if we have the cumulative distribution of $N_{0,1}$,

$$Z(t) = \int_{-\infty}^{t} N_{0,1}(u)\,du = \int_{-\infty}^{t} \frac{1}{\sqrt{2\pi}} e^{-u^2/2}\,du,$$

as a built-in function (or in ye olden days when I was young, as a table in a book), we can compute $P(c \le X \le d) = P((c - \mu)/\sigma \le Y \le (d - \mu)/\sigma) = Z((d - \mu)/\sigma) - Z((c - \mu)/\sigma)$.

The Gaussian distribution has the remarkable property that, if two independent random variables each follow a Gaussian, then their sum also follows a Gaussian.

Theorem 9.28. *Let X be a random variable whose density function is Gaussian with mean μ_x and variance σ_x^2, and let Y be an independent random variable whose density function is Gaussian with mean μ_y and variance σ_y^2. Then the density function of X + Y is Gaussian with mean $\mu_x + \mu_y$ and variance $\sigma_x^2 + \sigma_y^2$.*

One of the main reasons that the Gaussian distribution is important is because of the central limit theorem (Theorem 9.29).

Theorem 9.29 (central limit theorem.). *Let $X_1, \ldots, X_n$ be n independent random variables, each of which has mean μ and standard deviation σ. Let $Y = (X_1 + \ldots + X_n)/n$ be the average; thus, Y has mean μ and standard deviation $\sigma/\sqrt{n}$. If n is large, then the cumulative distribution of Y is very nearly equal to the cumulative distribution of the Gaussian with mean μ and standard deviation $\sigma/\sqrt{n}$.*

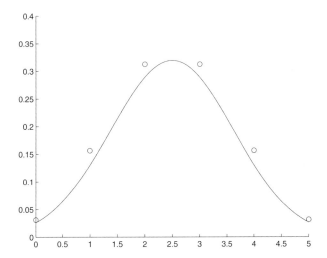

Figure 9.10. Gaussian density function. The dots are the binomial distribution $B_{5,0.5}$; and the connected curve is the scaled Gaussian, $N_{2.5,\sigma}/\sigma$, where $\sigma = \sqrt{5}/2$. For values of n larger than 5, the fit is even closer.

One particular and important example of a Gaussian distribution is the binomial distribution. Suppose $X_1, \ldots, X_n$ are independent and all have a Bernoulli distribution with parameter p. Then their sum $S = (X_1 + \ldots + X_n)$ has the binomial distribution $B_{n,p}$. The X_i have mean $\mu = p$ and standard deviation $\sigma = \sqrt{p(1-p)}$. Let $Y = S/n$ be the average of the X_i. By the central limit theorem, the cumulative distribution of Y is very nearly equal to the cumulative distribution of the Gaussian with mean μ and standard deviation $\sigma/\sqrt{n}$. Therefore, we can approximate the probability $P(c \le Y \le d)$ as the integral of $N_{\mu,\sigma}$ from c to d, and therefore, as discussed above, as $Z((d-\mu)/\sigma) - Z((c-\mu)/\sigma)$, where Z is the integral of the standard Gaussian, $Z(t) = \int_{-\infty}^{t} N_{0,1}(t)$.

Likewise, the binomial distribution $B_{n,p}$ is very well approximated by the Gaussian density $N(\mu, \sigma)$ with a scaling factor of σ, where $\sigma = \sqrt{p(1-p)n}$ (Figure 9.10). Specifically,[7] for any fixed p, for all k,

$$\left| B_{n,p}(k) - \frac{N_{np,\sigma}(k)}{\sigma} \right| \text{ is } O\left(\frac{1}{n}\right).$$

(For the unique value $p = 1/2$, this is $O(1/n^{3/2})$.)

Note that both of these approximations state only that the *difference* between the two functions is small. They do not state that the *ratio* is small, especially in the case where they are both close to zero. For example, in Section 9.6.3, on the Zipf function, we mentioned estimates for the values of $B_{n,p}(1)$

---

[7]Thanks to Gregory Lawler for these bounds.

and of $B_{n,p}(6 \cdot 10^6)$, where $p = 10^{-6}$ and $n = 10^8$. If we try to use the Gaussian as an estimate for the binomial distribution at these extreme values, we won't get an accurate answer.

9.9 MATLAB

The computation of the functions associated with discrete numerical random variables presents no particular new issues, and MATLAB provides few utilities particularly tailored to these computations. The function `nchoosek(N,K)` computes the binomial coefficient $N!/K!(N-K)!$.

```
>> nchoosek(5,2)
ans =
    10
>> nchoosek(10,5)
ans =
    252
```

For computations with continuous distributions, there are two important MATLAB functions: the general function for computing integrals and the specific function for the integral of the Gaussian.

A definite integral in MATLAB can be computed by using the function `quad(fn,a,b)`, where `fn` is a specification of the integrand, `a` is the lower bound, and `b` is the upper bound. There are a number of ways to write the integrand `fn`. The simplest is as follows: write a MATLAB expression $\alpha(v)$, where v corresponds to a vector of values of t $\langle t_1 = a, \ldots, t_n = b \rangle$ and $\alpha(v)$ corresponds to the vector $\langle f(t_1), \ldots, f(t_n) \rangle$. Then put $\alpha(v)$ between single quotation marks. The expression should contain only one variable, which is the variable being integrated over. (The support for functional programming in MATLAB is mediocre.)

For instance, you can evaluate the integral of $1/(1+t^2)$ from 1 to 2 as follows:

```
>> quad('1./(1+v.^2)',1,2)
ans =
    0.3218
```

The cumulative distribution function of the standard Gaussian $\int_{-\infty}^{x} N_{0,1}(t)dt$ is not an elementary function; that is, it is not expressible as a combination of polynomials, logs, exponentials, and trigonometric functions. However, a built-in MATLAB function `erf(x)` computes a closely related function called the *error function*, defined as

$$\operatorname{erf}(x) = \frac{2}{\sqrt{\pi}} \int_0^x e^{-t^2} dt.$$

Therefore the total weight of the standard Gaussian in an interval $[-x, x]$ symmetric around 0 can be computed as

$$\int_{-x}^{x} \frac{1}{\sqrt{2\pi}} e^{-u^2/2} du = \int_{-x/\sqrt{2}}^{x/\sqrt{2}} \frac{1}{\sqrt{\pi}} e^{-t^2} dt = \int_{0}^{x/\sqrt{2}} \frac{2}{\sqrt{\pi}} e^{-t^2} dt = \text{erf}\left(\frac{x}{\sqrt{2}}\right).$$

The first transformation follows from substituting $t = u/\sqrt{2}$. The second follows from the fact that the integrand is symmetric around 0.

The cumulative distribution of the standard Gaussian can be computed as

$$\int_{-\infty}^{x} \frac{1}{\sqrt{2\pi}} e^{-u^2/2} du = \int_{-\infty}^{x/\sqrt{2}} \frac{1}{\sqrt{\pi}} e^{-t^2} dt = \int_{-\infty}^{0} \frac{1}{\sqrt{\pi}} e^{-t^2} dt + \int_{0}^{x/\sqrt{2}} \frac{1}{\sqrt{\pi}} e^{-t^2} dt = \frac{1}{2}\left(1 + \text{erf}\left(\frac{x}{\sqrt{2}}\right)\right).$$

(The function erf does the right thing here for negative values of x.)

The MATLAB function `erfinv(y)` computes the inverse of erf. Therefore,

```
% To compute the probability that a normal distribution is within
% 1.6 standard deviations of the mean
>> erf(1.6/sqrt(2))
ans =
   0.8904


% To compute the probability that a normal distribution is less than
% mu + 1.6 sigma, where mu is the mean and sigma is the standard deviation
>> (1+erf(1.6/sqrt(2)))/2
ans =
   0.9452


% To find the value of d such that X is between mu-d*sigma and mu+d*sigma
% with probability 0.8
>> erfinv(0.8)*sqrt(2)
ans =
   1.2815
```

Exercises

Exercise 9.1. Let X be a random variable with values $-1, 2, 6$, such that $P(X = -1) = 0.2, P(X = 2) = 0.5, P(X = 6) = 0.3$. Compute $\text{Exp}(X), \text{Var}(X), \text{Std}(X)$.

Exercise 9.2. Let X be a random variable with values 0, 1, and 3, and let Y be a random variable with values $-1, 1, 2$, with the following joint distribution:

		-1	1	2
	0	0.12	0.08	0.10
X	1	0.20	0.04	0.25
	3	0.08	0.10	0.03

(a) Compute the marginal distributions.

(b) Are X and Y independent? Justify your answer.

(c) Compute $\text{Exp}(X)$ and $\text{Exp}(Y)$.

(d) Compute the distribution of $X + Y$.

(e) Compute $P(X \mid Y = 2)$ and $P(Y \mid X = 1)$.

Exercise 9.3. Let X be a random variable with values 0, 1, and 3, and let Y be a random variable with values $-1, 1, 2$. Suppose that $P(X = 1) = 0.5, P(X = 2) = 0.4$, and $P(X = 3) = 0.1$, with the following values of $P(Y \mid X)$:

		Y		
		-1	1	2
	0	0.5	0.3	0.2
X	1	0.2	0.7	0.1
	3	0.4	0.1	0.5

(a) Compute the joint distribution of X, Y.

(b) Compute the distribution of Y.

(c) Compute the corresponding table for $P(X \mid Y)$.

(d) Compute the distribution of $X + Y$.

(e) Compute $\text{Exp}(X)$, $\text{Exp}(Y)$, and $\text{Exp}(X + Y)$.

Problems

Problem 9.1. You may use MATLAB for this problem. A patient comes into a doctors office exhibiting two symptoms: $s1$ and $s2$. The doctor has two possible diagnoses: disease $d1$ or disease $d2$. Assume that, given the symptoms, the patient must have either $d1$ or $d2$, but cannot have both. The following probabilities are given:

$$
\begin{aligned}
P(s1 \mid d1) &= 0.8, \\
P(s1 \mid d2) &= 0.4, \\
P(s2 \mid d1) &= 0.2, \\
P(s2 \mid d2) &= 0.6, \\
P(d1) &= 0.003, \\
P(d2) &= 0.007.
\end{aligned}
$$

Assume that $s1$ and $s2$ are conditionally independent, given the disease.

(a) What are $P(d1\,|\,s1,s2)$ and $P(d2\,|\,s1,s2)$?

(b) The doctor has the choice of two treatments, $t1$ and $t2$. (It is not an option to do both.) Let c be the event that the patient is cured. The following probabilities are given:

$$P(c\,|\,d1,t1) = 0.8,$$
$$P(c\,|\,d2,t1) = 0.1,$$
$$P(c\,|\,d1,t2) = 0.3,$$
$$P(c\,|\,d2,t2) = 0.6.$$

Assume that event c is conditionally independent of the symptoms, given the disease and the treatment. What is $P(c\,|\,t1,s1,s2)$? What is $P(c\,|\,t2,s1,s2)$?

(c) Suppose that treatment $t1$ has a cost of \$1,000 and treatment $t2$ has a cost of \$500. If the patient has disease $d1$, then the value of being cured is \$20,000; if the patient has disease $d2$, then the value of being cured is \$15,000. Given that the patient is exhibiting symptoms $s1$ and $s2$, what is the expected value of applying $t1$? What is the expected value of applying $t2$?

(d) The doctor also has the option of ordering a test with a Boolean outcome. The test costs \$800. Logically, tests are like symptoms, so let event $s3$ be a positive result on this test. The following probabilities are given:

$$P(s3\,|\,d1) = 0.9,$$
$$P(s3\,|\,d2) = 0.1.$$

Assume that $s3$ is conditionally independent of $s1$ and $s2$, given the disease. Is it worthwhile ordering the test? What is the expected gain or cost from ordering the test?

Problem 9.2. You may use MATLAB for this problem. Continuing the example of the publisher and the reviewer discussed in Section 9.3.2, suppose that the publisher also has the option of consulting with two reviewers. Assume that the two reviewers follow the same probabilistic model, and that their reviews are conditionally independent, given the actual success or failure. Consider the following possible strategies:

1. Consult with one reviewer.

2. Consult with two reviewers. If both approve the manuscript, then publish, otherwise reject.

3. Consult with two reviewers. If either approves the manuscript, then publish, otherwise reject.

Suppose that a reviewer's fee is $500. Add these options to the decision tree
in Figure 9.3. What are the expected values of these strategies? Which is the
optimal strategy?

Problem 9.3. This problem is a continuation of Problem 9.2. If the publisher
has enough time, it may be possible to delay deciding whether to consult the
second reviewer until the opinion of the first is known. This allows two more
possible strategies:

- Consult reviewer A. If his opinion is favorable, consult reviewer B. If both
  are favorable, publish.

- Consult reviewer A. If his opinion is unfavorable, consult reviewer B. If
  either is favorable, publish.

(a) Add these to the decision tree in Problem 9.2. What are the expected
    values of these strategies? What is the optimal strategy?

(b) (This is difficult.) Present an argument that there are only four strategies
    worth considering: (1) not consulting with any reviewers, (2) consulting
    with one reviewer, and (3 and 4) the two conditional strategies described
    above. Your argument should be independent of the specific values of the
    probabilities, costs, and benefits involved; however, you should assume
    that reviewers A and B are equally accurate and equally expensive.

Problem 9.4. Let M be a matrix representing the joint distribution of two random variables X, Y, as in Table 9.1. Prove that if X and Y are independent, then
Rank$(M) = 1$.

Problem 9.5. Let X be a numerical random variable. A value q is defined as
a *median* of X if $P(X > q) \leq 1/2$ and $P(X < q) \leq 1/2$. (A random variable
may have more than one median.) Show that, for any random variable X with
finitely many values, any value of v that minimizes the expression Exp$(|X - v|)$
is a median of X. (This is stated without proof on p. 269. The statement is true
also for numerical random variables with infinitely many values, but the proof
is more difficult.)

   Hint: Consider how the value of Exp$(|X - v|)$ changes when v is moved
slightly in one direction or the other. Some experimentation with a simple example may be helpful.

Programming Assignments

Assignment 9.1. Random variable X takes on integer values from 1 to 500 and obeys an inverse power-law distribution with exponent 2.28. That is, $P(X = k) = \alpha / k^{2.28}$ for some constant α.

(a) Find α. (*Hint:* The probabilities must sum to 1.)

(b) Find $\text{Exp}(X)$.

(c) Find $\text{Std}(X)$.

(d) (This assignment is difficult; it requires some familiarity with infinite series.) Consider a variable Y that obeys the same powerlaw distribution as X but takes on integer values from 1 to ∞. Estimate the accuracy of the values you have computed in parts A, B, and C, as approximations for the corresponding values for Y.

Assignment 9.2. This assignment is a generalization of Problems 9.2 and 9.3. As in those problems, a publisher needs to decide whether to publish a book and has the option of consulting with a number of reviewers. The outcome of publishing is either Success, with a specified profit, or Fail with a specified loss. The outcome of not publishing is 0. Consulting with a reviewer costs a specified amount. A reviewer gives a Boolean answer, For or Against. Reviewers' opinions are conditionally independent, given the value of Success or Fail.

(a) Write a MATLAB function

```
PubValue(Profit,Loss,Fee,ProbSuc,ProbForSuc,ProbForFail,N)
```

that takes the following arguments:

> Profit = dollar profit if the book succeeds,
> Loss = dollar profit if the book fails,
> Fee = cost of hiring a reviewer,
> ProbSuc = $P(\text{Success})$,
> ProbForSuc = $P(\text{For} \mid \text{Success})$,
> ProbForFail = $P(\text{For} \mid \text{Fail})$,
> N = number of reviewers consulted; may be 0.

The function returns the expected profit to the publisher, assuming that N reviewers are consulted.

(b) Write a MATLAB function

```
OptimalN(Profit,Loss,Fee,ProbSuc,ProbForSuc,ProbForFail)
```

that returns a pair of values: the optimal number of reviewers to consult and the expected value.

Assignment 9.3. Consider the following simple model for disease diagnosis, with the following assumptions:

- There are N possible symptoms (including test results) which in this problem are taken, unrealistically, to be Boolean—that is, a patient either has the symptoms or doesn't; the test either succeeds or fails.

- There are M diseases under consideration.

- Symptoms are conditionally independent, given the disease.

- Any patient has exactly one diagnosis.

- There are Q different treatments. A patient can be given a single treatment. Similarly, assume, unrealistically, that the effectiveness of a treatment is Boolean; either a treatment entirely cures the disease or it is useless.

The *symptom matrix* is an $M \times N$ matrix S of probabilities: $S[I, J] = P(J \mid I)$, the probability that a patient exhibits symptom J, given that he has disease I.

(a) A *patient record* is a Boolean vector of length N indicating a patient's symptoms. Write a function `RecProb(R,S)` that takes as arguments a patient record R and a symptom matrix S and returns a vector D of length M such that $D[I] = P(R \mid I)$ for each disease I.

(b) The *frequency* vector is a vector F of length M such that $F[I]$ is the frequency of disease I in the population at large. Write a function `Diagnose(R,S,F)` that returns a vector D of length M such that $D[I] = P(I|R)$, the probability that a patient with symptoms R has disease I. Use Bayes' law. Include the normalizing factor.

(c) A *treatment efficacy* matrix is a $Q \times M$ matrix T, where $T[I, J]$ is the probability that treatment I will cure disease J. Assume that the event that I cures J is independent of the event that J manifests symptom K; that is, given that a patient has a particular disease, the effectiveness of the treatment is not affected by the particular symptoms he is manifesting. Write a function `Prognosis(R,S,F,T)` that returns a vector W of length Q, where $W[I]$ is the probability that a patient with symptoms R will be cured of his disease by treatment I.

(d) A *disease cost vector* is a vector C of length M indicating the cost of leaving disease I uncured. (Assume that this depends on the disease rather than on the symptoms.) A *treatment cost vector* is a vector $B[I]$ of length Q, indicating the cost of attempting treatment I. (Of course, for both of these, "cost" should be interpreted broadly as including all the undesirable consequences.) Write a function Benefit(R,S,F,T,C,B) that returns a vector A, where $A[I]$ is the expected benefit of applying treatment I to a patient with symptoms R. Note that the benefit of curing the disease applies only if the disease is cured, whereas the cost of the treatment applies whether the disease is cured or not.

Assignment 9.4. This part redoes parts A and B of Assignment 9.3 on the more realistic assumption that symptoms are numeric rather than Boolean, and the (common but often very unrealistic) assumption that each symptom is normally distributed. Specifically, assume that all symptoms are numeric, although diseases continue to be Boolean; that is, ignore the difference between a mild case and a severe case of the disease. Let S now be an $M \times N$ array, such that, for any disease I and symptom J, the measure x of symptom J given that the patient has disease I is normally distributed with mean $S[I,J]$ and standard deviation 1; that is, it follows the distribution $N_{S[I,J],\sigma}(x) = \exp(-(x - S[I,J])^2/2)/\sqrt{2\pi}$.

(a) Write a function RecProbN(R,P) that takes as arguments a patient record R and a symptom matrix S and returns a vector D of length M such that $D[I] = \tilde{P}(R|I)$ for each disease I.

(b) The *frequency* vector is a vector F of length M such that $F[I]$ is the frequency of disease I in the population at large. Write a function Diagnose(R,S,F) that returns a vector D of length M such that $D[I] = P(I|R)$, the probability that a patient with symptoms R has disease I.

Chapter 10

Markov Models

Markov models, also known as a *Markov chains* or *Markov processes*, are a type of probabilistic model that is very useful in the analysis of strings or sequences, such as text or time series. In a Markov process, time is viewed discretely; time instants are numbered 0, 1, 2, ... The process has a finite[1] number of *states;* at each instant, the process is in one particular state. Between time I and time $I + 1$, the process executes a *transition* from one state to another, and its destination is probabilistically determined.

The key property of a Markov process is that the probability of the transition depends only on the current state and not on any of the previous states. That is, metaphorically speaking, whenever we are in one state, we spin a wheel of fortune that tells where to go next. Each state has its own wheel, but the wheel at a given state remains constant over time, and the choice of where to move next depends only on the outcome of the wheel spin and on nothing else. This is known as the Markov condition. It is sometimes described by saying that the system is memoryless and time-invariant; what it does next depends only on where it is now and not on how it got there nor what the time is.

A simple example of a Markov model is the children's board game Chutes and Ladders (Davis and Chinn, 1985). In this game, the board has 100 squares in sequence; there are a number of "ladders," which take the player forward to a more advanced square, and a number of "chutes," which take the player back to an earlier square. Each player rolls a die and moves forward the number of squares shown on the roll. If he ends at the bottom of a ladder, he climbs it; if the player ends at the top of a chute, he slides down.

For a single player, the corresponding Markov model has a state for every possible resting square (i.e., square that is not the bottom of a ladder or the top of a chute.) Each state has six transitions, each with probability 1/6, corresponding to the six possible outcomes of the die roll. Figure 10.1 shows the Markov model corresponding to a simplified game in which there are only six resting squares and in which the die has only two outcomes, 1 or 2. Each of the

---

[1]We can also define Markov processes with infinitely many states; however, we will not consider these in this book.

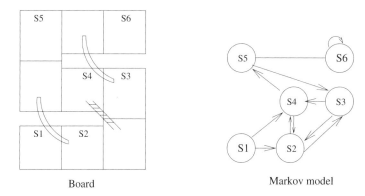

Board   Markov model

Figure 10.1. Markov model for Chutes and Ladders.

arcs has the label 1/2, except for the final self-loop from S6 to itself, which has label 1.

In a game on the same board with k players taking turns, a state is a $(k+1)$ tuple $S = \langle s_1, \ldots, s_k, p \rangle$ where s_i is the location of player i and p is an index from 1 to k indicating which player moves next. For example, if three players play the game in Figure 10.1 then one state is $\langle S5, S3, S3, 2 \rangle$. This has two transitions, each with probability 1/2: one to $\langle S5, S2, S3, 3 \rangle$ and the other to $\langle S5, S4, S3, 3 \rangle$.

We can describe a Markov model in terms of random variables. For each time instant, $t = 0, 1, 2, \ldots$, we define the random variable X_t to be the state of the system at time t. Thus, the domain of values for X_i is the set of states of the Markov model. The conditional probability $P(X_{t+1} = u \mid X_t = v)$ is the label on the arc in the model from u to v. The condition that the model is memoryless corresponds to the statement that X_{t+1} is conditionally independent of $X_1 \ldots X_{t-1}$, given X_t. The condition that the model is time-invariant corresponds to the statement that, for all times t and s,

$$P(X_{t+1} = u \mid X_t = v) = P(X_{s+1} = u \mid X_s = v).$$

Let us assign indexes $1, \ldots, k$ to the k states of the model. Then we can view the probability distribution of X_t as a k-dimensional vector $\vec{X}_t$. We can also characterize the Markov model by a $k \times k$ *transition* matrix M, where $M[i, j] = P(X_{t+1} = i \mid X_t = j)$. By the axioms of probability,

$$P(X_{t+1} = i) = \sum_{j=1}^{k} P(X_{t+1} = i, X_t = j) = \sum_{j=1}^{k} P(X_{t+1} = i \mid X_t = j) P(X_t = j).$$

This can be written as a matrix multiplication: $\vec{X}_{t+1} = M \cdot \vec{X}_t$.

For example, the Markov model in Figure 10.1 has the following transition matrix:

$$M = \begin{bmatrix} 0 & 0 & 0 & 0 & 0 & 0 \\ 1/2 & 0 & 1/2 & 1/2 & 0 & 0 \\ 0 & 1/2 & 0 & 0 & 1/2 & 0 \\ 1/2 & 1/2 & 1/2 & 0 & 0 & 0 \\ 0 & 0 & 0 & 1/2 & 0 & 0 \\ 0 & 0 & 0 & 0 & 1/2 & 1 \end{bmatrix}.$$

If $\vec{X}_0 = \langle 1, 0, 0, 0, 0, 0 \rangle$—that is, the player definitely starts at S1 at time 0—then

$$\vec{X}_1 = M \cdot \vec{X}_0 = \langle 0, 1/2, 0, 1/2, 0, 0 \rangle,$$

$$\vec{X}_2 = M \cdot \vec{X}_1 = \langle 0, 1/4, 1/4, 1/4, 1/4, 0 \rangle,$$

$$\vec{X}_3 = M \cdot \vec{X}_2 = \langle 0, 1/4, 1/4, 1/4, 1/8, 1/8 \rangle,$$

$$\vec{X}_4 = M \cdot \vec{X}_3 = \langle 0, 1/4, 3/16, 1/4, 1/8, 3/16 \rangle.$$

Note a few points about this analysis. First, we have $\vec{X}_3 = M \cdot \vec{X}_2 = M \cdot M \cdot \vec{X}_1 = M \cdot M \cdot M \cdot X_0 = M^3 \cdot \vec{X}_0$; so, in general, we have $\vec{X}_n = M^n \cdot X_0$.

Second, each column of matrix M adds up to 1. This corresponds to the fact that the transitions out of state j form a frame of discernment—exactly one transition occurs—so

$$\sum_{i=1}^{k} M_{i,j} = \sum_{i=1}^{k} P(X_{t+1} = i \mid X_t = j) = 1.$$

A matrix with this feature is called a *stochastic* matrix. As a consequence, the sum of the values of $M \cdot \vec{v}$ is equal to the sum of the values of $\vec{v}$ for any $\vec{v}$.

Third, the analysis here is oddly similar to Application 3.3, in Section 3.4.1, of population transfer between cities. In fact, it is exactly the same, with probability density here replacing population there. More precisely, that application was contrived as something that resembled a Markov process but did not require a probabilistic statement.

Here, we do not explore the mathematics of Markov models in depth; rather, we focus on two computer science applications of Markov models: the computation of PageRank for web search and the tagging of natural language text. First, we discuss the general issue of stationary distributions, which is needed for the analysis of PageRank.

10.1   Stationary Probability Distribution

Consider a Markov process with transition matrix M. A probability distribution $\vec{P}$ over the states of M is said to be *stationary* if $\vec{P} = M \cdot \vec{P}$. Thus, if $\vec{X}_t = \vec{P}$, then $\vec{X}_{t+1} = \vec{P}$; if we start with this probability distribution, then we stay with this probability distribution.

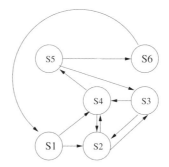

Figure 10.2. Modified Markov model.

Theorem 10.1. *Every finite Markov process has a stationary distribution. Equivalently, for every stochastic matrix M, there exists a probability distribution $\vec{P}$ such that $\vec{P} = M \cdot \vec{P}$.*

The example of the game Chutes and Ladders is, unfortunately, not a very interesting example of this theorem; the unique stationary distribution is $\langle 0, 0, 0, 0, 0, 1 \rangle$. We therefore change the example by removing the self-arc from S6 to itself and replacing it with an arc from S6 to S1; this means that when we finish the game, we start at the beginning again. Figure 10.2 shows the revised Markov model. In that case, the stationary distribution is $\langle 1/15, 4/15, 3/15, 4/15, 2/15, 1/15 \rangle$, which is easily checked. (For checking by hand, notice that the property $\vec{P} = M \cdot \vec{P}$ remains true if we multiply through by a scalar, so we can ignore the denominator 15.)

In fact, most Markov models have exactly one stationary distribution. Theorem 10.3 gives a condition that suffices to guarantee that the stationary distribution is unique. First, we must define the property *strongly connected*.

Definition 10.2. Let G be a directed graph. G is strongly connected, if, for any two vertices u and v in G, there is a path from u to v through G.

Theorem 10.3. *Let M be a finite Markov model. Let G be the graph whose vertices are the states of M and whose arcs are the transitions with nonzero probability. If G is strongly connected, then M has exactly one stationary distribution.*

Moreover, most Markov models have the property that, if we run them for a long enough time, they converge to the stationary distribution, no matter what the starting distribution. Theorem 10.5 gives a condition that ensures that the distribution converges. First, we present an explanation of the terms *periodic* and *aperiodic*.

Definition 10.4. Let G be a directed graph and let $k > 1$ be an integer. Vertex v in G is *periodic with period k* if every cycle from v to itself through G has a length divisible by k. G is *aperiodic* if no vertex in G is periodic for any period $k > 1$.

Theorem 10.5. *Let M be a finite Markov model. Let G be the graph whose vertices are the states of M and whose arcs are the transitions with nonzero probability. If G is strongly connected and aperiodic then, for any starting distribution $\vec{X}_0$, the sequence $\vec{X}_0, M\vec{X}_0, M^2\vec{X}_0, \dots$ converges to the stationary distribution.*

That is, if M is a finite Markov model, then however we start, if we run the model long enough, it converges to the stationary distribution $\vec{D}$. Thus, Markov models are memoryless in a second sense as well: over time, they entirely forget where they started from and converge to a distribution that depends only on the transition matrix.

For example, let M be the modified model in Figure 10.2. Then, as shown, the stationary distribution is $\vec{D} = \langle 1/15, 4/15, 3/15, 4/15, 2/15, 1/15 \rangle = \langle 0.0667, 0.2667, 0.2000, 0.2667, 0.1333, 0.0667 \rangle$. Let $\vec{X}_0 = \langle 1, 0, 0, 0, 0, 0 \rangle$ (i.e., at time 0 we are definitely at S1). Then

$$\vec{X}_4 = M^4 \cdot \vec{X}_0 = \langle 0.1250, 0.2500, 0.1875, 0.2500, 0.1250, 0.0625 \rangle,$$

$$\vec{X}_8 = M^8 \cdot \vec{X}_0 = \langle 0.0703, 0.2656, 0.1992, 0.2656, 0.1328, 0.0664 \rangle,$$

$$\vec{X}_{12} = M^{12} \cdot \vec{X}_0 = \langle 0.0669, 0.2666, 0.2000, 0.2666, 0.1333, 0.0667 \rangle,$$

$$|\vec{X}_{12} - \vec{D}| = 2.5 \cdot 10^{-4}.$$

10.1.1 Computing the Stationary Distribution

Given a transition matrix M, the stationary distribution $\vec{P}$ can be computed as follows. The vector $\vec{P}$ satisfies the equations $M\vec{P} = \vec{P}$ and $\sum_{I=1}^{n} \vec{P}[I] = 1$. We can rewrite the first equation as $(M - I_n) \cdot \vec{P} = \vec{0}$, where I_n is the identity matrix. For a Markov model with a unique stationary distribution, $M - I_n$ is always a matrix of rank $n - 1$. We can incorporate the final constraint $\sum_{I=1}^{N} \vec{P}[I] = 1$ by constructing the matrix Q whose first n rows are $M - I_n$ and whose last row is all 1s. This is an $(n + 1) \times n$ matrix of rank n. The vector $\vec{P}$ is the solution to the equation $Q \cdot \vec{P} = \langle 0, 0, \dots, 0, 1 \rangle$.

For example, suppose $M = [1/2, 2/3; 1/2, 1/3]$:

```
>> m=[1/2,2/3;1/2,1/3]
m =
     0.5000      0.6667
     0.5000      0.3333

>> q=m-eye(2)
q =
    -0.5000      0.6667
     0.5000     -0.6667
```

```
>> q(3,:)=[1,1]
q =
    -0.5000      0.6667
     0.5000     -0.6667
     1.0000      1.0000

>> q\[0;0;1]
ans =
     0.5714
     0.4286

>> m*ans
ans =
     0.5714
     0.4286
```

If the graph is large, there are more efficient, specialized algorithms.

10.2   PageRank and Link Analysis

The great Google empire began its existence as a project of Sergey Brin and
Lawrence Page, then two graduate students at Stanford, aimed at improving the
quality of search engines.[2] Their idea,[3] described in the now classic paper, "The
PageRank Citation Ranking" (Page, Brin, Motwani, and Winograd, 1998) was as
follows. The ranking of a webpage P to a query Q consists of two parts. First,
there is a *query-specific* measure of relevance—how well does this page match
the query? This is determined by the vector theory of documents or something
similar (see Assignment 2.1, Chapter 2). Second, there is a *query-independent*
measure of the quality of the page—how good a page is this, overall? This latter
measure, in the PageRank theory, is measured by using link analysis. Previous
to Google, it appears that search engines used primarily or exclusively a query-
specific measure, which could lead to anomalies. For example, Brin and Page
mention that, at one point, the top page returned by one of the search engines
to the query "Bill Clinton" was a page that said, simply, "Bill Clinton sucks."

---

[2]Readers who had not been using search engines prior to Google may not realize how thor-
oughly it revolutionized the field. The earliest web search engines were introduced in 1993, and
search engines such as Infoseek (launched 1994) and AltaVista (lauched 1995) soon became widely
used and were initially quite effective. By 1998, however, the quality of results returned by these
search engines was visibly deteriorating; the ranking algorithms simply could not keep up with
the exponential growth in the web. When Google was released, the improvement in quality was
startling.

[3]The ranking method now used by Google, although highly secret, is clearly much more complex
than the theory presented in this text. The current ranking system certainly makes heavy use of
user logs, which were barely available when Google was first launched. However, unquestionably,
link-based measures are still an important component of the current ranking method.

Why not? After all, two-thirds of the words on the page match words in the query; that would be hard to beat.

The first idea is that, if a page P is good, then many people will have linked to it. So one measure of quality would be just the number of inlinks to a page P. The second idea, however, says some links are more important than others. For example, consider these ideas:

1. Links from a different website, which indicate an outside interest, should count more than links from same domain, which just indicate the structure of the website.

2. Characteristics of anchor (e.g., font, boldface, position in the linking page) may indicate a judgment by the author of the linking page of relative importance of P.

3. A link from an important page is more important than a link from an unimportant page. This is circular, but the circularity can be resolved.

The PageRank algorithm works with idea (3). Suppose that each page P has an importance $I(P)$ computed as follows: First, every page has an inherent importance E (a constant) just because it is a webpage. Second, if page P has importance $I(P)$, then P contributes an indirect importance $F * I(P)$ that is shared among the pages to which P points. (F is another constant.) That is, let $O(P)$ be the number of outlinks from P. Then, if there is a link from P to Q, then P contributes $F \cdot I(P)/O(P)$ "units of importance" to Q.

(What happens if P has no outlinks, so that $O(P) = 0$? This actually turns out to create a little trouble for our model. For the time being, we assume that every page has at least one outlink, and we return to this problem in Section 10.2.2.)

We therefore have the following relation: for every page Q, if Q has inlinks $P_1, \ldots, P_m$ then

$$I(Q) = E + \frac{F \cdot I(P_1)}{O(P_1)} + \ldots + \frac{F \cdot I(P_m)}{O(P_m)}. \qquad (10.1)$$

Thus, we have a system of n linear equations in n unknowns, where n is the number of webpages.

We now make the following observation. Suppose we write down all the above equations for all the different pages Q on the web. Now we add all the equations. On the left, we have the sum of $I(Q)$ over all Q on the web; we call this sum S. On the right, we have N occurrences of E and, for every page P, O_P occurrences of $F \cdot I(P)/O_P$. Therefore, over all the equations, we have, for every page P, a total of $F \cdot I(P)$, and these add up to FS. Therefore, we have

$$S = NE + FS \quad \text{so} \quad F = 1 - (NE/S). \qquad (10.2)$$

Since the quantities $E, F, N,$ and S are all positive, it follows that $F < 1, E < S/N$.

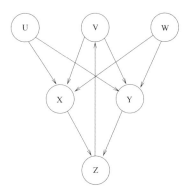

Figure 10.3. Link analysis for webpages.

We can normalize this as follows: let $J(Q) = I(Q)/S$ and $e = E/S$, then Equations (10.1) and (10.2), respectively, become

$$J(Q) = e + F \cdot (J(P_1)/O(P_1) + \ldots + J(P_m)/O(P_m)),$$

$$1 = Ne + F.$$

For example, suppose we have pages U,V,W,X,Y,Z linked as shown in Figure 10.3. Note that X and Y each have three inlinks, but two of these are from the "unimportant" pages U and W. Page Z has two inlinks and V has one, but these are from "important" pages. Let $e = 0.05$ and let $F=0.7$. We get the following equations (for simplicity, we write the page name rather than J(page name)):

```
U = 0.05
V = 0.05 + 0.7 * Z
W = 0.05
X = 0.05 + 0.7*(U/2+V/2+W/2)
Y = 0.05 + 0.7*(U/2+V/2+W/2)
Z = 0.05 + 0.7*(X+Y)
```

The solution to these equations is $U = 0.05$, $V = 0.256$, $W = 0.05$, $X = 0.175$, $Y = 0.175$, $Z = 0.295$.

10.2.1   The Markov Model

We can turn the above system of equations into a stochastic system of equations as follows. First, let $\vec{J}$ be the N-dimensional vector where $\vec{J}[Q] = J(Q)$; thus, $\sum_{Q=1}^{N} \vec{J}[Q] = 1$, so $\vec{J}$ is a probability distribution. Now, let M be the $N \times N$ matrix such that

$$M[P,Q] = \begin{cases} e + \frac{F}{O(Q)} & \text{if } Q \text{ points to } P, \\ e & \text{if } Q \text{ does not point to } P. \end{cases}$$

Then, for each column Q, $\sum_P M[P,Q] = \sum_{P=1}^{N} e + \sum_{P|Q\rightarrow P} F/O(Q) = Ne + F = 1$, so M is a stochastic matrix.

Moreover, for each row P,

$$\sum_{Q=1}^{N} M[P,Q]\vec{J}[Q] = \sum_{Q=1}^{N} e\vec{J}[Q] + \sum_{Q\rightarrow P} \frac{F}{O(Q)}\vec{J}[Q]$$

$$= e \cdot \sum_{Q=1}^{N} \vec{J}[Q] + \sum_{Q\rightarrow P} \frac{F}{O(Q)} \cdot J(Q)$$

$$= e + \sum_{Q\rightarrow P} F \cdot \frac{J(Q)}{O(Q)} = \vec{J}[P]. \qquad (10.3)$$

In other words, $M\vec{J} = \vec{J}$, so $\vec{J}$ is the stationary distribution for M.

Matrix M then corresponds to the Markov model where there is a link from every page Q to every page P (including self-link). The probability of the link is e if Q does not point to P and is $e + F/O(Q)$ if Q does point to P. We can view this in terms of the following stochastic model: imagine that we are browsing the web for a very long time, and each time we read a page P, we decide where to go next by using the following procedure.

1. We flip a weighted coin that comes up heads with probability F and tails with probability $1 - F$.

2. If the coin comes up heads, we pick an outlink from P at random and follow it. (Again, we assume that every page has at least one outlink.)

3. If the coin comes up tails, we pick a page at random in the web and go there.

In this example, suppose $F = 0.7$, and $e = (1 - 0.7)/6 = 0.05$, so the matrix is

$$M = \begin{bmatrix} 0.05 & 0.05 & 0.05 & 0.05 & 0.05 & 0.05 \\ 0.05 & 0.05 & 0.05 & 0.05 & 0.05 & 0.75 \\ 0.05 & 0.05 & 0.05 & 0.05 & 0.05 & 0.05 \\ 0.40 & 0.40 & 0.40 & 0.05 & 0.05 & 0.05 \\ 0.40 & 0.40 & 0.40 & 0.05 & 0.05 & 0.05 \\ 0.05 & 0.05 & 0.05 & 0.75 & 0.75 & 0.05 \end{bmatrix}.$$

Since the graph is fully connected, Theorem 10.3 applies, and there exists a unique stationary distribution.

From a computational point of view, this would at first seem to be a step in the wrong direction. We have turned an extremely sparse system of equations into an entirely dense one; moreover, with N equal to several billion, the transition matrix cannot even be generated or stored, let alone used in calculation. However, it turns out that, because of the special form of this matrix, fast iterative algorithms are known for computing the stationary distribution. See Langville and Meyer (2006).

Also note that the connection to Markov processes is actually bogus; the justification of the link analysis does not actually require any reference to any kind of random event or measure of uncertainty. The reasons to present this in terms of Markov models are first, because the mathematics of stochastic matrices has largely been developed in the context of Markov models; and second, because the stochastic model is a useful metaphor in thinking about the behavior of PageRank and in devising alternative measures.

In understanding the behavior of the algorithm, it is helpful to consider its behavior for different values of the parameter F. If $F = 0$, then the links are ignored, and PageRank is just the uniform distribution $I(Q) = 1/N$.

If $F = \epsilon$ for small ϵ (specifically $1/\epsilon$ is much larger than the number of inlinks of any page), then the all pages have PageRank closed to $1/N$. In this case,

$$I(P) = (1/N)\left(1 + \epsilon \sum_{Q \to P} 1/O(Q)\right) + O(\epsilon^2/N).$$

If $F = 1$, the method blows up. There still exists a stationary distribution, but there may be more than one, and iterative algorithms may not converge to any stationary distribution.

If F is close to 1, the system is unstable (i.e., small changes to structure may make large changes to the solution). The iterative algorithms converge only slowly.

The experiments used in the original PageRank paper used $F = 0.85$.

10.2.2   Pages with No Outlinks

Of course, in the actual web, many pages (in fact, the majority of pages—again, a Zipf distribution) have no outlinks. In that case, the above formulas in Section 10.2.1 break down because the denominator $O(P)$ is equal to zero. A number of solutions have been proposed, including the following.

Solution 1. The simplest solution is to add a self-loop from each page to itself. Thus, every page has at least one outlink. The problem is that this rewards a page for not having any outlinks. In our random model, if we are randomly at a page with no outlink, we flip the coin until it comes up tails, at which point we jump at random. With $F = 0.85$, this means that once we reach a page, we will generally stay there through about six more, so a page with no outlinks will be ranked as about seven times as important as another page with the same real inlinks but many outlinks.

Solution 2. The original PageRank paper proposes the following method:

Step 1. We prune all pages with no outlinks, then prune all pages that had onlinks only to the pages we just pruned, and keep on doing this until all pages have an outlink.

Step 2. We compute PageRank over this set;

Step 3. We reinstate the pages we pruned and let importance propagate, using Equation (10.1).

Solution 3. The rank computed in Solution 2 seems rather arbitrary. Langville and Meyer (2006) propose a method with a more principled formulation. When we randomly come to a page with no outlinks, we should simply skip flipping the control coin, and just jump randomly to any page in the web. That seems logical. Equation (10.1) is still satisfied; E is the probability that a nonlink-based jump from somewhere in the web reaches page P. The stochastic matrix becomes more complicated because it now contains three kinds of values:

- $M[P,Q] = 1/N$ if Q has no outlinks,
- $M[P,Q] = (1-F)/N$ if Q has some outlinks, but not to P,
- $M[P,Q] = (1-F)/N + F/O(Q)$ if Q links to P.

Further, more complex methods of dealing with "dangling nodes" are studied in Langville and Meyer (2006).

10.2.3   Nonuniform Variants

It is not necessary for the "inherent" importance of all pages P to be the same value E; we can have any distribution $E(P)$, representing some other evaluation of inherant importance. For example, we could have $E(P)$ ranked higher for pages on a .edu site, or for the Yahoo homepage, or for our own page, and so on. We just change E to $E(Q)$ in Equation (10.1).

Likewise, there may be a reason to think that some outlinks are better than others (e.g., font, font size, or links to a different domain are more important than links within a domain). We can assign $W(Q,P)$ to be the weight of the link from Q to P however we want; the only constraints are that the weights are nonnegative and that the weights of the outlinks from Q add up to 1. We replace $1/O(Q)$ by $W(Q,P)$ in Equation (10.3).

10.3   Hidden Markov Models and the K-Gram Model

A number of problems in natural language analysis can be formulated in the following terms: We are given a *text*, which is viewed as a string of *elements*. The problem is to assign one *tag* out of a collection of tags to each of the elements. Examples of such problems are the following.

- The elements are words and the tags are parts of speech. In applications, "parts of speech" here often involves finer distinctions than the usual linguistic categories. For instance, the "named entity" recognition problem involves categorizing proper nouns as personal names, organizations, place names, and so on.

- The elements are words and the tags are the meanings of the words (lexical disambiguation).

- The elements are phonemes, as analyzed in a recording of speech, and the tags are words.

In the statistical, or supervised learning, approach to this problem, we are given a large *training corpus* of correctly tagged text. The objective, then, is to extract patterns from the training corpus that can be applied to label the new text.

The same kind of analysis applies in domains other than natural language analysis. For example, a computer system may be trying to analyze the development over time of some kind of process, based on a sequence of observations that partially characterize the state. Here the elements are the observations and the tags are the underlying state.

10.3.1   The Probabilistic Model

In a probabilistic model of tagging, we formulate the problem as looking for the *most likely* sequence of tags for the given sequence of elements. Formally, let n be the number of elements in the sentence to be tagged. Let E_i be the random variable for the ith element, let T_i be the random variable for the ith tag, and let e_i be the actual value of the ith element. The problem is to find the sequence of tag values $t_1, \ldots, t_N$ such that $P(T_1 = t_1, \ldots, T_n = t_n \mid E_1 = e_1, \ldots, E_N = e_n)$ is as large as possible. We will abbreviate the event $T_i = t_i$ as $\bar{t}_i$ and $E_i = e_i$ as $\bar{e}_i$.

For instance, in tagging the sentence "I can fish" by part of speech, we want to arrive somehow at the estimate that $P(T_1{=}\text{pronoun}, T_2{=}\text{modal}, T_3{=}\text{verb} \mid E_1{=}\text{"I,"} E_2 = \text{"can,"} E_2 = \text{"fish"})$ is large; that $P(T_1{=}\text{pronoun}, T_2{=}\text{verb}, T_3{=}\text{noun} \mid E_1{=}\text{"I,"} E_2 = \text{"can,"} E_2 = \text{"fish"})$ (meaning "I put fish into cans") is small; and that the probability for any other sequence of three tags is tiny or zero.

In principle, what we are after is, considering all the times that the sentence "I can fish" might be written or spoken, determining what fraction of the time the intended meaning is the one corresponding to ⟨pronoun modal verb⟩ and what fraction of the time it is the one corresponding to ⟨pronoun verb noun⟩. However, there is obviously no way to directly measure this for most sentences.

So we have to estimate this probability in terms of numbers that can actually be measured in a training corpus. As is done throughout applied probability, to do this, we make independence assumptions that are violently and obviously false. We proceed as follows.

We first apply Bayes' law to derive

$$P(\bar{t}_1 \ldots \bar{t}_n \,|\, \bar{e}_1 \ldots \bar{e}_n) = P(\bar{e}_1 \ldots \bar{e}_n \,|\, \bar{t}_1 \ldots \bar{t}_n) \cdot P(\bar{t}_1 \ldots \bar{t}_n) / P(\bar{e}_1 \ldots \bar{e}_n). \tag{10.4}$$

However, the denominator above $P(\bar{e}_1, \ldots, \bar{e}_n)$ is just a normalizing factor, which depends only on the sequence of elements, which is given. As this is the same for all choices of tags, we can ignore it in comparing one sequence of tags to another.

So we have re-expressed the problem as finding the sequence of tags that maximizes the product $P(\bar{t}_1, \ldots, \bar{t}_n) \cdot P(\bar{e}_1, \ldots, \bar{e}_n \,|\, \bar{t}_1, \ldots, \bar{t}_n)$. The first factor here is the inherent likelihood of the sequence of tags, determined even before we have seen what the elements are. The second is the likelihood that a particular sequence of tags $\langle t_1, \ldots, t_n \rangle$ will be instantiated as the sequence of elements $\langle e_1, \ldots, e_n \rangle$. For instance, in the speech understanding problem, the first term is the probability that a speaker will speak a given sentence, sometimes called the linguistic model; the second is the probability, given the sentence, that a speaker would pronounce it in the way that has been heard, sometimes called the phonetic model. (The interpretation of this product in the part-of-speech problem is not as natural.) We now have to estimate these two terms by using independence assumptions.

We can use the rule for the probability of a conjunction to expand the first factor of the right-hand term in Equation (10.4) as follows:

$$P(\bar{e}_1, \ldots, \bar{e}_n | \bar{t}_1, \ldots, \bar{t}_n) = P(\bar{e}_1 | \bar{t}_1, \ldots, \bar{t}_n) \cdot P(\bar{e}_2 | \bar{e}_1, \bar{t}_1, \ldots, \bar{t}_n) \cdot, \ldots, \cdot P(\bar{e}_n | \bar{e}_1, \ldots, \bar{e}_{n-1}, \bar{t}_1, \ldots, \bar{t}_n).$$
$$\tag{10.5}$$

We now make the independence assumption that $\bar{e}_i$ depends only on $\bar{t}_i$; specifically, that $\bar{e}_i$ is conditionally independent of $\bar{t}_j$ for $j \neq i$ and of $\bar{e}_j$ for $j < i$. We also make the "time invariance" assumption that the probability that, for example, "pronoun" is instantiated as "he" is the same whether we are talking about the first word in the sentence or about the fifth.[4]

Having made this assumption, we can rewrite the right-hand side of Equation (10.5) in the form $P(\bar{e}_1 \,|\, \bar{t}_1) \cdot P(\bar{e}_2 \,|\, \bar{t}_2) \cdot \ldots \cdot P(\bar{e}_n \,|\, \bar{t}_n)$. We can now use Bayes' law again, and write this in the form $P(\bar{e}_i \,|\, \bar{t}_i) = P(\bar{t}_i \,|\, \bar{e}_i) P(\bar{e}_i) / P(\bar{t}_i)$. But again, the factors $P(\bar{e}_i)$ do not depend on the choice of $\bar{t}_i$, so they can be ignored.

What is gained by this second turning around? After all whether we use $P(\bar{e}_i \,|\, \bar{t}_i)$ or $P(\bar{t}_i \,|\, \bar{e}_i) / P(\bar{t}_i)$, we will end up estimating it, for each element, in

---

[4]To express this condition mathematically would require making our notation substantially more complicated. This is common in write-ups of complex probabilistic models such as this; substantial parts of the theory are left either stated only in English, or left entirely implicit, to be understood by the reader.

terms of the fraction of instances of t_i in the corpus that are also instances of e_i. The answer is that, since $P(\bar{t}_i \mid \bar{e}_i)$ is the more natural quantity, there may be other sources of information that give good evidence about this, but not about $P(\bar{e}_i \mid \bar{t}_i)$. For instance, we may have a dictionary that shows a word has possible parts of speech that do not appear in the training corpus. In that case, we can assign $P(\bar{t}_i \mid \bar{e}_i)$ some default estimate on the strength of the dictionary entry.)

We now turn to the factor $P(\bar{t}_1, \ldots, \bar{t}_n)$ in Equation (10.4). Again, using the rule for conjunctions, we can write this in the form

$$P(\bar{t}_1, \ldots, \bar{t}_n) = P(\bar{t}_1) \cdot P(\bar{t}_2 \mid \bar{t}_1) \cdot P(\bar{t}_3 \mid \bar{t}_1, \bar{t}_2) \cdot \ldots \cdot P(\bar{t}_n \mid \bar{t}_1, \ldots, \bar{t}_{n-1}). \qquad (10.6)$$

We now make the second independence assumption, known as the k-gram assumption, for some small integer k: the dependence of the ith tag on all the previous tags in fact reduces to its dependence on the $k-1$ previous tags. More precisely stated, T_i is conditionally independent of $T_1, \ldots, T_{i-k}$ given $T_{i+1-k}, \ldots,$ T_{i-1}. Thus, $P(\bar{t}_i \mid \bar{t}_1, \ldots, \bar{t}_{i-1}) = P(\bar{t}_i \mid \bar{t}_{i+1-k}, \ldots, \bar{t}_{i-1})$.

From this point forward in this section, to simplify the notation, we consider the case $k = 3$, which is the most commonly used value; however, it should be easy to see how the formulas can be modified for other values of k. The trigram assumption ($k = 3$) states that T_i is conditionally independent of $T_1, \ldots,$ T_{i-3}, given T_{i-2}, T_{i-1}. Thus, $P(\bar{t}_i \mid \bar{t}_1, \ldots, \bar{t}_{i-1}) = P(\bar{t}_i \mid \bar{t}_{i-2}, \bar{t}_{i-1})$. Again, we make the time invariance assumption that this probability is the same for all values of $i > k$.

To get more uniform notation, we use the convention that there are two random variables, T_{-1} and T_0, that have the special tag *S* (start) with probability 1. We can then write Equation (10.6) in the form

$$P(\bar{t}_1, \ldots, \bar{t}_n) = \Pi_{i=1}^{n} P(\bar{t}_i \mid \bar{t}_{i-2}, \bar{t}_{i-1}).$$

Putting all this, we can rewrite Equation (10.4) as

$$P(\bar{t}_1, \ldots, \bar{t}_n \mid \bar{e}_1, \ldots, \bar{e}_n) = \alpha \cdot \Pi_{i=1}^{n} \frac{P(\bar{t}_i \mid \bar{e}_i) \cdot P(\bar{t}_i \mid \bar{t}_{i-2}, \bar{t}_{i-1})}{P(\bar{t}_i)}, \qquad (10.7)$$

where α is a normalizing factor.

10.3.2   Hidden Markov Models

Based on these independence assumptions, we can now view sentences as the output of the following random process. The process outputs a sequence of ⟨tag,element⟩ pairs. The *state* is the pair ⟨t_{i-2}, t_{i-1}⟩ of the last two tags that it has output (in the general k-gram model, this is the tuple of the last $k-1$ tags it has output). If the process is in state ⟨t_{i-2}, t_{i-1}⟩, then it first picks a new tag t_i to output, with probability distribution $P(\bar{t}_i \mid \bar{t}_{i-2}, \bar{t}_{i-1})$, and then, based on that tag, picks an element e_i with probability distribution $P(\bar{e}_i \mid \bar{t}_i)$. The new state is $\langle t_{i-1}, t_i \rangle$. The problem we have to solve is that we are given the sequence of elements $\langle e_1, \ldots, e_n \rangle$ and we wish to find the sequence of tags $\langle t_1, \ldots, t_n \rangle$ for which the posterior probability $P(\bar{t}_1, \ldots, \bar{t}_n \mid \bar{e}_1, \ldots, \bar{e}_n)$ is maximal. Note that finding the sequence of states is equivalent to finding the sequence of tags; if the process goes from state $\langle x, y \rangle$ to $\langle y, z \rangle$, then the tag output is z.

This is an instance of a *hidden Markov model (HMM)*, which is a modified Markov model with the following properties:

- There is a particular starting state s_0.

- There may be multiple arcs between any two vertices. (For that reason, the transition matrix representation cannot be used.)

- Each arc is labeled by an element.

- The arc $A(u, e, v)$ from state u to state v labeled e is labeled by the probability $P(A(u, e, v) \mid u)$, the probability, given that we are in state u at time t, that we will proceed to traverse arc $A(u, e, v)$. These probabilities on the outarcs from u form a probability distribution.

As the process moves on a path through the model, it outputs the elements that label the arcs it traverses. The problem is, given a sequence of output elements, what is the most probable path that outputs that sequence? The model is called "hidden" because all we see is the sequence of elements, not the sequence of states.

We proceed as follows:

$$P(S_0 = s_0, S_1 = s_2, \ldots, S_n = s_n \mid S_0 = s_0, \bar{e}_1, \ldots, \bar{e}_n)$$
$$= \frac{P(S_0 = s_0, \bar{e}_1, S_1 = s_1, \bar{e}_2, S_2 = s_2, , \ldots, S_n = s_n, \bar{e}_n, S_{n+1} = s_{n+1} \mid S_0 = s_0)}{P(\bar{e}_1, \ldots, \bar{e}_n)}$$
$$= \alpha \cdot \Pi_{i=1}^{n} P(A(s_{i-1}, e_i, s_i) \mid s_{i-1}),$$

where α is a normalizing factor depending only on the e_i and not on the s_i.

In the case of the k-gram model, as we have seen in Equation (10.7),

$$P(A(\langle t_{i-2}, t_{i-1} \rangle, e_i, \langle t_{i-1}, t_i \rangle)) = \alpha \frac{P(\bar{t}_i \mid \bar{e}_i) \cdot P(\bar{t}_i \mid t_{i-2}, t_{i-1})}{P(\bar{t}_i)}.$$

For convenience in computation, we use the negative logarithm of the probability of the path; maximizing the probability is equivalent to minimizing the negative logarithm. This procedure replaces maximizing a product by minimizing a sum, which is a more standard formulation, and avoids the risk of underflow. Discarding the normalization factor α, which does not affect the choice of maximum, our task is now to maximize the expression

$$-\log_2\left(\Pi_{i=1}^{n}\frac{P(\bar{t}_i \mid \bar{e}_i) \cdot P(\bar{t}_i \mid \bar{t}_{i-2}, \bar{t}_{i-1})}{P(\bar{t}_i)}\right) = \sum_{i=1}^{n} -\log_2\left(\frac{P(\bar{t}_i \mid \bar{e}_i) \cdot P(\bar{t}_i \mid \bar{t}_{i-2}, \bar{t}_{i-1})}{P(\bar{t}_i)}\right). \quad (10.8)$$

10.3.3   The Viterbi Algorithm

In the analysis in Section 10.3.2, we have reduced the problem of finding the most probable sequence of tags to the following graph algorithm problem. Given

1. a directed multigraph M (i.e., a graph with multiple arcs between vertices), where each arc is labeled with an element and a numeric weight;

2. a particular starting vertex $S0$ in M;

3. a string W of N elements;

To find the path through M that traverses arcs labeled with the elements in S in sequence such that the sum of the weights is maximal.

Since there are, in general, exponentially many such paths, we might naively suppose that the problem is intractable. In fact, however, it is easily reduced to the well-known single-source shortest-paths graph problem. Thus, we construct a graph G with the following constraints:

- The vertices in G are all pairs $\langle V, I \rangle$, where V is a vertex of M and I is an integer from 1 to N, plus a final vertex END.

- There is an edge from $\langle U, I-1 \rangle$ to $\langle V, I \rangle$ in G only if there is an edge from U to V in M whose label is $S[I]$.

- There is an edge of weight 0 from every vertex $\langle V, N \rangle$ to END.

Clearly, any path through M starting in $S0$ that outputs W corresponds to a path through G from $\langle S0, 0 \rangle$ to END.

Because the graph G is "layered"—that is, all arcs go from a vertex in the Ith layer to one in the $(I+1)$th layer—the algorithm to solve this is particularly simple. The cost of the shortest path ending at a vertex V in layer $I+1$ is the minimum over all vertices U in layer I of the cost of the shortest path to U plus the cost of the arc from U to V. Details of the algorithm, known as Viterbi's algorithm, are given in Algorithm 10.1, and an example is given in Section 10.3.4. Some easy further optimizations can be carried out. In particular, as we generate states at the Ith level, we can check that they have an outarc labeled $W[I]$ and you put them on a list; that way, on the next iteration, we can just go through the list of states that satisfy the two tests, $E = W[I]$ and Total$[U, I-1] \neq \infty$, which may be much smaller than the set of all states. Also, once all the values of Total at level $I+1$ have been calculated, the values at level I can be overwritten, saving some space.

---

function Viterbi
    (**in** W : string of elements of length N;
        $M = \langle SS, AA \rangle$: directed multigraph.
            Each arc A in AA has the form $\langle$ tail, element, head $\rangle$;
        $C(A)$; numerical cost of arcs in AA;
        $S0$: starting state in M)
    return path of minimal total cost through M starting in $S0$ and outputting W;

/* The array Total(S, I) is the total cost of the shortest path through M from $S0$ to S
    traversing the first I elements of W.
The array Back(S, I) is the state preceding S on this path.
The graph G discussed in the text is left implicit, rather than constructed explicitly. */

{ Construct arrays Total(S, I) and Back(S, I) where $S \in SS$ and $I \in 0 \ldots N$;
  Initialize Total to ∞;
  Total$[S0, 0] = 0$;

  for $(I \leftarrow 1 \ldots N)$ {
      for (each arc $A = \langle U, E, V \rangle$) in AA {
        if $(E = W[I]$ and Total$[U, I-1] \neq \infty)$ {
          NewWeight $\leftarrow$ Total$[U, I-1] + C(A)$;
          if NewWeight < Total(V, I) {
            Total$(V, I) \leftarrow$ NewWeight;
            Back$(V, I) \leftarrow U$;
          } } } }

/* Find the best path by tracing back pointers from the best final state. */
  Best $\leftarrow$ the state S for which Total(S, N) is minimal;
  Path $\leftarrow \langle$Best$\rangle$;
  for $(I \leftarrow N \ldots 1)$ {
    Best $\leftarrow$ Back(Best, I);
    Path $\leftarrow$ Best, Path;
    }

return Path;
}

Algorithm 10.1. Viterbi algorithm.

10.3.4   Part of Speech Tagging

Consider the following toy example: Fish can swim fast. Note that:

- "Fish" and "swim" can be nouns or verbs (go fishing).

- "Can" can be a modal, a noun, or a verb (pack stuff into cans)

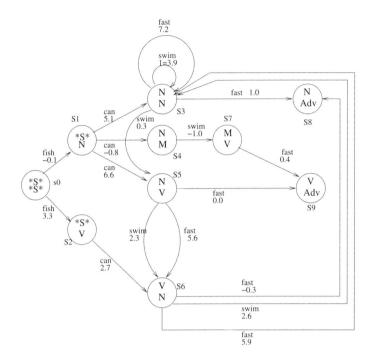

Figure 10.4. Markov model for part of speech tagging for "Fish can swim fast."

- "Fast" can be an adverb ("he works fast"), an adjective ("a fast horse"), a noun ("the fast of Yom Kippur"), or a verb ("Jews fast on Yom Kippur"). However, to simplify this example, we consider only the possibilities adverb and noun for "fast."

We use a trigram model here, and we suppose that we have the probabilities shown in Table 10.1. Only the probabilities relevant to the sample sentence are shown. Probabilities not shown are either zero or are irrelevant (i.e., they appear only in terms where they are multiplied by zero).

Figure 10.4 shows the Markov model corresponding to Table 10.1. The weights on an arc from state $\langle t_1, t_2 \rangle$ to $\langle t_2, t_3 \rangle$, labeled e, is equal to

$$-\log_2\left(\frac{P(\bar{e}_i \mid \bar{t}_i) \cdot P(\bar{t}_i \mid \bar{t}_{i+1-k}, \bar{t}_{i-1})}{P(\bar{t}_i)}\right).$$

For example, the arc labeled "can" from state S1 to S4 has weight

$$-\log_2(P(\text{M} \mid \text{"can"}) \cdot P(\text{M} \mid SN)/P(\text{M})) = -\log_2(0.9 \cdot 0.4/0.1) = -1.8480.$$

The numbers on the arcs in Figure 10.4 are negative logs of unnormalized probabilities, and thus their interpretation is a little abstruse. If we consider the

Relevant absolute tag probabilities:
$P(N) = 0.3$,
$P(V) = 0.2$,
$P(M) = 0.1$,
$P(Adv) = 0.1$.

Tag given element probabilities:
$P(N \mid \text{"fish"}) = 0.8$,
$P(V \mid \text{"fish"}) = 0.2$,
$P(M \mid \text{"can"}) = 0.9$,
$P(N \mid \text{"can"}) = 0.09$,
$P(V \mid \text{"can"}) = 0.01$,
$P(N \mid \text{"swim"}) = 0.2$,
$P(V \mid \text{"swim"}) = 0.8$,
$P(Adv \mid \text{"fast"}) = 0.5$,
$P(N \mid \text{"fast"}) = 0.02$.

Relevant transitional probabilities:
$P(N \mid *S**S*) = 0.4$,
$P(V \mid *S**S*) = 0.1$,
$P(M \mid *S*N) = 0.2$,
$P(V \mid *S*N) = 0.2$,
$P(N \mid *S*N) = 0.1$,
$P(N \mid *S*V) = 0.5$,
$P(N \mid MV) = 0.35$,
$P(Adv \mid MV) = 0.15$,
$P(V \mid NM) = 0.5$,
$P(M \mid NN) = 0.2$,
$P(V \mid NN) = 0.2$,
$P(N \mid NN) = 0.1$,
$P(Adv \mid NN) = 0.1$,
$P(N \mid NV) = 0.3$,
$P(Adv \mid NV) = 0.2$,
$P(N \mid VN) = 0.25$,
$P(Adv \mid VN) = 0.25$.

Table 10.1. Example of part of speech tagging for "Fish can swim fast."

At the start
Total(S0,0) = 0.

After 1 word

Total(S1,1) = Total(S0,0) + C(S0, "fish," S1) =	−0.1,	Back(S1,1) = S0
Total(S2,1) = Total(S0,0) + C(S0, "fish," S2) =	3.3,	Back(S2,1) = S0

After 2 words

Total(S3,2) = Total(S1,1) + C(S1, "can," S3) =	5.0,	Back(S3,2) = S1
Total(S4,2) = Total(S1,1) + C(S1, "can," S4) =	−0.9,	Back(S4,2) = S1
Total(S5,2) = Total(S1,1) + C(S1, "can," S5) =	6.5,	Back(S5,2) = S1
Total(S6,2) = Total(S2,1) + C(S2, "can," S6) =	6.0,	Back(S6,2) = S2

After 3 words

Total(S3,3) = min(Total(S3,2) + C(S3, "swim," S3), Total(S6,2) + C(S6, "swim," S3)) =	8.6,	Back(S3,3) = S6
Total(S5,3) = Total(S3,2) + C(S3, "swim," S5) =	5.3,	Back(S5,3) = S3
Total(S6,3) = Total(S5,2) + C(S5, "swim," S6) =	8.8,	Back(S6,3) = S5
Total(S7,3) = Total(S4,2) + C(S4, "swim," S7) =	−1.9,	Back(S7,3) = S4

After 4 words

Total(S3,4) = min(Total(S3,3) + C(S3, "fast," S3), Total(S6,3) + C(S6, "fast," S3)) =	14.7,	Back(S3,4) = S6
Total(S6,4) = Total(S5,3) + C(S5, fast," S6) =	10.9,	Back(S6,4) = S5
Total(S8,4) = min(Total(S3,3) + C(S3, "fast," S8), Total(S6,3) + C(S6, "fast," S8)) =	8.5,	Back(S3,4) = S6
Total(S9,4) = min(Total(S7,3) + C(S7, "fast," S9), Total(S5,3) + C(S5, "fast," S9)) =	−1.5,	Back(S9,4) = S7

Table 10.2. Execution of Viterbi's algorithm.

probability that a given arc will be traversed, given the starting state and the element, for two different arcs, then the ratio of those probabilities is 2^{L2-L1}, where $L1$ and $L2$ are the two labels. For example, the arc labeled "swim" from $S4$ to $S7$ is -1.0, and the arc labeled "can" from $S2$ to $S6$ is 2.7. Therefore,

$$\frac{P(T_3 = V \mid T_1 = N, T_2 = M, E_3 = \text{"swim"})}{P(T_2 = N \mid T_0 = *S*, T_1 = V, E_2 = \text{"can"})} = 2^{3.7} = 13.$$

The execution of Viterbi's algorithm proceeds as shown in Table 10.2. Infinite values are not shown. The result of Table 10.2 is that S9 is the winning state with a score of -1.5. Tracing backward from S9, the winning tagging is found to be $\langle$ N, M, V, Adv $\rangle$, as we would expect.

10.3.5 The Sparse Data Problem and Smoothing

To compute the product in Equation (10.8), we need three types of probabilities: $P(t_i \mid t_{i+1-k}, \ldots, t_{i-1})$, $P(t_i \mid e_i)$, and $P(t_i)$. This is where our training corpus comes in; we estimate each of these quantities by the frequency in the corpus:

$$P(t_i \mid t_{i-2}, t_{i-1}) = \text{Freq}_C(t_i \mid t_{i-2}, t_{i-1}),$$
$$P(t_i \mid e_i) = \text{Freq}_C(t_i \mid e_i),$$
$$P(t_i) = \text{Freq}_C(t_i),$$

where $\text{Freq}_C(\cdot)$ means the frequency in the training corpus C.

Unfortunately, there is often a problem with the first of these quantities in particular; namely, that some possible k-tuples of tags $t_{i+1-k}, \ldots, t_i$ may actually never occur in the corpus C. This is unlikely in the case where the tags are parts of speech, but is altogether likely in the case where the tags are words. (As discussed in Section 9.6.3, the frequency of trigrams in text follows a Zipf distribution with a very long tail.)

We have seen this kind of problem before, with naive Bayes classifiers (Section 8.11), and the solution we use here is a similar kind of smoothing. We estimate the probability of the k-gram as a weighted sum of the unigram, the bigram, the trigram, all the way to the $(k-1)$ gram. For example, with $k = 3$, we estimate

$$P(\bar{t}_i \mid \bar{t}_{i-2}, \bar{t}_{i-1}) = \alpha_1 \text{Freq}_C(\bar{t}_i \mid \bar{t}_{i-2}, \bar{t}_{i-1}) + \alpha_2 \text{Freq}_C(\bar{t}_i \mid \bar{t}_{i-1}) + \alpha_3 \text{Freq}_C(\bar{t}_i),$$

where $\alpha_1, \alpha_2, \alpha_3$ are weights (e.g., 0.6, 0.3, 0.1). For example, suppose that in speech understanding we encounter a sequence of phonemes that sounds like "burnt strawberry jelly." We do not want to reject this interpretation as having probability zero just because we have not previously encountered the trigram "burnt strawberry jelly" in our corpus. Rather, we estimate the probability of seeing "jelly" following "burnt strawberry" as 0.7 times the frequency with which "burnt strawberry" is followed by "jelly" in the corpus, plus 0.2 times the frequency with which "strawberry" is followed by "jelly" in the corpus, plus 0.1 times the frequency of "jelly" in the corpus. Thus, we give preference to trigrams that actually occur in the corpus, while admitting the possibility of seeing new trigrams and even new bigrams.

Exercises

Use MATLAB for all of these exercises.

Exercise 10.1. Consider the Markov model shown in Figure 10.5.

(a) What is the transition matrix for the model?

(b) Assume that at time 0 the probability of state A is 1. What is the probability distribution at time 1? At time 2? At time 10?

(c) What is the stationary probability distribution for the model?

Exercise 10.2. (This exercise is time consuming, but possibly fun.) Construct the Markov model for a player's position at the end of each roll of the dice on the standard Monopoly board, keeping in mind (a) that if you land on "Go to jail" you go to jail; (b) that if you land on Community Chest or Chance, with some probability you will get a card sending you somewhere else (e.g., nearest railroad, St. Charles Place, Go). Ignore the rule that you go to jail if you roll three doubles in a row, and assume that the player always pays to get out of jail. Use MATLAB to compute the stationary distribution for the model. Can you glean any useful strategic hints?

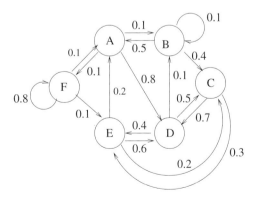

Figure 10.5. Markov model for Exercise 10.1.

Exercise 10.3.

(a) Redo the PageRank computation related to Figure 10.3 on page 306 by using $F = 0.1$, $E = 0.15$. How do the new results compare to the old ones? Give an intuitive explanation.

(b) Modify the graph in Figure 10.3 by adding arcs from U to W and from W to U. Compute the PageRank in the new graph, by using $F = 0.7$, $e = 0.05$. How do the results compare to the PageRank in the original graph? Give an intuitive explanation.

Exercise 10.4. Compute the absolute probabilities of all the possible taggings of the sentence "Fish can swim fast," by using the model in Section 10.3.4.

Problems

Problem 10.1. The *one-dimensional random walk with absorption* is the following Markov model:

- There are $2N + 1$ states, labeled $-N, \ldots, -1, 0, 1, \ldots, N$.

- If $K \neq -N, N$ then there are two transitions from K, one to $K - 1$ and one to $K + 1$, each with probability $1/2$.

- The only transition from N and from $-N$ is to itself, with probability 1.

Figure 10.6 illustrates this model with $N = 3$.

(a) There is more than one stationary distribution for the random walk. Characterize the space of stationary distributions. (You should be able to solve this problem without using MATLAB.)

(b) Generate the transition matrix for the case $N = 50$.

(c) Let $N = 50$. Assume that at time 0 the system is in state 0 with probability 1. Compute the distribution at time 50. Plot the distribution. What does it look like? Explain this outcome.

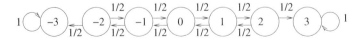

Figure 10.6. Random walk with $N = 3$.

(d) Let $N = 50$. Clearly, any random walker will either find his way to state 50 or to state -50 and then will stay there forever. We can define a function $f(I)$, which is the probability that a walker will end up in state 50 given that at some earlier time he is at state I. The values $f(-50), \ldots, f(0), \ldots,$ $f(50)$ satisfy a system of linear equations. Describe the system of linear equations and solve them, either by using MATLAB or by inspection.

Problem 10.2. Suppose that a Markov model has two different stationary distributions $\vec{p}$ and $\vec{q}$. Show that, for any t such that $0 < t < 1$, $t \cdot \vec{p} + (1 - t)\vec{q}$ is also a stationary distribution. Note that there are two things you have to show: first, that this is stationary, and second, that it is a legitimate distribution.

Problem 10.3. Construct an example of a Markov model that has only one stationary distribution, but where there are starting distributions that do not converge to the stationary distribution. (*Hint:* There is an example with only two states.)

Programming Assignments

Assignment 10.1.

(a) Write a function PageRank(A,E) that computes page rank based on links, as described in Section 10.2. The input parameter A is an $N \times N$ adjacency matrix; $A(I, J) = 1$ if page I has a link to page J. The input parameter E corresponds to the probabilistic parameter e as described.

Ignore any self-loops. Treat a page with no outlinks as if it had an outlink to every other page, as in Solution 3, Section 10.2.2.

(b) Choose some specific collection of interlinked documents, such as the Wikipedia pages for US presidents, or some collection of messages on a chat board. Construct the link graph between the documents, and compute the PageRank. Order the documents by PageRank. Does the ordering correspond well to your own sense of the relative importance of these documents?

Chapter 11

Confidence Intervals

11.1 The Basic Formula for Confidence Intervals

Suppose that we carry out a poll of 1,000 randomly chosen Americans, asking each of them whether they prefer apple pie or blueberry pie. The results are 574 prefer apple, and 426 prefer blueberry. What can we say about the fraction of the entire population that prefers apple pie? It would not be reasonable to conclude with certainty that exactly 57.4% of the population prefers apple. It seems safe, however, to conclude with near certainty that the true fraction is larger than 10% and smaller than 90%. What we would like to assert, clearly, is a statement of the form, "The fraction of the population that prefers apple pie is probably somewhere close to 57.4%," with some numbers attached to the vague terms "probably" and "somewhere close."

For example, "The probability is at least 95% that the fraction of people who prefer apple pie is between 54% and 60%" would be more descriptive. Symbolically, let f be the true fraction of the population and let $\bar{f} = 0.574$ be the fraction in the poll. Then the claim is $P(0.54 \leq f \leq 0.60 \,|\, \bar{f} = 0.574) \geq 0.95$. The interval $[0.54, 0.60]$ is called the *confidence interval for f at the 95% level.* (Sometimes it is called the confidence interval at the 5% level. Since no one is ever interested in intervals where the probability is less than $1/2$, this does not give rise to ambiguity.)

The actual calculation is quite simple, but the justification is complicated. In fact, the same calculation can be justified in two quite different ways, one corresponding to the likelihood theory of probability and one corresponding to the frequentist theory. Since 99% of the time, all you need is to do the calculation (and 99% of the people who do these calculations neither understand the justification nor care about it), we will begin here how to do the calculation, and we point out some important features of it. The justifications are discussed in optional Sections 11.3 and 11.4.

As long as the sample size n is large, the sample fraction $\bar{f}$ is not close to 0 or to 1, and we are not interested in probabilities *very* close to 1 (that is, probabilities such as $1 - 2^{-n}$), then we can consider that the true fraction f to follow a

Gaussian distribution mean $\bar{f}$ and standard deviation $\sigma = \sqrt{\bar{f}(1-\bar{f})/n}$. Specifically, let $E(x)$ be the integral of the standard Gaussian: $E(x) = \int_{-\infty}^{x} N_{0,1}(t)dt$. For any q that is substantially less than $\min(f, 1-f)/\sigma$, the probability that f lies within $q \cdot \sigma$ of $\bar{f}$ is

$$P(\bar{f} - q \cdot \sigma \le f \le \bar{f} + q \cdot \sigma) = \int_{\bar{f}-q\cdot\sigma}^{\bar{f}+q\cdot\sigma} N_{\bar{f},\sigma}(t)dt = \int_{-q}^{q} N_{0,1}(x)dx = 2E(q) - 1.$$

Therefore, to get a confidence interval at level p, look up $E^{-1}((p+1)/2)$ in a statistics table, or compute the MATLAB expression `erfinv(p)*sqrt(2)` (see Section 9.9), and that is the value of q. The confidence interval at level p is then $[\bar{f} - q \cdot \sigma, \bar{f} + q \cdot \sigma]$.

For example, suppose for the same American pie poll, we have $n = 1000$ and $\bar{f} = 0.574$, and we want to get a confidence interval at the 95% level. We compute $\sigma = \sqrt{\bar{f}(1-\bar{f})/n} = 0.0156$. We have $(1+p)/2 = 0.975$ and we can find out that $q = E^{-1}(0.975) = 1.96$. Therefore the 95% confidence interval is

$$[0.574 - 1.96 \cdot 0.0156, \ 0.574 + 1.96 \cdot 0.0156] = [0.5434, 0.6046].$$

This is the "$\pm 3\%$ margin of error" that is always reported together with political polls.

If we want to do a back-of-the envelope estimate, and we don't have the inverse Gaussian integral handy, we can do the following: first, $\sqrt{\bar{f}(1-\bar{f})}$ is usually about $1/2$ for most reasonable values of $\bar{f}$; if $\bar{f}$ is between 0.2 and 0.8, then $\sqrt{\bar{f}(1-\bar{f})}$ is between 0.4 and 0.5. So σ is generally $1/2\sqrt{n}$. Second, we really need to remember only three values of q:

- A 5% confidence interval is 2 standard deviations (q=2).

- A 1% confidence interval is 2.6 standard deviations (q=2.6).

- A 0.1% confidence interval is 3.3 standard deviations (q=3.3).

So as long as we know how to estimate $\sqrt{n}$, we're all set.

Four general points about sampling and confidence intervals should be kept in mind. First, the standard deviation σ is proportional to $1/\sqrt{n}$. Therefore, to reduce the width of the confidence interval by a factor of k, it is necessary to increase the sample size by k^2. That is the reason that every political poll has a sample size of about 1,000 and a 3% margin of error. Reducing the margin of error to 1% would require sampling 9,000 people, and would therefore cost nine times as much to carry out; it's not worthwhile, especially considering that this is a political poll, and the numbers will be all different next week anyway. (Self-selecting polls, such as Internet votes, can have much higher values of n, but gauging the accuracy of such a poll is a very different question.)

Second, the confidence interval does not at all depend on the size of the population, as long as the population is large enough that we can validly approximate sampling without replacement by sampling with replacement. (If we sample *with* replacement, then the same confidence intervals apply, even if the actual population is *smaller* than the sample; that is, if we ask everyone the same question several times.) A sample of size 1,000 has a ±3% confidence interval at the 95% level whether the population is ten thousand or ten billion.

Third, you are now in a position to be as annoyed as I always am when you hear a TV pundit, in one sentence, state responsibly that the poll he or she has quoted has a margin of error of 3%, and in the next sentence, speculate on the reason it went from 55% yesterday to 54% today. For a poll with $n = 1,000$ and $\bar{f} = 0.55$, the interval $[0.54, 0.56]$ has a confidence of 0.495. So there is a better than 50% chance that two polls of 1,000 people will differ by 1% in one direction or the other purely by chance, even if they are conducted in the same way and taken at the same time. There is no need to look for any other reason.

The fourth point is the most important, even though it runs the risk of sounding like the villainous lawyers for PG&E in *Erin Brockovich*. If we set our significance level at 99%, then, in one out of every 100 experiments, we will get results that pass the test just due to chance. If we are mining medical data looking for a cause of a disease, and we have records of 1,000 different possible influences, then we will probably find 10 that are significant at the 99% level and 1 that is significant at the 99.9% level, even if they are all actually independent of the disease. If we run a statistical test on our particular city and find that the incidence of bone cancer is above the national average with a confidence level of 99.9%, then we have reason to suspect that there may well be an environmental cause. But if someone in each of the 30,000 cities in the United States runs such a test, then we have to expect that 30 of these will come up positive even if there is no environmental cause anywhere. The moral is that we have to be very careful in drawing conclusions purely from statistical evidence.[1]

11.2   Application: Evaluating a Classifier

One of the most common uses of confidence intervals in computer science is to evaluate the quality of a classifier. As discussed in Section 8.11, a *classifier* is a program that determines whether a given instance belongs to a given category (e.g., whether an email message is spam). Most classifiers are constructed at least in part by applying supervised machine learning (ML) techniques to a corpus of labeled examples. Suppose that we have such a classifier; how do we measure how well it works?

---

[1]This kind of problem has recently become a serious concern in scientific research. See, for example, the article "Why Most Published Research Findings Are False" (Ioannidis, 2005).

First, we must decide on a measure of "goodness." For our discussion here, we take that to be the *overall accuracy* of the classifier—that is, the fraction of instances that are correctly classified, in either direction.

Using the standard approach, we begin by randomly separating the labeled corpus into a *training set R* and a *test set E*. We then use the training set R as input to the ML algorithm A, which outputs a classifier A_R. We run the classifier A_R on the test set E; determine the accuracy of A_R over E; and calculate a confidence interval $[L, U]$ at level p based on the size of E.

We can now make the following claim: Let Ω be the probability space of all instances. Assume that the corpus C is a random sample of Ω; this is a large assumption, which is generally difficult to justify but works out well enough. In that case, the test set and the training set are nearly independent[2] random samples of Ω, since they were chosen randomly from the corpus. Therefore, the event "instance I is in E" is independent of the event "A_R labels I correctly" relative to the probability distribution Ω. Therefore, the confidence interval we have calculated is valid.

Why should we not run the ML algorithm on the entire corpus, and then test it on the test set—or, for that matter, test it on the entire corpus? The answer is that, if $R \subset E$ or $R = E$, then A_R is no longer independent of E because A has used E in computing the classifier. In the extreme case, suppose that the classifier output by A_R consists of a pair: a general rule to be used on examples that are not in R; and a hash table with the answer recorded for all the instances in R.[3] In that case, if $E \subset R$, then the accuracy of A_R on E will be 1; however, this is no reason to think that this is a good estimate of the accuracy of A_R over Ω.

In practice, of course, machine learning algorithms generally do better the larger the training set, and it seems like a pity to throw away useful training data. Therefore, what is often done in practice is that, after the classifier has been evaluated, the learning algorithm is rerun on all the data, and one assumes that the new classifier, based on more data, is at least as good as the classifier that was evaluated. (For some learning algorithms, it is possible to prove that, on average, they do better given more data, and for some others, it is currently just a plausible assumption.)

One issue that requires care relates to the problem of multiple experiments discussed at the end of Section 11.1. In general, there are many different machine learning techniques that can be used for any given application, and each general technique can be used with a variety of options, parameter settings, feature spaces, and so on. Suppose that we divide the corpus into a training

---

[2] They are not quite independent because they are mutually exclusive.

[3] This may seem like an absurd cheat, but if the distribution over examples follows a power law, so that a large fraction of the instances encountered come from a small number of cases, it may be perfectly reasonable (see Section 9.6.3).

set R and a test set E of size 1,000; we run a hundred different machine learning techniques over R, we evaluate all the output classifiers using E, and we
get an accuracy for each one. Let us say that these range from 0.64 to 0.72.
We now, naturally, take the classifier that did best—call it CBest—and we publish a paper that claims that the accuracy of CBest has a confidence interval
[0.69,0.75] at the 95% confidence level. This claim, however, is not justified. It
is quite possible that the classifiers are all of *equal* quality; they each have a 0.68
chance of being correct on any given item. If the classifiers are independent of
one another,[4] then with 100 classifiers, we are likely to get one of accuracy 0.72
just by chance. (It is, however, proper to publish a paper that describes the entire procedure of running 100 classifiers, and in that context, we may certainly
say that CBest had an accuracy of 0.72.)

To get a proper evaluation of the classifier, we need a new test set. This
test set has to be independent of both sets R and E because both of these have
been used in formulating CBest—R to train it and E to select it. Another way
of viewing this is that the overall procedure, "Run these 100 ML techniques
on R; test them on E; choose the best," is itself a meta-ML procedure that
takes $R \cup E$ as input and generates the classifier CBest. E cannot be used to
evaluate CBest because it was part of the input to the program that generated
CBest.

Therefore, if a scientist anticipates that he or she will be testing multiple
classifiers and choosing the best, and if there is plenty of labeled data, then a
common practice is to divide the corpus of data into *three* parts: the training
set R, the test set E, and the validation set V. The scientist then trains each of
the machine learning algorithms on R; tests using V; chooses the best CBest;
and then reevaluates CBest by using the test set E. This evaluation of CBest is
valid. However, if this reevaluation does not produce satisfying results, it is not
permissible to then test the *second-best* on V and see whether it does better on
E and then publish that, if that seems better, because now the same test set has
again been used both for selection and for evaluation.

A very common problem, and very difficult problem in evaluating classifiers, is that labeled data are often limited and expensive to obtain. If we have
only 100 items of labeled data, then we cannot very well spare 30 of those for
testing purposes; in any case, a test set of size 30 has a 95% confidence interval
of ±18%, which is so large as to be useless. One technique that is often used in
this case is *cross-validation*, as shown in Algorithm 11.1. The data are divided
ten different ways into a training set and a validation set, the classifier is trained
on the training set and evaluated on the validation set, and the average accuracy is returned. However, it is difficult to analyze this technique statistically
or to determine valid confidence intervals for the result because the different
iterations are anything but independent.

---

[4]They almost certainly are not, but for the sake of argument, let's say they are.

```
function CrossValidate(in D: labeled data; M: classification algorithm)
                      return estimate of accuracy

{ Partition the data D into ten subsets S_1 ... S_10;
  for (i ← 1 ... 10)
      R ← D \ S_i;
      C ← M(R); /* Compute classifier */
      A_i ← accuracy of C on S_i;
  endfor
  return average of the A_i.
}
```

Algorithm 11.1. Cross-validation algorithm.

11.3 Bayesian Statistical Inference (Optional)

The derivation and interpretation of confidence intervals in the likelihood (Bayesian) interpretation of probability theory is quite straightforward. Let's return to the American pie poll described in Section 11.1. Let F be a random variable whose value is the fraction of the whole population that prefers apple pie. This is "random" in the sense of likelihood judgment; we don't know what the likelihood is, so there is some chance of it having any value between 0 and 1. If the population count is known to be M, then the value of F must actually have the form i/M, but for simplicity we approximate this as a continuous probability density. Let X be a random variable, where $X = \bar{f}$ is the event that, in a random sample of size n, the fraction that prefers apple pie is $\bar{f}$. What we want to know is the probability distribution of $\tilde{P}(F = t \mid X = 0.574)$ over all values of t.

As is so often done, we start with Bayes' law and write

$$\tilde{P}(F = t \mid X = 0.574) = P(X = 0.574 \mid F = t)\tilde{P}(F = t)/P(X = 0.574).$$

We know how to compute the first expression $P(X = 0.574 \mid F = t)$, this is given by the binomial distribution as $B_{1000,t}(574)$. The denominator is just a constant normalizing factor, independent of t. But what is $\tilde{P}(F = t)$?

$\tilde{P}(F = t)$ is called the *prior;* it expresses our preconceptions about the relative likelihood of different values of the fraction of the population that prefers apple pie *before we see the results of the poll*. In contrast, $\tilde{P}(F = t \mid X = 0.574)$, our judgment *after* we have seen the poll, is called the *posterior* probability. We first consider the simplest approach to the prior, which is, in fact, often used, and then we return to the question of how reasonable that is. The simplest solution is to assume that F is uniformly distributed, and thus $\tilde{P}(F = t)$ is a constant. In that case, $\tilde{P}(F = t \mid X = 0.574) = \alpha \cdot B_{1000,t}(574)$, where α is a normalizing constant. This is not actually equal to the binomial distribution, which would be $B_{1000,0.574}(1000t)$, but, like the binomial distribution, it is accurately approximated by a Gaussian distribution with mean $\mu = 0.574$ and standard deviation

$$\sigma = \sqrt{0.574 \cdot (1 - 0.574)/1000} = 0.0156.$$

So, as in our previous discussion, for any q, the probability that f is between $\bar{f} - q\sigma$ and $\bar{f} + q\sigma$ is given by $2E(q) - 1$.

The advantage of the Bayesian approach is that it answers the question that we want answered: namely, what, probably, is the fraction of the population that prefers apple pie? (As we see in Section 11.4, the sample space approach refuses to answer this question.)

The weak point of the Bayesian approach is the arbitrary choice of prior. Why should we assume that F is uniformly distributed? One answer that is often given is that once you have enough evidence, the priors don't matter. For certain kinds of priors, that is demonstrably true. For example, suppose that, rather than apple pie, the poll is measuring approval ratings for the president. One might think that positing a uniform distribution on approval ratings is ignoring history. We know that approval ratings for the president are generally somewhere between 35% and 65% and almost never lower than 20% or higher than 80%. Perhaps we should use a prior that reflects that knowledge; for example, set $\bar{P}(F = t)$ to be 1.5 for t between 0.2 and 0.8 and 0.25 for $t < 0.2$ and for $t > 0.8$. But when we have done all the computation, we find that the change in the confidence interval is infinitesimal because the posterior probability is tiny that the approval rating is outside the range $[0.2, 0.8]$. (On the other hand, if our poll does return a result of 20%, then this choice of priors will make a difference.)

But if we instead use a radically different model for the prior, then the choice of prior may make a difference. Returning to the pie example, suppose we choose the following probabilistic model as a prior. We define a Boolean random variable X_i that expresses the pie preferences of person i; further, we assume that each X_i follows a Bernoulli distribution with its own parameter p_i, the X_i's are all independent, and the p_is are themselves uniformly distributed over $[0, 1]$. In this model, the prior on F is Gaussian, with norm 0.5 and standard deviation $3.5 \cdot 10^{-5}$. We will need a vast poll to get a posterior probability that is not very close to 0.5. To put it more simply, in this model, when we poll our 1,000 subjects, we get information about their personal values of p_i, but we get no information about anyone else's value of p_i because our model says that these are all independent. We need to poll essentially everyone to get an accurate view of F.

Now, one may say that on that model, the history of opinion polls in this country is inexpressibly unlikely, which is true; and based on that fact, a sensible statistician will not choose that prior. However, from a logical standpoint

this is not altogether satisfying. After all, all the other polls are also just posterior information; according to a pure Bayesian theory, the proper way to consider them is to combine them with the priors that existed before ever seeing
any polls. Simply *rejecting* a probabilistic model that makes the posterior data
unlikely is essentially the hypothesis testing approach (see Section 11.5), which
is generally frowned on in Bayesian theories.

Moreover, in some cases, we actually *do* want to use a prior that is something like this last example. For instance, consider again the problem of determining whether our municipality has unusually high rates of cancer. In that
case, a reasonable prior might be that, in each municipality, there is some underlying probability of getting cancer, and that all these probabilities are uniformly distributed and independent from one municipality to the next.

11.4   Confidence Intervals in the Frequentist Viewpoint (Optional)

The interpretation of confidence intervals in the frequentist (sample space)
view of probability is much more roundabout. In this viewpoint, the statement,
"Based on this poll, the probability is .95 that between 54% and 60% of people
prefer apple pie" is entirely meaningless, since it does not refer to any sample
space. Instead, the meaning of confidence intervals is a statement about the
procedure used to compute them.

Definition 11.1. The confidence interval at level p for a sample of size n is
given by a pair of monotonically nondecreasing functions, $L_n(\bar{f})$ and $U_n(\bar{f})$,
with the following property. Consider any property Φ whose true frequency
in the space Ω is f. Let X be the random variable whose value is the number of elements with property Φ in a random sample of size n from Ω. Then
$P(L_n(X) \le f \le U_n(X)) \ge p$.

That is, if we take a random sample of size n, find the fraction $\bar{f}$ of elements
in the sample that are Φ, and compute the values $L_n(\bar{f})$ and $U_n(\bar{f})$, then, with
probability at least p, $L_n(\bar{f})$ will be less than the true frequency f and $U_n(\bar{f})$
will be greater than f.

The value of any particular confidence interval is just the application of
this procedure to $\bar{f}$; for example, when $\bar{f} = 0.574$, then $L_{1000}(\bar{f}) = 0.54$ and
$U_{1000}(\bar{f}) = 0.60$.

Here the sample space Ω is the set of random samples of size n. Note that
this does not make any claim whatsoever about this particular poll, or even
about polls where $\bar{f} = 0.574$. It could be that all the poll questions we ever
ask are actually lopsided by at least 80% to 20% in the population, so that the
only times we get $\bar{f} = 0.574$ are when we have, by incredibly bad luck, chosen a

wildly unrepresentative sample. In that case, whenever we compute the confidence interval for $\bar{f} = 0.574$, it *never* includes the true value of f, which is either less than 0.2 or greater than 0.8. That would not at all affect the correctness of the claim being made about the confidence interval. This claim does not say that the confidence interval for $\bar{f} = 0.574$ usually—or ever—contains the true value. What it says is that—whatever poll is being carried out—for *most* random samples, this *procedure* will give valid bounds on the true fraction. If $f = 0.2$ then we will only very rarely get the value $\bar{f} = 0.574$.

The condition that $L(\bar{f})$ and $U(\bar{f})$ are monotonically nondecreasing is needed for the following reason. Suppose that we have come up with procedures L and U that satisfy Definition 11.1. Now, suppose some troublemaker defines procedures L' and U' as follows:

$$\text{if } \bar{f} \neq 0.574, \quad \text{then} \quad L'(\bar{f}) = L(\bar{f}) \quad \text{and} \quad U'(\bar{f}) = U(\bar{f}),$$
$$\text{if } \bar{f} = 0.574, \quad \text{then} \quad L'(\bar{f}) = 0.1 \quad \text{and} \quad U'(\bar{f}) = 0.2.$$

Then L' and U' satisfy the condition $P(L'(X) \leq f \leq U'(X)) \geq p'$, where p' is very slightly less than p, because the odds of getting a sample fraction of *exactly* 0.574 is small even for the case $f = 0.574$. The condition that L and U are monotonically increasing excludes this kind of shenanigans.

We can go about constructing a confidence interval procedure as follows. Suppose we can find monotonically increasing functions $Q(f)$, $R(f)$ such that $P(Q(f) \leq X \leq R(f)) \geq p$ for every value of f. This is not quite what we are looking for because the roles of f and X are reversed. Now, we let $U(X) = Q^{-1}(X)$ and $L(X) = R^{-1}(X)$. Since Q is monotonically increasing, $f \leq U(X)$ if and only if $Q(f) \leq Q(Q^{-1}(X)) = X$; and likewise, $f \geq L(X)$ if and only if $R(f) \geq X$. Therefore,

$$P(L(X) \leq f \leq U(X)) = P(R^{-1}(X) \leq f \leq Q^{-1}(X)) = P(Q(f) \leq X \leq R(f)) \geq p,$$

which is what we wanted.

How can we find these functions Q and R? There is a range of acceptable possibilities, but the simplest thing is to choose $Q(t)$ such that, with probability $(1-p)/2$, X is in the interval $[0, Q(f)]$, and with probability $(1-p)/2$, X is in the interval $[R(f), 1]$. Thus, with probability p, X is in neither of those intervals, and thus X is in the interval $[Q(f), R(f)]$. Now, nX is distributed according to the binomial distribution $B_{n,f}$. Let $C_{n,f}$ be the cumulative distribution corresponding to the binomial distribution. Then we have

$$(1-p)/2 = P(0 \leq X \leq Q(f)) = C_{n,f}(n \cdot Q(f)),$$

so

$$Q(f) = (1/n) \cdot C_{n,f}^{-1}((1-p)/2);$$

and

$$(1-p)/2 = P(R(f) \leq X \leq 1) = 1 - P(0 \leq X \leq R(f)) = 1 - C_{n,f}(n \cdot R(f)),$$

so

$$R(f) = (1/n) \cdot C_{n,f}^{-1}(1 - ((1-p)/2)) = (1/n) \cdot C_{n,f}^{-1}((1+p)/2).$$

If n is large, f is not close to either 0 or 1, and p is not very close to 1, then we can make this task much simpler by using the Gaussian distribution. Specifically, let $\sigma = \sqrt{f(1-f)/n}$, which is the standard deviation of $\bar{f}$. Let $N_{f,\sigma}(t)$ be the Gaussian with mean f and standard deviation σ, and let $E_{f,\sigma}(x)$ be the corresponding cumulative distribution. By the central limit theorem, the cumulative distribution of the Gaussian approximates the cumulative distribution of the binomial, so for any k, as long as k is substantially less than $\min(f/\sigma, (1-f)/\sigma)$, we have

$$p = P(f - k\sigma \leq X \leq f + k\sigma) \approx E_{f,\sigma}(f+k\sigma) - E_{f,\sigma}(f-k\sigma) = E_{0,1}(k) - E_{0,1}(-k) = 2E_{0,1}(k) - 1.$$

so $k = E_{0,1}^{-1}(1+p)/2$. Using that value of f, we have $Q(f) = f - k\sigma$ and $R(f) = f + k\sigma$. Finally, if n is large, so that $k\sigma \ll f$, then Q and R are very nearly inverses of one another, and we can approximate σ as $\sigma' = \sqrt{\bar{f}(1-\bar{f})/n}$, so $L(X) = R^{-1}(X) \approx Q(X) = f - k\sigma'$ and $U(X) = f + k\sigma'$. The method for computing the confidence interval is exactly the same as in the Bayesian approach,[5] even though the interpretation is very different.

The advantage of the frequentist approach is that we no longer have to make any assumptions about the prior distribution of f. The answer to the frequentist question is valid whatever the prior distribution; it is valid even if, as many frequentists would claim, the idea of the prior distribution of f is meaningless. The disadvantage of this approach is that it no longer answers the question in which we are interested: "What is the actual fraction of people who prefer apple pie?"; it makes a claim about a certain statistical procedure. It is a consequence that if we follow the procedure and treat the confidence interval *as if it were* an actual statement about f, then most of the time we will be all right; but that seems like a very indirect, not to say obscure, approach.

More important, for many purposes, such as computing an expected value or to calculate another probability, it is convenient to have a probability distribution over f. Now, it may be possible to rephrase these purposes as well in the probability distribution over the sample rather than over f itself, but this is always indirect, and generally difficult and problematic.

---

[5]The Bayesian and the frequentist confident intervals are very nearly the same for large values of n and for $\bar{f}$ not close to 0 or 1; for small values of n or extreme probabilities, they can be quite different.

11.5 Hypothesis Testing and Statistical Significance

A *hypothesis testing method* is a method that uses statistical characteristics of a body of data to accept or reject a particular hypothesis about the processes that generated that data.

In this section, we discuss the traditional statistical approach to hypothesis testing, which is commonly used in statistical data analysis. Like the frequentist definition of confidence intervals, this approaches the issue from a rather backward point of view. In this approach, the objective of a successful (e.g., publishable) statistical analysis is formulated as *rejecting* a hypothesis, known as the *null hypothesis*, about the process that generated the data. Thus, the null hypothesis is the hypothesis that we are rooting against, so to speak. To decide whether to reject the null hypothesis, the statistician uses a *statistical test*, which is a Boolean function computed from the data. The hypothesis can be rejected if the outcome of the test would be very unlikely if the hypothesis were true.

Some examples of datasets, hypotheses, and tests follow.

Example 11.2. The pie preferences of Americans are being tested. The dataset consists of the results of the poll. What we want to prove is that most Americans prefer apple pie to blueberry pie. The null hypothesis is that the dataset is the outcome of a uniform distribution over all samples of 1,000 Americans, and that the fraction of Americans who prefer apple pie is at most 0.5. The statistical test is the proposition, "$0.53 \leq \text{Freq}_D(\text{Apple})$," or at least 53% of the people sampled prefer apple pie.

Example 11.3. A drug is being tested for its effectiveness against a disease. Thirty patients who have the disease are given the drug, and 30 are given a placebo. The dataset is the record of which patients had recovered by the end of the week. The hypothesis we want to prove is that the drug has some effectiveness. The null hypothesis is, therefore, that it is at best useless. Specifically, the hypothesis posts that among patients who receive the drug, recovery is a Bernoulli process with parameter a; that among the patients who receive the placebo, recovery is a Bernoulli process with parameter b; and that $a > b$, which we can write as

$$P(\text{Recover} \mid \text{Drug}) > P(\text{Recover} \mid \text{Placebo}).$$

The test T is the Boolean function

$$\text{Freq}_D(\text{Recover} \mid \text{Drug}) - \text{Freq}_D(\text{Recover} \mid \text{Placebo}) > 0.2.$$

Example 11.4. We wish to show that playing Mozart to fetuses in the womb has some effect (positive or negative) on their IQs. Dataset D records how many

hours of Mozart the fetus was exposed to and what the child's IQ was at the age
of 6 months. The null hypothesis is that the two variables are the outcome
of independent random variables. The statistical test is that the correlation
between hours of Mozart and IQ score is greater than 0.2.

A test T gives good evidence for rejecting hypothesis H if positive outcomes
for T are unlikely for any process Q satisfying H. Formally, we posit the following definition.

Definition 11.5. Let H, the *null hypothesis*, be a set of random processes, each
of which generates values in a set Ω. For each process $Q \in H$, let P_Q be the
probability distribution associated with the outcome of Q. Let T, the *statistical
test*, be a subset of Ω (the set of samples for which the test succeeds). Let D,
the *dataset*, be an element of Ω. Then *H is rejected at significance level s, where
$s > 0$, using T as applied to D* if $D \in T$ and $P_Q(T) \leq s$ for all $Q \in H$.

For instance, in Example 11.2, Ω is the set of all samples of size 1,000 taken
from all Americans. For any sample $s \in \Omega$, let $A(s)$ be the fraction that prefer
apple pie. The null hypothesis H is the set of all binomial distributions $B_{p,1000}$,
where $p \leq 0.5$. The test $T = \{s \in \Omega \mid A(s) > 0.53\}$. The dataset D is the particular
sample; note that $A(D) = 0.574$, so $D \in T$. Clearly, over the processes $Q \in H$,
$P_Q(T)$ reaches its maximum when $p = 0.5$, and for that particular Q, $P_Q(T) =
0.0275$. Therefore, the null hypothesis can be rejected at the 2.75% level.

There is an obvious problem with Definition 11.5. Nothing in the definition prevents us from choosing T so $P_Q(T)$ is unlikely, whether Q satisfies the
hypothesis or not. For instance, in the pie test, we could choose as our test
$T = \{s \in \Omega \mid A(s) = 0.574\}$; again, $D \in T$. This test on exact equality has a probability of no more than about 1/30 whatever the actual fraction, and therefore it
can be used to reject any hypothesis at all at the 0.03 significance level.

The problem is that there are two kinds of errors that can be made in hypothesis testing. The first is to reject the null hypothesis when it is in fact true;
this is known as a *type 1* error. The second is to fail to reject it when it is in fact
false; this is known as a *type 2* error. In general, in choosing a test for a particular hypothesis, there is a trade-off between the two types of errors: to make
the probability of a type 1 error small, we must accept a large probability of a
type 2 error, and vice versa. Definition 11.5 ensures that it is unlikely that we
will make a type 1 error, but it says nothing about type 2 errors.

There is a reason for this asymmetry; guarding against type 2 errors is problematic. We need to establish that if the null hypothesis is false, then it is unlikely that the test will succeed. But we cannot speak of the probability of the
event T on the assumption that the sample was *not* generated by process Q.
We could require that the probability of T is small, given any process Q outside
H, but in general, the set of processes outside H is unmanageable and useless.

In Example 11.2, for instance, the process Q that generates D with probability 1 is certainly outside of H, since it is not a Bernoulli process of any kind, and $P(T \mid Q) = 1$.

This problem can be solved as follows:

1. Consider only a fixed space of hypotheses Ω. For instance, assume that the data were generated by a binomial distribution $B_{n,q}$ for some value of q.

2. Let the *alternative hypothesis* $H_a = \Omega \setminus H$. Define the *power of test T over H_a* as the maximum over all tests $Q \in H_a$ of $P(\neg T \mid Q)$, the probability that a type 2 failure will not occur, given that the data are generated by Q.

3. Consider only tests of a particular form; generally either $f(D) \geq l$ (a *one-sided* test) or $l \leq f(D) \leq u$ (a *two-sided test*) for constants l and u. (The function $f(D)$ is called a *statistic* of the data D.)

4. Require that the test T being applied satisfy the following: Of all tests T' of the form in (3) that reject the hypothesis H at level α, T should be the test that achieves the maximum power over H_a.

Thus, the test T is very unlikely to be satisfied if the data were generated by a process Q satisfying the null hypothesis, but it is not inherently unlikely, in the sense that it is likely for at least some processes satisfying the alternative hypothesis.

In the apple pie scenario (Example 11.2), if H is the set of all binomial distributions $B_{n,p}$ such that $p \leq 0.5$, then H_a is the set of all $B_{n,p}$ such that $p > 0.5$. Consider again the defective test T, "Reject H on sample s if and only if $A(s) = 0.574$"; thus, $D \in T$. Over Q in H, the maximum likelihood of $P(T \mid Q)$ is achieved when $p = 0.5$ at

$$B_{1000,0.5}(0.574) \approx N_{500,15.811}(574)/15.811 = N_{0,1}(4.6802)/15.811 = 4.42 \cdot 10^{-7},$$

so a type 1 error is extremely unlikely. However, the maximum likelihood of $P(T \mid Q)$ for $Q \in H_a$ is achieved when $p = 0.574$, and is only $B_{1000,0.574}(574) \approx N_{0,1}(0)/15.637 = 0.0155$.

By contrast, consider the test T', "Reject H on sample s if $A(s) \geq 0.574$." Over Q in H, the maximum likelihood of $P(T' \mid Q)$ is likewise achieved when $p = 0.5$; its value is only very slightly larger than $P(T \mid Q)$. However, the maximum likelihood of $P(T' \mid Q)$ is achieved with $p = 1$ and is equal to 1.

Note that the restriction (1) above was implicitly imposed in the Bayesian analysis of confidence intervals in Section 11.3. We considered a prior probability distribution over the fraction f; we did not assign any probability to the possibility that the sample was not generated by random selection.

11.6   Statistical Inference and ESP

The conflict between the frequentist and the likelihood interpretations of probability appeared in a front-page *New York Times* article on January 5, 2011. The circumstances concerned a number of experiments performed by Prof. Daryl Bem to detect precognition. For example, in one experiment, first subjects were presented with a list of 48 words to study; second, they were given a memory test on these words; and third, they were given a subset of 24 of the words and asked to categorize them. (In general, people remember a list of words better if they have gone through the work of categorizing them than if they have just tried to memorize them as a list.) In Bem's experiments, he found that subjects did better on the memory test on words that they spent time categorizing *after* the test was complete. Using a standard statistical measure, he determined that the null hypothesis—that the difference was entirely due to chance—could be rejected at a high level of significance.

A number of critics (e.g., Rouder and Morey, 2011) argued that, from a Bayesian point of view, it would be necessary to consider the prior probability of the hypothesis that the subjects were benefiting from extrasensory perception (ESP), which, they stated, was very small. On this view, even though some of Bem's experiments substantially raised the posterior estimate of the reality of ESP, they did not raise it to the point where reasonable people would actually believe it. The critics did not, however, give any guidance on the difficult question of how one could reasonably estimate the prior probability that ESP was possible: 10^{-8}, 10^{-50}, $10^{-10^{50}}$?

Exercises

Exercise 11.1. Using pencil and paper but no computer, estimate confidence intervals for polls over samples of sizes $N = 400$, $N = 10,000$, $N = 1,000,000$; $\bar{f} = 0.2$ and $\bar{f} = 0.5$; and confidence levels of 0.99 and 0.999. (Consider all combinations; thus your answer should have 12 parts.)

Exercise 11.2. Using MATLAB, compute confidence intervals for polls over samples of sizes $N = 750$, $N = 3,500$, $N = 1,250,000$; $\bar{f} = 0.28$ and $\bar{f} = 0.65$; and confidence levels of 0.86 and $1 - 10^{-5}$. (As in Exercise 11.2, your answer should have 12 parts.)

Problems

Problem 11.1. As mentioned on page 323, the simple formula for confidence interval given in Section 11.1 breaks down if N is small, $\bar{f}$ is close to 1, or a very high degree of confidence is required.

(a) Consider the problem of computing a 80% confidence interval, under the frequentist definition, for the situation in which you poll eight people and seven respond that they prefer apple pie. Compute a confidence interval (there is more than one possible answer). Use the exact discrete distribution; do not approximate it by a continuous one.

(b) (This part is difficult; it can be done by using MATLAB, but it requires some care to avoid running into overflow problems.) Compute a 99% confidence interval for the situation in which you poll 1,000 people and 995 prefer apple pie.

Problem 11.2. Repeat Problem 11.1, but use instead the Bayesian interpretation of confidence intervals, assuming a prior that is a uniform distribution over the true fraction f. Again, part (b) is difficult.

Problem 11.3. The Bayesian treatment of hypothesis testing is conceptually simpler than the frequentist idea of statistical significance; one simply computes the posterior probability that the hypothesis is true.

Suppose you are flipping a coin. You are considering three hypotheses:

(a) The coin is weighted so it comes up heads 1/4 of the time and tails 3/4 of the time.

(b) The coin is fair.

(c) The coin is weighted so it comes up heads 3/4 of the time and tails 1/4 of the time.

Assume that the prior probability of each of these hypotheses is 1/3. You now flip the coin ten times and get seven heads. What are the probabilities of the three hypotheses given the results of the coin flips?

Problem 11.4. Starting with the same priors as in Problem 11.3, suppose you flip the coin 100 times and get 70 heads. Estimate the posterior probabilities of the three hypotheses.

Problem 11.5. Suppose that, as in Problem 11.3, you are flipping a coin. Let X be the bias of the coin (e.g., if $X = 1/4$, then the coin comes up heads 1/4 of the time). Suppose that your prior on X is the uniform distribution between 0 and 1. You now flip the coin ten times and it comes up heads seven times; call these data D. What is the posterior probability density of X; that is, $\bar{P}(X|D)$? Use MATLAB to plot this curve.

Chapter 12

Monte Carlo Methods

Broadly speaking, a Monte Carlo method is one that uses random choices to solve a problem. In this chapter, we look at a particular class of methods that work by generating a large number of values within some probability space and doing a simple calculation over them. In Section 12.9, we sketch some further kinds of probabilistic algorithms and applications.

12.1 Finding Area

Suppose that you want to find the area of the region defined by the inequalities $0 \leq x \leq 10; 0 \leq y \leq 10; -0.1 \leq x \cdot \sin(\pi x) \cdot y \cdot \sin(\pi y) \leq 0.1$. It is not possible to compute this as a closed form analytic expression, and it is difficult to use deterministic methods of numerical integration.

There is, however, a very simple Monte Carlo method. Let R be the region satisfying the given inequalities, and let Q be the square $[0, 10] \times [0, 10]$. For some large N, pick N points uniformly over Q; that is, choose x- and y-coordinates each uniformly distributed between 0 and 10. For each such point, calculate whether the point is in Q by checking whether the inequality $-0.1 \leq x \cdot \sin(\pi x) \cdot y \cdot \sin(\pi y) \leq 0.1$ is satisfied. Let K be the number of sample points in R, and let $p = K/N$. Since $R \subset Q$, the probability that any given sample point x is in R is just Area(R)/Area(Q). Therefore, K follows the binomial distribution $B_{n,p}$, where $p = $ Area(R)/Area(Q) = Area(R)/100. We can therefore estimate Area(R) = $100p$ with a standard deviation of $50\sqrt{p(1-p)/N}$. In an experiment with $N = 10,000$, there were 566 points that satisfied the constraint, so the area can be estimated at 5.66, with a 95% confidence interval $[5.22, 6.10]$.

As another example, Figure 12.1 shows the case where R is the unit circle and Q is the square $[-1, 1] \times [-1, 1]$. Of 100 points chosen at random in Q, 78 are inside the circle, so the estimated area is Area(Q) $\cdot 78/100 = 3.12$, with the 95% confidence interval $[2.64, 3.44]$. Since the true answer is $\pi = 3.14$, we were luckier in this instance than we deserved to be. An experiment with $N = 10,000$ gives a value of 3.13.

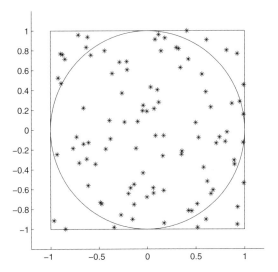

Figure 12.1. Monte Carlo estimate of the area of a circle.

Why not pick points systematically rather than randomly? Actually, for estimating the area of a two-dimensional region Q, systematically checking grid points *is*, in fact, a reasonable alternative to Monte Carlo methods. Pick a value of $\epsilon > 0$; construct a grid of $\epsilon \times \epsilon$ within the square Q (note that this has $N = \text{Area}(Q)/\epsilon^2$ points); count the number K of grid points in R; and estimate $\text{Area}(R)$ by $K\epsilon^2$. It is easily shown that, for any fixed R, the maximal error in this approach is $\Theta(\epsilon) = \Theta(1/\sqrt{N})$—thus, comparable to the error in Monte Carlo search—and the result is guaranteed, rather than merely probabilistic.

There are two drawbacks, however. The first is that the result is asymptotic and only order of magnitude. The error in the estimate is, in general, proportional to the circumference of R, so unless we can bound that in some way, there is no way to know how large to choose N to get a given accuracy. If R has a sufficiently convoluted shape, it may be necessary to choose a very large value of N to get an accurate answer. Furthermore, if we don't know anything about R, then there is the fear that our systematic method of picking points may be related to the shape of R in a way that throws off the estimate.[1] For instance in the (contrived) example at the beginning of this chapter, if we choose as grid the points with integer coefficients in 1, …, 10, all of them satisfy the inequality, so we would estimate $\text{Area}(R)$ at 100. With a Monte Carlo search, by contrast, the accuracy of the estimate is given by the standard confidence

---

[1] Unless R is a pathological set, such as the set of all points with rational coordinates—technically, if R is bounded and either topologically open or topologically closed—then it is always the case that any *sufficiently* fine uniform grid will give an accurate estimate.

interval, and depends on N and Area(R), but not on any other characteristic of R.

The second drawback is that using grid points is only competitive in two dimensions. In k-dimensions, the error in the grid estimate is still $O(1/\epsilon)$, but this is now $N^{-1/k}$, whereas the error in the Monte Carlo estimate is $O(N^{-1/2})$ regardless of the dimension.[2] In large number of dimensions, the grid method is hopeless; for example, in 100-dimensional space, a $2 \times 2 \times \ldots \times 2$ grid has 2^{100} points. But the Monte Carlo method doesn't care; it has exactly the same accuracy even if the dimensionality of the space is much greater than the number of sample points.

There is a catch, however: we need to be able to find an enveloping region Q that satisfies the following conditions:

- we can compute Area(Q) accurately;

- we can generate points uniformly within Q;

- $R \subset Q$;

- Area(Q)/Area(R) is not very large.

In large dimensional space, this can be difficult, even if R is a quite reasonable shape. For example, suppose that R is the 100-dimensional sphere, and we try to estimate its volume by using the cube $[-1,1]^{100}$ as Q. It worked well in two dimensions, but in 100 dimensions, the ratio Area(R)/Area(Q) = $\pi^{50}/(2^{100} \cdot 50!) \approx 10^{-68}$. So we would have to generate 10^{68} points at random in the cube before finding a single one that fell in the sphere. If we want to do Monte Carlo simulation in a large dimensional space, therefore, it is necessary to have techniques for finding regions that fit the target closely and generating points uniformly in these.

12.2   Generating Distributions

In Monte Carlo search, it is often important to generate a *nonuniform* distribution. The built-in MATLAB rand function, however, is designed to generate a uniform distribution over [0,1]. It is therefore often necessary to turn this uniform distribution into a nonuniform distribution. To generate a uniform distribution over the finite set $\{0 \ldots n-1\}$, use floor(n*rand).

To generate a finite, nonuniform distribution with probabilities $p_1, \ldots, p_k$, partition the unit interval into subintervals of length $p_1, \ldots, p_k$. Let $b_0 = 0$; $b_1 = p_1$; $b_2 = p_1 + p_2$; $b_3 = p_1 + p_2 + p_3$; $\ldots$; $b_k = 1$. If random variable U is uniformly distributed between 0 and 1, then the event $b_{i-1} \leq U \leq b_i$ has probability p_i. Thus, to generate this distribution, we precompute the values $b_0, \ldots, b_k$;

---

[2]In *one* dimension, the grid estimate, with an error of $O(1/N)$, beats Monte Carlo.

we execute rand; and find the subinterval that contains the value returned by rand.

To turn a uniform distribution into a continuous, nonuniform distribution, we use the inverse of the cumulative probability. (The finite case described in the previous paragraph is actually a special case of this.) Suppose that X is a random variable with cumulative probability function $F(t)$; that is $P(X \le t) = F(t)$. Let $u = F(t)$; then $u = F(t) = P(X \le t) = P(X \le F^{-1}(u)) = P(F(X) \le u)$. Thus, $F(X)$ is uniformly distributed; so $X = F^{-1}(U)$, where U is uniformly distributed.

For example, suppose we want to generate points according to the density function $f(t) = e^{-t}, t \ge 0$. The cumulative probability function is $\int_0^x e^{-t}dt = 1 - e^{-x}$. If $y = 1 - e^{-x}$, then $x = -\ln(1 - y)$. So we generate values of a random variable Y uniformly between 0 and 1, compute $X = -\ln(1 - Y)$, and then the pdf is $\tilde{P}(X = t) = e^{-t}$.

Likewise, if we want to generate points with the distribution $f(t) = 1/t^2$, $t \ge 1$, then the cumulative function is $1 - 1/x$. The inverse is the function $x = 1/(1 - y)$. So, if Y has a uniform distribution from 0 to 1, then $X = 1/(1 - Y)$ has the pdf $\tilde{P}(X = t) = 1/t^2$.

Computing the inverse of the cumulative function for the Gaussian is not so simple. Luckily, MATLAB has a built-in function randn, which generates random variables according to the standard normal distribution $N_{0,1}$.

12.3   Counting

Essentially the same technique that we have used for estimating area can be used to estimate the size of a finite set R. To estimate R, find a set Q such that

- $R \subset Q$;

- $|Q|$ is easily computed;

- it is easy to generate random elements uniformly in Q;

- $|R|/|Q|$ is not very small.

In this case, we generate a sample of N points uniformly in Q; let K be the number of points in the sample that lie in R; let $p = K/N$; and estimate $|R|$ as $p|Q|$, with a standard deviation of $\sqrt{p(1-p)/N}$, where $p = K/N$.

For instance, suppose we want to estimate the number of integer solutions $\langle x, y, z \rangle$ to the constraints

$$0 \le x \le W, \quad 0 \le y \le W, \quad 0 \le z \le W,$$
$$x^2 + x + y^2 + y + z^2 + z + 1 \text{ is a prime number,}$$

for various values of W. This can be estimated as follows:

- Generate N random triples $\langle x, y, z \rangle$ of integers between 0 and W, using independent, uniform distributions for each.

- Count the number K of triples satisfying the condition that $x^2 + x + y^2 + y + z^2 + z + 1$ is a prime number. Let $p = K/N$.

- Since there are $(W + 1)^3$ triples in total, estimate the total number of triples as $p \cdot (W + 1)^3$, with a standard deviation of $\sqrt{p(1-p)/N}(W+1)^3$.

12.4 Counting Solutions to a DNF Formula (Optional)

An interesting example of a counting problem that can be solved by using Monte Carlo methods is finding the number of solutions to a formula in disjunctive normal form (DNF).[3] A propositional formula ϕ is in DNF if it is expressed as the disjunction of a collection of conjunctions of literals, where a literal is either a propositional atom or its negation. For example, the following formula is in DNF:

$$(A \wedge B \wedge \neg C \wedge \neg D) \vee (\neg A \wedge \neg B) \vee (\neg B \wedge C). \tag{12.1}$$

Let $\mu_1, \ldots, \mu_k$ be the conjunctions in ϕ. Let m be the total number of propositional atoms in ϕ. For example, in Formula (12.1), $m = 4$ (A, B, C, and D), $k = 3$, and $\mu_1 = A \wedge B \wedge \neg C \wedge \neg D$, $\mu_2 = \neg A \wedge \neg B$, $\mu_3 = \neg B \wedge C$.

A valuation—that is, an assignment of T or F to each atoms—satisfies a formula ϕ in DNF if it satisfies *all* of the literals in *at least one* of the μ_i. For example, the valuations that satisfy Formula (12.1) are the following.

$$\langle A = T, B = T, C = F, D = F \rangle \quad \text{satisfies} \quad \mu_1;$$
$$\langle A = T, B = F, C = T, D = T \rangle \quad \text{satisfies} \quad \mu_3;$$
$$\langle A = T, B = F, C = T, D = F \rangle \quad \text{satisfies} \quad \mu_3;$$
$$\langle A = F, B = F, C = T, D = T \rangle \quad \text{satisfies both} \quad \mu_2 \quad \text{and} \quad \mu_3;$$
$$\langle A = F, B = F, C = T, D = F \rangle \quad \text{satisfies both} \quad \mu_2 \quad \text{and} \quad \mu_3;$$
$$\langle A = F, B = F, C = F, D = T \rangle \quad \text{satisfies} \quad \mu_2;$$
$$\langle A = F, B = F, C = F, D = F \rangle \quad \text{satisfies} \quad \mu_2.$$

So there are seven valuations that satisfy Formula (12.1).

The abstract problem of counting the solutions to a DNF formula has a number of important practical applications. For example, if we have a network in which individual edges can fail with a specified probability, and we want to know the probability that two nodes will be able to communicate, then that problem can be expressed in terms of the number of solutions to a particular DNF formula. However, the problem is what is called "#P-complete"

---

[3]This algorithm is from Karp, Luby, and Madras (1989).

(read "sharp P complete"). The class of #P-complete problems bears the same relation to the space of counting problems as the relation of the class of NP-complete problems to the space of decision problems; as far as anyone knows, no polynomial time-exact solution exists. However, there is an efficient Monte Carlo algorithm that gives a probabilistic, approximate solution for counting the solutions to a DNF formula.

The obvious way to construct a Monte Carlo algorithm here would be to take the target set R as the set of all valuations satisfying the formula, and the enveloping set Q to be the set of all 2^m valuations over the m atoms. That does not work, however; there are cases—moreover, cases of practical importance—in which $|R|$ is exponentially small compared to $|Q|$. We have to be more clever.

Instead, we take the enveloping set Q to be the set of all pairs $\langle s, i \rangle$ where i is an index between 1 and k and s is a valuation satisfying μ_i. We then take the set R to be the subset of S that is the set of all pairs $\langle s, i \rangle$ where s is any valuation satisfying ϕ and i is the smallest index for which s satisfies μ_i.

For example, in Formula (12.1), the enveloping set Q and the target set R are

$$
\begin{aligned}
Q = \{ &\langle\langle A = T, B = T, C = F, D = F\rangle, 1\rangle, \quad R = \{ \langle\langle A = T, B = T, C = F, D = F\rangle, 1\rangle, \\
&\langle\langle A = T, B = F, C = T, D = T\rangle, 3\rangle, \quad\quad \langle\langle A = T, B = F, C = T, D = T\rangle, 3\rangle, \\
&\langle\langle A = T, B = F, C = T, D = F\rangle, 3\rangle, \quad\quad \langle\langle A = T, B = F, C = T, D = F\rangle, 3\rangle, \\
&\langle\langle A = F, B = F, C = T, D = T\rangle, 2\rangle, \quad\quad \langle\langle A = F, B = F, C = T, D = T\rangle, 2\rangle, \\
&\langle\langle A = F, B = F, C = T, D = T\rangle, 3\rangle, \quad\quad \langle\langle A = F, B = F, C = T, D = F\rangle, 2\rangle, \\
&\langle\langle A = F, B = F, C = T, D = F\rangle, 2\rangle, \quad\quad \langle\langle A = F, B = F, C = F, D = T\rangle, 2\rangle, \\
&\langle\langle A = F, B = F, C = T, D = F\rangle, 3\rangle, \quad\quad \langle\langle A = F, B = F, C = F, D = F\rangle, 2\rangle\}. \\
&\langle\langle A = F, B = F, C = F, D = T\rangle, 2\rangle, \\
&\langle\langle A = F, B = F, C = F, D = F\rangle, 2\rangle\};
\end{aligned}
$$

In this case, R contains all the pairs in Q except the two pairs, $\langle\langle A = F, B = F, C = T, D = T\rangle, 3\rangle$ and $\langle\langle A = F, B = F, C = T, D = F\rangle, 3\rangle$, because those valuations satisfy μ_2 as well. Note the following consequences:

- Any valuation that satisfies ϕ appears exactly once in R; namely, associated with the first μ_i that it satisfies. Therefore, $|R|$ is the number of valuations satisfying ϕ.

- Any valuation appears at most k times in Q. Therefore $|R|/|Q| \geq 1/k$.

- If we have chosen a pair $\langle s, i \rangle$ from Q, we can check whether it is in R by looping through the conjunctions $\mu_1, \ldots, \mu_{i-1}$ and seeing whether s satisfies any of these. This can be done in time $O(|\phi|)$.

What remains is to come up with a method for picking elements of Q uniformly. That is done by first choosing i, and then choosing s. We next see how these are done in the opposite order:

Suppose we have chosen a particular conjunction μ_i. This has the form $\mu_i = \lambda_1 \wedge \ldots \wedge \lambda_{r_i}$ for some value of r_i. So the atoms in $\lambda_1, \ldots, \lambda_{r_i}$ must have the values indicated by the λs, and the other atoms can have any combination of values. Therefore, there are 2^{m-r_i} different valuations satisfying μ_i; we can pick one of these at random just by giving all the atoms in $\lambda_1, \ldots, \lambda_r$ the right value and assigning the other atoms values T or F with probability $1/2$.

For example, in Formula (12.1), if we choose $i = 3$, then $\mu_3 = \neg B \wedge C$, so $r = 2$. So there are $2^{4-2} = 4$ valuations satisfying μ_3. We can select a random valuation by choosing $B = F$ and $C = T$, picking $A = T$ or $A = F$ randomly with probability $1/2$, and picking $D = T$ or $D = F$ randomly with probability $1/2$.

Thus, the number of pairs in Q with index i is 2^{m-r_i}, so the total number of pairs in Q is $|Q| = \sum_i^k 2^{m-r_i}$. Therefore, if we want to pick pairs in Q uniformly, we should pick index i with probability $p_i = 2^{m-r_i}/|Q|$.

For example, in Formula (12.1), we have $m = 4, r_1 = 4, r_2 = 2, r_3 = 2$, so $|Q| = 2^0 + 2^2 + 2^2 = 9$ and $p_1 = 2^0/|Q| = 1/9$, $p_2 = 4/9$; $p_3 = 4/9$. Thus, there is a $1/9$ probability that we will choose $i = 1$, and therefore a $1/9$ probability that we will choose the single valuation satisfying μ_1; there is a $4/9$ probability that we will choose $i = 2$, and, having done so, a $1/4$ probability of choosing each of the four valuations satisfying μ_2 for a total probability of $1/9$ for each of these; and likewise for $i = 3$. Therefore, each element of Q has a $1/9$ probability of being selected, as promised. Putting all this together, we get Algorithm 12.1, CountDNF.

12.5   Sums, Expected Values, and Integrals

If Q is a set and f is a numeric function over Q, we can estimate $\sum_{x \in Q}$ as $(|Q|/|S|) \cdot \sum_{x \in S} f(x)$, where S is a random sample of Q.

If Q is a numerical random variable, then we can estimate $\text{Exp}(Q)$ by taking a sample of Q of size n and computing the average over those samples. We can estimate $\text{Std}(Q)$ by computing the standard deviation of the sample.

If f is an integrable function over $\mathbb{R}^k$ and Q is a region in $\mathbb{R}^k$, then we can estimate $\int_Q f(t)dt$ as $\sum_{x \in S} f(x) \cdot \text{Area}(Q)/|S|$, where S is a random sample of points in Q. The method of computing the area of region R in Section 12.1 is actually a special case of this, where Q is the enveloping region and $f(x)$ is the characteristic function of R—that is, $f(x) = 1$ for $x \in R$ and 0 for $x \notin R$.

The accuracy of all these estimates is generally proportional to $1/\sqrt{n}$, where n is the sample size. The constant of proportionality depends on the characteristics of the example.

```
function CountDNF(in φ: A formula in DNF; n: number of Monte Carlo iterations)
                    return An estimate of the number of valuations satisfying φ

/* Preprocessing */
{ m ← the number of propositional atoms in φ;
  μ₁ ... μₖ ← the conjuncts in φ;
  for (i = 1 ... k) rᵢ ← the number of literals in μᵢ;
  for (i = 1 ... k) Cᵢ ← 2^(m−rᵢ);
  |Q| = Σ_{i=1}^{k} Cᵢ;
  for (i = 1 ... k) pᵢ = Cᵢ/|Q|;

/* Monte Carlo loop */
  Count ← 0
  for (j ← 1 ... n)
      pick i at random, following distribution pᵢ;
      construct a random valuation s satisfying μᵢ
          by setting all the atoms in μᵢ as specified there
          and setting all the other atoms to be T/F with probability 1/2;
      SinR ← true;
      for (w ← 1 ... i − 1)
          if s satisfies μ_w then { SinR ← false; exitloop } endif;
      endfor
      if SinR then Count ← Count + 1 endif
  endfor

  return Count ∗ |Q|/N;
}
```

Algorithm 12.1. Monte Carlo algorithm to count the number of solutions to a formula in DNF.

A couple of caveats should be mentioned, though. The technique does not work in the case where there is a very small region where f is immensely larger than everywhere else. Specifically, for computing a sum over a finite set Q, suppose that there is a subset Z of Q such that

- $|Z|/|Q| < \epsilon$;

- the average value of $f(x)$ over Z is at least on the order of $1/\epsilon$ times the average value of $f(x)$ over $Q \setminus Z$—therefore, the sum of f over Z is at least comparable to the sum of f over $Q \setminus Z$.

Then if the size of the sample is $O(1/\epsilon)$, the sample will probably not contain any elements of Z. The average size of elements in the sample will be about the average size on $Q \setminus Z$.

For all of these, if there is a subset B of relative size ϵ over which f is very large $(\Omega(1/\epsilon))$, then n has to be at least several times $1/\epsilon$ in order to be confident that the sample contains a representative number of elements from B. If f follows a power law distribution, where the top few elements have most of the total f, then a random sample is likely to be very inaccurate. You can be sure that this condition does not arise if either

(a) the ratio between $\max_{x \in Q} f(x)$ and $\min_{x \in Q} f(x)$ is not very large; or

(b) the space of y such that $f(y)$ is close to $\max_{x \in Q} f(x)$ is not very small.

These conditions are made precise in the literature on Monte Carlo methods (see, e.g., Fishman 1996).

Moreover, to ensure that the estimate of $\mathrm{Exp}(Q)$ approaches its true value with probability 1 as the sample size goes to infinity, $\mathrm{Var}(X)$ must be finite. To assure that this holds for $\int_Q f(t)dt$, the integral $\int_Q f^2(t)dt$ must be finite.

Finally, if f is a well-behaved function, then deterministic methods of numerical integration give answers that are much more accurate than Monte Carlo methods for a given amount of computation. Monte Carlo methods for integration win out for functions that, even though they are integrable, are not well-behaved.

12.6   Probabilistic Problems

Not surprisingly, Monte Carlo methods can be very effective at solving complex probabilistic problems. For example, suppose that we are interested in a particular form of solitaire, and we want to know how frequently we end up winning with some specified strategy. It would almost certainly be hopeless to try to do an exact combinatorial analysis. However, Monte Carlo testing is comparatively easy: program up the game and the strategy, generate 10,000 shuffles of the deck at random, and test.

One of the nice features of this example is that you can be sure at the start that the probability is neither very close to 0 nor very close to 1 because forms of solitaire in which the player nearly always wins or nearly always loses usually do not survive.

A Monte Carlo search can be applied in the same way to multiplayer card games. Suppose that we are playing 5-card draw in poker against three other players. We want to know what the probability is that we are holding the winning hand. The Monte Carlo solution is simple: we carry out N random deals of the remaining cards to the other players, and we check the fraction in which we have the winning hand.

Moreover, the strategy is easy to adapt to a new game. Suppose one of the other players calls for some ridiculous variant: two cards are visible, red jacks are wild, 2s can count as aces, and if the total pointcount in your hand is exactly equal to 43, then that beats anything. We could work for a week trying to come up with an exact analysis of the combinations. But the Monte Carlo approach can be up and running in minutes, as long as two conditions hold:

- We can easily generate random hands that satisfy the known information. In poker, ignoring the information from the bets, this is just a matter of randomly dealing the invisible cards.

- We can easily calculate whether one hand beats another.

Likewise, bridge-playing programs work by doing a Monte Carlo search over distributions of the unseen cards and finding the move that most often leads to a successful outcome (Ginsberg, 1999).

The same approach applies outside the game context, in all kinds of complex or dynamic probabilistic situations. In financial applications, for example, the value of a portfolio of investments may be affected by a series of external events with complex dependencies. In such cases, it may be much more effective to run a Monte Carlo simulation, generating a large random collection of scenarios, than to attempt an analytic analysis.

Problems involving continuous distributions can be addressed in this way as well. Suppose, for example, that X is a random variable with distribution $N_{5,0.2}$, and Y is a random variable uniformly distributed over $[3,7]$. What is $P(X \cdot Y > 20)$? Another problem involves a collection of n points in the plane, where the x-coordinate is drawn from X and the y-coordinate is drawn from Y. If we draw an edge between every pair of points that are less than d apart, what is the probability that the resultant graph is connected? For the first problem, with enough hard work we can probably formulate and evaluate a definite integral. For the second, the exact analysis is certainly very difficult, and quite possibly hopeless. In this case, or similar cases, if a high degree of accuracy is not required, then a Monte Carlo evaluation can be quickly programmed up and executed, and with very high probability it will give a reasonably accurate answer.

12.7   Resampling

A method of statistical analysis that has recently become popular is called *resampling* or *nonparametric statistics*. In resampling, statistical analysis is carried out by doing Monte Carlo tests over subsamples of the data or other collections directly derived from the data. Resampling has two advantages over traditional statistics. First, many traditional statistical methods require strong assumptions about the processes that generated the data (e.g., that the data were generated by a normal distribution), but resampling methods make no such assumptions. Second, traditional statistical analysis often requires complex mathematical analysis and the use of obscure special functions; in contrast, resampling techniques are much simpler to learn, to program, and to apply.

For example,[4] suppose that we want to determine whether a drug is more effective than a placebo. Suppose we have measured the improvement on a scale from 0 to 100 for a number of trials of both drugs, and the results are

> Placebo: 54 51 58 44 55 52 42 47 58 46
> Drug: 54 73 53 70 73 68 52 66 65

Eyeballing these data, it certainly *looks* like the drug is effective. The placebo has numbers in the 40s and 50s; the drug has numbers in the 50s, 60s, and 70s. The average for the drug is 63.78, and the average for the placebo is 50.70; the difference between the averages is thus 13.08. But is this actually a significant difference? We carry out the following experiment: We take the same data points, but we shuffle the labels placebo and drug randomly. For instance, one reshuffling might be

> 54-P 51-D 58-D 44-P 55-D 52-P 42-P 47-P 58-P 46-D
> 54-P 73-D 53-D 70-D 73-D 68-P 52-P 66-P 65-D

Here, the average for label D is 60.44, and the average for label P is 53.7. Testing 10,000 different relabelings, the difference between the average for label D and the average for label P was 13.08 or greater in only 8 cases. That establishes with a high degree of confidence that the drug *is* indeed effective; if the choice between drug and placebo were actually independent of the improvement measure, then the chances of getting a difference with these data that is as large as 13.08 seems to be less than 0.001.

Note that the structure here is rather different than in our previous examples. Roughly speaking, the target set R here is the set of labelings where the difference in mean is at least 13.08 and the enveloping set Q is the set of all relabelings. But here we are *not* interested in an accurate measure of R; all we want to do is to establish that it is much smaller than Q.

We next want to get a confidence interval for the *magnitude* of the difference between the drug and the placebo (the previous experiment established only the *existence* of a difference.) The method used is called bootstrapping. We assume that our data for the drug are a random sample of size nine out of a large space of results for the drug. What is that large space? The simplest assumption is that it is *exactly the same as the data we have*, repeated a zillion times. A sample from this space is just a sample of size nine from the drug data

---

[4]These examples are taken from Shasha and Wilson (2008).

with replacement; that is, some values may appear multiple times and some values may appear zero times. So, to find a 90% confidence interval, we take 10,000 pairs of ⟨ a sample from the drug data and a sample from the placebo data ⟩, we compute the difference of the means for each such pair, sort all these differences, and find the 5% and the 95% percentile levels on all these differences. For these particular data, we find a confidence interval of [7.5, 18.1].

Be careful not to get confused about the two different kinds of samples here. At the object level, there are the samples of drug data (size nine) and of placebo data (size ten). At the meta-level, there are the bootstrap samples—there are 10,000 of these, each of which is a pair of a hypothetical sample of drug data and a sample of placebo data. The samples discussed in Chapter 11 correspond to the object-level samples here. In particular, the width of the confidence interval is determined by the size of the object level samples (nine and ten) and *not* by the size of the bootstrap sample. As we make the bootstrap sample larger, the confidence interval does not shrink—we just get a better estimate.

12.8   Pseudorandom Numbers

A limitation on Monte Carlo methods is that a function such as MATLAB's `rand` is not a truly random process; it is a pseudorandom number generator. These generally work by taking an arbitrary starting value x called the *seed*, and then returning `f(x)`, `f(f(x))`, `f(f(f(x)))`, ... for some carefully selected generator function `f`. Therefore, after a certain point they cycle.

The use of pseudorandom numbers in a Monte Carlo search can give rise to two kinds of problems:

1. There could be some relation between the generator function `f` and the phenomenon being studied, so that the phenomenon is not independent of the random number stream.

2. If the Monte Carlo search involves a very large sample, then the cyclic period of the random number generator can be reached. This problem places a limit on the precision of results attainable through a Monte Carlo search, particularly because we need to generate a sample whose size is proportional to the inverse square of the desired accuracy.

Both of these problems can arise in practice. See Fishman (1996, Chapter 7) for an extensive discussion.

Random sequences that come from hardware random number generators, which are based on physical random processes, do not suffer from these problems.

12.9 Other Probabilistic Algorithms

Monte Carlo methods were pioneered for solving physics problems that were too difficult to be solved analytically; much of the subsequent development has likewise been aimed at these kind of problems. (Indeed, the method was first extensively used in the Manhattan Project, a program that built the first atomic bomb.)

Many successful search techniques, such as simulated annealing, WALKSAT, genetic algorithms, and random restart, use an element of random choice to escape local minima or wander around local plateaus. See Russell and Norvig (2009) for an introduction to this active area of research.

In algorithmic analysis and theory of computation, if D is a decision problem (i.e., a problem with a yes or no answer), then a probabilistic algorithm A for D is one that makes a number of random choices and has the following properties.

- If the answer to D is true then A always returns true.

- If the answer to D is false, then there is at least a 50% chance that A will return false, where the probability is over the sample space of random choices made.

Therefore, if we run algorithm A k times and it returns false on any of those runs, the answer must be false. Moreover, if the answer is false, then the probability is at most 2^{-k} that A will return true every time. Therefore, if we have run A k times and have gotten true every time, then we can be reasonably (although never entirely) sure that the answer is indeed true.

12.10 MATLAB

The function `rand` with no arguments generates a random number uniformly between 0 and 1. The function call `rand(n)` generates an $n \times n$ matrix of random numbers, independently chosen uniformly between 0 and 1. The function call `rand(m,n)` generates an $m \times n$ matrix of random numbers.

The function `randn` generates a random number distributed according to the standard Gaussian distribution $N_{0,1}$. The function calls `randn(n)` and `randn(m,n)` generate matrices.

```
>> rand
ans =
    0.3816

>> rand
ans =
    0.7655
```

```
>> rand(1,7)
ans =
    0.7952    0.1869    0.4898    0.4456    0.6463    0.7094
0.7547

>> rand(4)
ans =
    0.2760    0.1190    0.5853    0.5060
    0.6797    0.4984    0.2238    0.6991
    0.6551    0.9597    0.7513    0.8909
    0.1626    0.3404    0.2551    0.9593

% The function randn generates random numbers according to the normal
% distribution with mean 0 and standard deviation 1.

>> randn
ans =
    0.0859

>> randn
ans =
   -1.4916

>> randn(1,7)
ans =
   -0.7423   -1.0616    2.3505   -0.6156    0.7481   -0.1924
0.8886
```

It is often useful to be able to get the same sequence of random numbers; one example is debugging. To do this, carry out the following operations.

- Create a stream s of random numbers, as shown in the following code.

- Make this the default stream. Thus, this is the stream that functions such as rand will consult.

- To restart the same sequence over again, call reset(s).

- To go back to an intermediate state of a random stream, save the state by saving s.State in a variable, and restore it by assigning to s.State the variable where you have saved it.

```
>> s=RandStream('mt19937ar')         % Create a random stream
s =
mt19937ar random stream
          Seed: 0
       RandnAlg: Ziggurat

>> RandStream.setDefaultStream(s)   % Make it the default stream

>> rand(1,7)                        % Generate some random numbers
```

```
ans =
    0.8147    0.9058    0.1270    0.9134    0.6324    0.0975    0.2785

>> rand(1,5)                    % Generate more random numbers
ans =
    0.5469    0.9575    0.9649    0.1576    0.9706

>> state2=s.State;              % Save the current state in state2

>> rand(1,6)                    % Generate more random numbers
ans =
    0.9572    0.4854    0.8003    0.1419    0.4218    0.9157

>> reset(s)                     % Go back to the beginning

>> rand(1,7)
ans =
    0.8147    0.9058    0.1270    0.9134    0.6324    0.0975    0.2785

>> s.State=state2;              % Jump ahead to state2.

>> rand(1,6)
ans =
    0.9572    0.4854    0.8003    0.1419    0.4218    0.9157
```

Exercises

Use MATLAB for all of these exercises.

Exercise 12.1. Use a Monte Carlo method to estimate the size of the ellipse $x^2 + xy + y^2 \le 1$. *Note:* This ellipse lies inside the rectangle $[-2,2] \times [-2,2]$. Give the 95% confidence interval for your answer.

Exercise 12.2. An integer is *square-free* if its prime factorization consists of distinct primes. For example, $35 = 5 \cdot 7$ is square-free; $12 = 2 \cdot 2 \cdot 3$ is not square free. Use a Monte Carlo method to estimate what fraction of integers less than 1,000,000 are square-free. The MATLAB function factor(N) computes the prime factorization of integer N. Give the 95% confidence interval for your answer.

Exercise 12.3. Use a Monte Carlo method to estimate the probability that a random string of 200 integers between 0 and 9 contains a consecutive subsequence of five increasing values (e.g., 2,3,6,8,9). Give the 95% confidence interval for your answer.

Problems

Problem 12.1. (This problem refers to Assignment 12.2 at the end of this chapter). In Assignment 12.2, we write the main body of the routine as version 1:

```
for I=1:N1
  use RandomWord Q times to generate a set S of word ranks;
  for I=1:N2
     count = count + (RandomWord() in S);
  end
end
```

Two other possible versions would use only one level of looping but the same total number of iterations of the inner loop. Version 2 is

```
use RandomWord Q times to generate a set S of word ranks;
for I=1:(N1*N2)
     count = count + (RandomWord() in S);
  end
end
```

and version 3 is

```
for I=1:(N1*N2)
     use RandomWord Q times to generate a set S of word ranks;
     count = count + (RandomWord() in S);
  end
end
```

(a) Discuss the pros and cons of these three versions in terms of running time and accuracy.

(b) Suppose that it is necessary to assert a confidence interval on the answer p that corresponds to a sample of size $N1 * N2$. Which of these versions should be used? Why?

Programming Assignments

Assignment 12.1. Implement Algorithm 12.1, described in Section 12.4. That is, write a function CountDNF(F). The input parameter F is an $m \times n$ array, where n is the number of propositional atoms and m is the number of conjunctions. F[i,j]=1 if j is a conjunct in μ_i; F[i,j]=-1 if ¬j is a conjunct in μ_i; and F[i,j]=0 if j does not appear in μ_i. Thus, the input matrix for Formula (12.1) would be

$$\begin{bmatrix} 1 & 1 & -1 & -1 \\ -1 & -1 & 0 & 0 \\ 0 & -1 & 1 & 0 \end{bmatrix}.$$

The function should return an estimate of the number of valuations satisfying the formula.

Assignment 12.2. The values for p in Table 9.4 were generated by using a Monte Carlo simulation. Write a function `AlreadySeen(Q)` that carries out a comparable calculation; namely, given a subfunction `RandomWord()` that generates random word ranks, and the number of words already seen, `Q`, compute the probability that the new word you see is one you have already seen.

Step 1. Write the function `RandomWord()` to follow the distribution used in Table 9.4; for $I = 1, \ldots, 100,000$, the probability that `RandomWord` returns I is $1/(I \cdot H(100,000))$. Here the normalizing factor $H(N)$ is the harmonic function $\sum_{I=1}^{N} 1/I$.

Hint: Use the technique described in Section 12.2 for generating finite distributions. Note that $P(\texttt{RandomWord} \leq K) = H(K)/H(100,000)$. A brute force approach is simply to compute $P(\texttt{RandomWord} \leq K)$ for $K = 1, \ldots,$ 100,000, save the results in an array, and then use a binary search on the array to find the proper subinterval. A more clever way that is both more time efficient and space efficient is to precompute these values up to $K = $ 1,000 and then to use the fact that, for $K > 1,000$, to sufficient accuracy, $H(K) \approx \ln(K) + \gamma$, where γ is Euler's constant, 0.5772 (see Assignment 2.2). In either case, the precomputation should be done *once* and saved in a global variable; it should not be done each time `RandomWord` is called.

Step 2. Implement the following pseudocode:

```
function P = AlreadySeen(Q)
  count = 0;
  for I=1:N1
    use RandomWord Q times to generate a set S of word ranks;
    for I=1:N2
      count = count + (RandomWord() in S);
    end
  end
  P=count/(N1*N2);
end
```

Here `N1` and `N2` are reasonably large parameters that govern the Monte Carlo search.

Assignment 12.3. Aces Up Solitaire is played with a standard deck of 52 cards and has the following rules.

```
deal four cards face up on the table;
the remaining 48 cards are the deck (face down);
repeat until (the deck is empty) {
  while (two or more visible cards have the same suit)
    discard all but the highest card of that suit;
  if (there is any empty space)
      move the top card from any other pile into that space;  ***
    else deal one card from the deck face-up onto each pile;
 }
```

The player wins if, at the end, only the aces remain and the other 48 cards have been discarded.

Note that at the step marked by three asterisks above, the player can make a choice; hence there is some element of strategy. Consider the following four strategies for executing this step.

1. Choose randomly among the remaining piles with at least two cards.

2. Take a card from the deepest pile.

3. Take a card from the shallowest pile with at least two cards.

4. If there are any moves that will allow you to make a discard, then choose randomly among such moves. If not, choose randomly among the piles with at least two cards.

Use the Monte Carlo search for each of these strategies:

(a) Compute the probability of winning.

(b) Compute the expected number of cards discarded.

Assignment 12.4. Write a program that uses a Monte Carlo search to estimate the probability you hold a winning hand in poker, assuming that

• every player has been dealt five cards,

• you can see all your own cards,

• you can see K of each of your opponents' cards.

That is, write a function WinningPokerHand(My,Theirs), where

• My is a 2×5 array. My[1,J] is the number value of the Jth card. My[2,J] is a number from 1 to 4, indicating the suit of the Jth card (clubs, diamonds, hearts, spades).

- **Theirs** is a three-dimensional $N \times 2 \times K$ array indicating the visible cards of your opponents. N is the number of opponents. **Theirs[I,1,J]** and **Theirs[I,2,J]** are the number and suit of the Jth visible card of Ith opponent. The ace should be input as 1, although, of course, it can count as either 1 or 14.

For example, suppose that you hold the three of clubs, the three of spades, the queen of diamonds, the queen of spades, and the ten of clubs. You have two opponents, and $K = 3$. The visible cards of opponent 1 are the ace of spades, the jack of spades, and the four of spades. The visible cards of opponent 2 are the king of diamonds, the king of spades, and the six of hearts. Then the input parameters are

$$\text{My} = \begin{bmatrix} 3 & 3 & 12 & 12 & 10 \\ 1 & 4 & 2 & 4 & 1 \end{bmatrix}, \quad \text{Theirs} = \begin{bmatrix} \begin{bmatrix} 1 & 11 & 4 \\ 4 & 4 & 4 \end{bmatrix} & \begin{bmatrix} 13 & 13 & 6 \\ 2 & 4 & 3 \end{bmatrix} \end{bmatrix}.$$

Carry out a Monte Carlo search by randomly dealing the hidden cards to the opponents from the rest of the deck, and calculating whether you have the winning hand. The function should return an estimate of the probability that you have a winning hand.

Assignment 12.5. Use a Monte Carlo search to solve the following variant of the problem discussed at the end of Section 12.6. Write a function `CloselyConnectedPoints(N,D)` to estimate the probability that N points chosen at random and connected with an arc of length D form a connected graph. Specifically:

(a) Choose N points randomly, with the X- and Y-coordinates generated independently by the standard Gaussian $N_{0,1}$.

(b) Let G be the graph whose vertices are the points in (A) and where there is an arc from U to V if $d(U,V) < D$.

Then what is the probability that G is a connected graph?

Draw a plot of the probability versus D for $N = 100$, over a range of value in which D goes from close to 0 to close to 1.

Draw a plot of the probability versus N for $D = 0.1$ for $N = 2$ to a large enough value such that the probability is near 1.

Chapter 13

Information and Entropy

The two central concepts in information theory are the *information* of an event and the *entropy* of a random variable.

13.1   Information

The *information* of a probabilistic event E, denoted $\text{Inf}(E)$, measures the amount of information that we gain when we learn E, starting from scratch (i.e., from some presumed body of background knowledge). $\text{Inf}(E)$ is measured in number of bits. As in probabilistic notation, "E, F" represents the event that both E and F are true; thus, $\text{Inf}(E, F)$ is the information gained by learning both E and F. The conditional information $\text{Inf}(E \mid F)$ is the amount of information that we gain if we learn E after we have already learned F. Therefore, $\text{Inf}(E \mid F) = \text{Inf}(E, F) - \text{Inf}(F)$.

The information of E is related to its probability: $P(E) = 2^{-\text{Inf}(E)} = 1/2^{\text{Inf}(E)}$, or equivalently, $\text{Inf}(E) = -\log(P(E)) = \log(1/P(E))$. Likewise, $\text{Inf}(E \mid F) = -\log(P(E \mid F))$. (Throughout this chapter, $\log(x)$ means $\log_2(x)$.) Note that, since $P(E) \leq 1$, $\log(P(E)) \leq 0$, so $\text{Inf}(E) \geq 0$; finding out the outcome of an event never constitutes a loss of information.

The intuition here is as follows. Suppose we have k independent events, each of which has probability $1/2$. Then, clearly, communicating the outcome of all these events requires k bits; thus, the sequence of k outcomes constitutes k bits of information. The probability of any particular sequence of outcomes is 2^{-k}. So, for a random string S of k bits, $P(S) = 2^{-\text{Inf}(S)}$ and $\text{Inf}(S) = -\log(P(S))$. The natural generalization is that, for any event E, $\text{Inf}(E) = -\log(P(E))$.

That last step may seem like a leap. We can elaborate it as follows: suppose that the information function $\text{Inf}(E)$ satisfies the following properties.

1. If E is an event of probability $1/2$, then $\text{Inf}(E) = 1$ bit.

2. If $P(E) = P(F)$, then $\text{Inf}(E) = \text{Inf}(F)$; that is, the information of an event is purely a function of its probability.

3. $\text{Inf}(E \,|\, F) = \text{Inf}(E, F) - \text{Inf}(F)$, as discussed above.

4. If E and F are independent events, then $\text{Inf}(E \,|\, F) = \text{Inf}(E)$. The argument is as follows. If E and F are independent, then knowing F gives no information about E. Therefore, learning E after we know F gives you as much new information as learning E before we learned F.

5. The entropy is a continuous function of the probability distribution.

Given these five premises, we can prove that $\text{Inf}(E) = -\log(P(E))$.

Proof: By premises (3) and (4), if E and F are independent, then $\text{Inf}(E, F) = \text{Inf}(F) + \text{Inf}(E \,|\, F) = \text{Inf}(F) + \text{Inf}(E)$.

Let k be a large number. Let $F_1, \ldots, F_k$ be independent events, each of probability $2^{-1/k}$. Then $P(F_1, \ldots, F_k) = 1/2$, so $\text{Inf}(F_1, \ldots, F_k) = 1$. But then, since $F_1, \ldots, F_k$ are independent, $1 = \text{Inf}(F_1, \ldots, F_k) = \text{Inf}(F_1) + \ldots + \text{Inf}(F_k)$. Since the Fs all have equal probability, this sum equals $k \cdot \text{Inf}(F_1)$. So $\text{Inf}(F_1) = 1/k$. Now, let E be any event and let $q = \lfloor k \cdot -\log(P(E)) \rfloor$, so $\log(P(E)) \approx -q/k$. Then $P(F_1, \ldots, F_q) = 2^{-q/k} \approx 2^{\log(P(E))} = P(E)$.

By the continuity premise (4), $\text{Inf}(P(E)) \approx \text{Inf}(F_1, \ldots, F_q) = q/k \approx -\log(P(E))$. In the limit, as k goes to infinity, the approximation converges to an equality. $\square$

We give another argument at the end of Section 13.4, justifying the definition of $\text{Inf}(E)$ as $-\log(P(E))$.

For convenient reference, we restate the above additivity properties of information as Theorem 13.1.

Theorem 13.1.

(a) *For any two events E and F,* $\text{Inf}(E, F) = \text{Inf}(F) + \text{Inf}(E \,|\, F)$.

(b) *If E and F are independent, then* $\text{Inf}(E, F) = \text{Inf}(E) + \text{Inf}(F)$.

(c) *If $E_1, \ldots, E_k$ are independent and have the same distribution, then* $\text{Inf}(E_1, \ldots, E_k) = k \cdot \text{Inf}(E_1)$.

Note that each of these rules is just the logarithm of the corresponding rule for probabilities:

(a) $P(E, F) = P(F) \cdot P(E \,|\, F)$.

(b) If E and F are independent, then $P(E, F) = P(E) \cdot P(F)$.

(c) If $E_1, \ldots, E_k$ are independent and identically distributed, then $P(E1, \ldots, E_k) = (P(E_1))^k$.

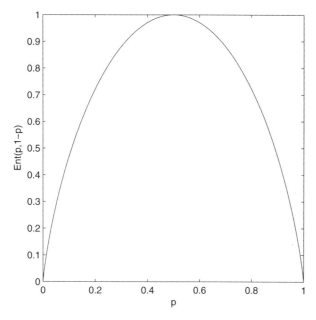

Figure 13.1. Entropy function.

13.2   Entropy

The *entropy* of random variable X, denoted $\mathrm{Ent}(X)$, is the expected value of its
information: $\mathrm{Ent}(X) = \mathrm{Exp}(\mathrm{Inf}(X))$. Thus, if X has distribution $\langle p_1, p_2, \ldots, p_k \rangle$,
then

$$\mathrm{Ent}(X) = -(p_1 \log(p_1) + \ldots + p_k \log(p_k)).$$

By convention, if $p_i = 0$, we take $p_i \log(p_i)$ to be 0.

Since the entropy depends only on the probability distribution, it is com-
mon to write $\mathrm{Ent}(p_1 \ldots p_k) = -(p_1 \log(p_1) + \ldots + p_k \log(p_k))$ as the entropy of a
random variable with distribution $\langle p_1, \ldots, p_k \rangle$.

If X is a Boolean variable with $P(X = t) = p$ and $P(X = f) = 1 - p$, then
$\mathrm{Ent}(X) = \mathrm{Ent}(p, 1-p) = -(p \log(p) + (1-p) \log(1-p))$. This is shown as a function of p in Figure 13.1. For $p = 0$ and $p = 1$ we have $\mathrm{Ent}(X) = 0$. (If $p = 0$, then
the event $X = t$ has infinite information but zero chance of occurring; and the
event $X = f$ will certainly occur, but carries zero information.) The function
reaches a maximum of 1 at $p = 1/2$. If $p = \epsilon$ or $p = 1 - \epsilon$, where ϵ is small, then
$\mathrm{Ent}(X) \approx -\epsilon \log \epsilon$.

If X is a random variable with k values, each of probability $1/k$, then $\mathrm{Ent}(X) =
\mathrm{Ent}(1/k \ldots 1/k) = \log(k)$. It is a fact that if Y is any random variable with k val-
ues, then $\mathrm{Ent}(Y) \leq \mathrm{Ent}(X) = \log(k)$ (Section 13.6).

13.3   Conditional Entropy and Mutual Information

Let E be an event and let X be a random variable. The *entropy of X given E* is the entropy of X once we know E:

$$\text{Ent}(X \mid E) = -\sum_x P(X = x \mid E) \log(P(X = x \mid E)).$$

We can consider $X \mid E$ to be a random variable over the range of X defined by the equation $P((X \mid E) = x) = P(X = x \mid E)$. In that case, $\text{Ent}(X \mid E)$ is simply the entropy of the random variable $X \mid E$.

Let Y be a second random variable. The *conditional entropy of X given Y*, $\text{CEnt}(X \mid Y)$, is the expected value, taken over all the possible values y of Y of [the entropy of X given $Y = y$], or

$$\text{CEnt}(X \mid Y) = \sum_y P(Y = y) \cdot \text{Ent}(X \mid Y = y)$$

$$= -\sum_{x,y} P(Y = y) \cdot P(X = x \mid Y = y) \log(P(X = x \mid Y = y)).$$

In other words, imagine that we currently don't know the value of either X or Y, but we are about to find out the value of Y. The conditional entropy measures what we can expect the entropy of X will be, on average, after we find out Y.

Note that the entropy conditioned on an event $\text{Ent}(X \mid E)$ is indeed an entropy of the random variable $X|E$. The "conditional entropy" of variable X conditioned on variable Y, $\text{CEnt}(X \mid Y)$ is not actually the entropy of any random variable, which is why we use a different symbol.

Theorem 13.2. *For any two random variables X and Y,* $\text{Ent}(X, Y) = \text{Ent}(Y) + \text{CEnt}(X \mid Y)$. *If X and Y are independent,* $\text{CEnt}(X \mid Y) = \text{Ent}(X)$ *and* $\text{Ent}(X, Y) = \text{Ent}(X) + \text{Ent}(Y)$.

Proof: The first statement follows from Theorem 13.1 and the additivity of expected value. The second statement follows from the definition of conditional entropy. □

Example 13.3. Suppose that X has two values $\{f, t\}$, with $P(X = f) = P(X = t) = 1/2$. Suppose also that Y is independent of X; it has values $\{r, w, b\}$, with $P(Y = r) = 1/2$, $P(Y = w) = 1/4$, and $P(Y = b) = 1/4$. Then

$$\text{Ent}(X) = -(1/2 \cdot \log(1/2) + 1/2 \cdot \log(1/2)) = 1,$$
$$\text{Ent}(Y) = -(1/2 \cdot \log(1/2) + 1/4 \cdot \log(1/4) + 1/4 \cdot \log(1/4)) = 3/2.$$

The joint distribution of X, Y is then

$$P(X = f, Y = r) = 1/4, \quad P(X = f, Y = w) = 1/8, \quad P(X = f, Y = b) = 1/8,$$
$$P(X = t, Y = r) = 1/4, \quad P(X = t, Y = w) = 1/8, \quad P(X = t, Y = b) = 1/8.$$

Therefore,

$$\begin{aligned}
\mathrm{Ent}(X,Y) &= -(1/4 \cdot \log(1/4) + 1/8 \cdot \log(1/8) + 1/8 \cdot \log(1/8) + 1/4 \cdot \log(1/4) \\
&\quad + 1/8 \cdot \log(1/8) + 1/8 \cdot \log(1/8)) \\
&= 5/2 \\
&= \mathrm{Ent}(X) + \mathrm{Ent}(Y).
\end{aligned}$$

Example 13.4. Let W and Z be random variables with the following joint distribution:

$$P(W = f, Z = r) = 0.4, \quad P(W = f, Z = w) = 0.05, \quad P(W = f, Z = b) = 0.05,$$
$$P(W = t, Z = r) = 0.1, \quad P(W = t, Z = w) = 0.1, \quad P(W = t, Z = b) = 0.3.$$

Thus, W and Z are not independent: if $W = f$, then probably $Z = r$; if $W = t$, then probably $Z = b$. The conditional probabilities are

$$P(Z = r \mid W = f) = 0.8, \quad P(Z = w \mid W = f) = 0.1, \quad P(Z = b \mid W = f) = 0.1,$$
$$P(Z = r \mid W = t) = 0.2, \quad P(Z = w \mid W = t) = 0.2, \quad P(Z = b \mid W = t) = 0.6.$$

So we have

$$\begin{aligned}
\mathrm{Ent}(W) &= \mathrm{Ent}(1/2, 1/2) = 1, \\
\mathrm{Ent}(Z \mid W = f) &= \mathrm{Ent}(0.8, 0.1, 0.1) = 0.9219, \\
\mathrm{Ent}(Z \mid W = t) &= \mathrm{Ent}(0.2, 0.2, 0.6) = 1.3710, \\
\mathrm{CEnt}(Z \mid W) &= P(W = f) \cdot \mathrm{Ent}(Z \mid W = f) + P(W = t) \cdot \mathrm{Ent}(Z \mid W = t) \\
&= 0.5 \cdot 0.9219 + 0.5 \cdot 1.3710 \\
&= 1.1464, \\
\mathrm{Ent}(W, Z) &= \mathrm{Ent}(0.4, 0.05, 0.05, 0.1, 0.1, 0.3) = 2.1464.
\end{aligned}$$

Note that $\mathrm{Ent}(W, Z) = \mathrm{Ent}(W) + \mathrm{CEnt}(Z \mid W)$.

It is a remarkable fact that the conditional entropy is always less than or equal to the entropy, as stated in Theorem 13.5.

Theorem 13.5. *For any two random variables X and Y, $\mathrm{CEnt}(X \mid Y) \leq \mathrm{Ent}(X)$, with equality holding if and only if X and Y are independent.*

What this theorem means is that, on average, finding out Y decreases our uncertainty about X unless the two are independent, in which case it leaves the uncertainty unchanged. We omit the proof of Theorem 13.5.

It is certainly true, however, that finding out a particular *event* can increase our uncertainty. If we see that the clock reads 2:45, then our entropy on the time of day is low; if we then observe that the clock is stopped, though, the entropy goes way up again. But that can happen only because it was originally unlikely that the clock would be stopped; had we known from the first that the clock might well be stopped, then our original probability distribution on the time would have been much smoother and the original entropy much higher. In either case, finding out the reverse fact, that the clock is running, will lower our entropy on the time; if we had originally thought it likely that the clock might be stopped, finding out that it is running will lower the entropy by a good deal. What Theorem 13.5 states is that, on average, finding out whether or not the clock is running cannot increase our entropy on the time of day—it can only fail to decrease our entropy if the random variables *Status of Clock* and *Time of Day* are independent (which they probably are).

Theorem 13.2 has the following interesting consequence. We have that $\text{Ent}(X,Y) = \text{Ent}(Y) + \text{CEnt}(X|Y)$. By symmetry, it is equally true that $\text{Ent}(X,Y) = \text{Ent}(X) + \text{CEnt}(Y|X)$. It follows, therefore, that

$$\text{Ent}(Y) - \text{CEnt}(Y|X) = \text{Ent}(X) - \text{CEnt}(X|Y). \tag{13.1}$$

This quantity is known as the *mutual information* of X and Y, denoted $\text{MInf}(X,Y)$. Since entropy is a measure of uncertainty, this measures how much less uncertain we are about Y once we have learned X; in other words, how much (in bits) we learn about Y when we find out the value of X. Equation (13.1) states that this is equal to the amount we learn about X when we find out the value of Y. By Theorem 13.5, this is always nonnegative, and it is zero only if X and Y are independent.

In Example 13.4, we have

$$\text{Ent}(Z) = \text{Ent}(0.5, 0.15, 0.35) = 1.4406),$$

so

$$\text{MInf}(W,Z) = \text{Ent}(Z) - \text{CEnt}(Z|W) = 1.4406 - 1.1464 = 0.2942.$$

Therefore, if we originally know the distribution of Z and then find out the value of W, we have gained 0.2942 bits of information about the value of Z.

13.4   Coding

The entropy of a probability distribution is a critical number in calculating how efficiently a string generated by a random process can be transmitted. This problem and other problems relating to the transmission of information were indeed the original motivation for the invention of information theory by Claude Shannon in the late 1940s and early 1950s (Luenberger 2006).

We address here the problem of encoding a finite string of characters as a string of bits. We assume that the character string is drawn from a finite alphabet, denoted $\mathscr{A}$.

Definition 13.6. Let $\mathcal{A}$ be a finite alphabet. Then the set of all strings of characters in $\mathcal{A}$ of length exactly n is denoted $\mathcal{A}^n$. The set of all finite strings is denoted $\mathcal{A}^\omega$.

In particular, we are interested in the alphabet of bits $\{0,1\}$, which we denote $\mathcal{B}$.

Definition 13.7. Let $\mathcal{A}$ be an alphabet. A *coding scheme for* $\mathcal{A}$ is an injection from $\mathcal{A}^\omega$ into $\mathcal{B}^\omega$.

That is, a coding scheme $\Gamma(s)$ is a function that maps every string over $\mathcal{A}$ in an unambiguous way; if $x, y \in \mathcal{A}$ and $x \neq y$, then $\Gamma(x) \neq \Gamma(y)$.

Suppose that we need to communicate very long strings over $\mathcal{A}$, and that these strings are generated by choosing each character in the string independently according to a distribution given by random variable X. We want to design a coding scheme Γ that encodes these strings efficiently in terms of the lengths of the bit strings output. We measure the efficiency of Γ by fixing a large value of n, asking what is the expected number of bits in $\Gamma(s)$, where s is a string of length n generated by this random process, and dividing that expected number by n. We then let n go to infinity; this is the asymptotic efficiency of Γ, denoted $\mathrm{BpC}(\Gamma, X)$, where BpC is bits per character. (If the limit as n goes to infinity does not exist, then $\mathrm{BpC}(\Gamma, X)$ is undefined.)

Definition 13.8. Let X be a random variable over $\mathcal{A}$. Let $X_1, \ldots, X_n$ be independent random variables with the same distribution as X. Then X^n will denote the joint distribution of $X_1, \ldots, X_n$.

Thus, X^n is a random variable over $\mathcal{A}^n$ corresponding to choosing each character in the string independently according to X.

Definition 13.9. Let X be a probability distribution over $\mathcal{A}$ and let Γ be a coding scheme for X. The *bit per character ratio* for Γ over X is defined as

$$\mathrm{BpC}(\Gamma, X) = \lim_{n \to \infty} \mathrm{Exp}(|\Gamma(X^n)|)/n,$$

assuming that this limit exists.

Example 13.10. Let $\mathcal{A} = \{a, b, c, d\}$ and let X have the distribution

$$P(X = a) = 3/4,$$
$$P(X = b) = 1/8,$$
$$P(X = c) = 1/16,$$
$$P(X = d) = 1/16.$$

Let Γ_1 be defined as follows. Do a character-by-character encoding, where $a \to 00$, $b \to 01$, $c \to 10$, and $d \to 11$. For example,

$$\Gamma_1(\text{aacaaadbaab}) = 0000100000001101000001.$$

For readability, we henceforth add commas, which aren't actually in the bit string; thus, the previous string will be written 00,00,10,00,00,00,11,01,00,00,01. Then for any string $s \in \mathscr{A}^\omega$, $|\Gamma_1(s)| = 2|s|$, where $|s|$ denotes the length of s. In particular, for any $s \in \mathscr{A}^n$, $|\Gamma_1(s)| = 2n$. Therefore, $\mathrm{BpC}(\Gamma_1, X) = 2$; this holds for any distribution X.

Example 13.11. Let $\mathscr{A}$ and X be as in Example 13.10. Let Γ_2 be defined as follows: Do a character-by-character encoding, where $a \to 0$, $b \to 10$, $c \to 110$ $d \to 111$. For example, $\Gamma_2(\text{aacaaadbaab}) = 0,0,110,0,0,0,111,10,0,0,10$.

Consider forming an n-character string in $\mathscr{A}^n$, and let $X_1, X_2, \ldots, X_n$ be random variables, where the value of X_i is the ith character in the string. Then $|\Gamma_2(X^n)| = |\Gamma_2(X_1)| + \ldots + |\Gamma_2(X_n)|$, so $\mathrm{Exp}(|\Gamma_2(X^n)|) = \mathrm{Exp}(|\Gamma_2(X_1)|) + \ldots + \mathrm{Exp}(|\Gamma_2(X_n)|) = n \cdot \mathrm{Exp}(|\Gamma_2(X)|)$, since all the X_i have the same distribution as X. But

$$\mathrm{Exp}(|\Gamma(X)|) = P(X = `a') \cdot |\Gamma_2(`a')| + P(X = `b') \cdot |\Gamma_2(`b')| + P(X = `c') \cdot |\Gamma_2(`c')| + P(X = `d') \cdot |\Gamma_2(`d')|$$
$$= (3/4) \cdot 1 + (1/8) \cdot 2 + (1/16) \cdot 3 + (1/16) \cdot 3 = 11/8 = 1.375.$$

Therefore, $\mathrm{Exp}(|\Gamma_2(X^n)|) = n \cdot \mathrm{Exp}(|\Gamma_2(X)|) = 1.375n$, so $\mathrm{BpC}(\Gamma_2, X) = 1.375$. Note that this is substantially smaller than $\mathrm{BpC}(\Gamma_1, X) = 2$.

It is not obvious on the face of it that Γ_2 is a coding scheme at all because it is not obvious that there could not be two alphabetic strings with the same encoding (once the commas are removed). However, we can prove that encodings are unique because this scheme is *prefix-free*; that is, no code for one character is a prefix for another. That being the case, the bit string can be read left to right, "peeling off" the code for each character at a time. For example, to decode the string "0100111101100," we observe that

- The first '0' is a code for 'a' and does not start the code for any other character.

- The '1' at index 2 is not in itself a code for anything.

- The '10' at index 2–3 is a code for 'b' and does not start the code for any other character.

- The '0' at index 4 is a code for 'a' and does not start the code for any other character.

- The '1' at index 5 and the '11' at index 5–6 are not in themselves codes for anything. The '111' at indices 5–7 is the code for 'd'.

```
R--->a  {0}
  |
  |->*--->b  {10}
      |
      |->*--->c  {110}
          |
          |->d  {111}


R  :  Root.
*  :  Interior  node.
Horizontal  link:  0.
Vertical  elbow:  1.
```

Figure 13.2. Binary tree for code Γ_2.

Continuing on in this way, we can recover the entire string.

We can generalize this observation in Definition 13.12 and Theorem 13.15.

Definition 13.12. A set S of bit strings is *prefix-free* if no element in S is a prefix of any other element.

A prefix-free code corresponds to a set of leaves in a binary tree (see Figures 13.2 and 13.3).

Definition 13.13. A *character code* over $\mathscr{A}$ is an injection from $\mathscr{A}$ to $\mathscr{B}^\omega$.

A character code Γ over $\mathscr{A}$ is *prefix-free* if $\Gamma(\mathscr{A})$ is a prefix-free set.

Definition 13.14. A function Γ from $\mathscr{A}^\omega$ is a *simple code* if

- Γ is a character code over $\mathscr{A}$;

- for any string $s = \langle s[1], \ldots, s[k] \rangle$, $\Gamma(s)$ is the concatenation of $\Gamma(s[1]), \ldots, \Gamma(s[k])$.

Theorem 13.15. *If Γ is a simple code that is prefix-free over $\mathscr{A}$, then Γ is an unambiguous coding scheme over $\mathscr{A}^\omega$.*

Proof: Using the left-to-right decoding method described in Example 13.11, it is clear that any bit string can be decoded in at most one way.   □

Example 13.16. Let $\mathscr{A}$ and X be as in Example 13.10. Let $\Gamma_3(s)$ be defined as follows: break s up into two-character blocks; if s is odd, there will be a final one-character block. We then can do the following block-by-block encoding

```
R--->aa {0}
   |
   |->*--->*--->ab {100}
        |      |
        |      |->ba {101}
        |
        |->*--->*--->ac {1100}
             |      |
             |      |->ad {1101}
             |
             |->*--->ca {1110}
                  |
                  |->*--->da {11110}
                       |
                       |->*--->*--->bb {1111100}
                            |      |
                            |      |->*--->bc {11111010}
                            |           |
                            |           |->cb {11111011}
                            |
                            |->*--->*--->bd {11111100}
                                 |      |
                                 |      |->db {11111101}
                                 |
                                 |->*--->*--->cc {111111100}
                                      |      |
                                      |      |->cd {111111101}
                                      |
                                      |->*--->dc {111111110}
                                           |
                                           |->*--->dd {1111111110}
                                                |
                                                |--->*--->a Final {111111111100}
                                                     |      |
                                                     |      |->b Final {111111111101}
                                                     |
                                                     |->*--->c Final {111111111110}
                                                          |
                                                          |->d Final {111111111111}
R : Root. \\
\* : Interior node. \\
Horizontal link: 0. \\
Vertical elbow: 1.
```

Figure 13.3. Binary tree for code Γ_3.

(the corresponding binary tree is shown in Figure 13.3):

$aa \to 0$	$ab \to 100$	$ac \to 1100$	$ad \to 1101$
$ba \to 101$	$bb \to 1111100$	$bc \to 11111010$	$bd \to 11111100$
$ca \to 1110$	$cb \to 11111011$	$cc \to 111111100$	$cd \to 111111101$
$da \to 11110$	$db \to 11111101$	$dc \to 111111110$	$dd \to 1111111110$

Final $a \to 111111111100$ Final $b \to 111111111101$ Final $c \to 111111111110$ Final $d \to 111111111111$

In a string of length n, there will be $\lfloor n/2 \rfloor$ blocks of length 2 and possibly one block of length 1. We can ignore the final block, in the limit as $n \to \infty$. The expected total length of the nonfinal blocks is

$$(n/2) \cdot [P(`aa') \cdot |\Gamma(`aa')| + P(`ab') \cdot |\Gamma(`ab')| + \ldots + P(`dd') \cdot |\Gamma(`dd')|]$$

$$= (n/2)[(9/16) \cdot 1 + (3/32) \cdot 3 + \ldots + (1/256) \cdot 10]$$

$$= 1.2129n$$

(if you work it out). So $\mathrm{BpC}(\Gamma_3) = 1.2129$. Note that we have achieved a still greater compression.

How efficient can we make these codes? The answer, as we shall prove next, is that we can get codes whose BpC is arbitrarily close to the entropy, and that we cannot do better than the entropy. In Examples 13.10–13.16, the entropy is $-((3/4)\log(3/4) + (1/8)\log(1/8) + (1/16)\log(1/16) + (1/16)\log(1/16) = 1.1863$, so the code in Example 13.16 is within 2.5% of optimal.

Examples 13.11 and 13.16 illustrate two techniques used to make the naive code of Example 13.10 more efficient. First, we assign shorter codes to more common characters and longer codes to less common ones, lowering the overall average. Second, we group the characters into blocks; this allows us to carry out the first technique with more delicate discrimination.

We now prove three theorems that show the effectiveness of these techniques. Theorem 13.17 states that by applying the first technique, we can get a simple prefix-free code with a BpC that is at most the entropy plus 1. Theorem 13.20 states that by applying both techniques, we can get a nonsimple code with a BpC that is arbitrarily close to the entropy. Theorem 13.22 states that no coding scheme of any kind can have a BpC that is less than the entropy.

Theorem 13.17. *Let $\mathcal{A}$ be an alphabet and let X be a random variable over $\mathcal{A}$. There exists a simple code Γ for $\mathcal{A}$ such that* $\mathrm{BpC}(\Gamma, X) \leq \mathrm{Ent}(X) + 1$.

Proof: We present Algorithm 13.1, which constructs such a code. As in Figures 13.2 and 13.3, we assign a leaf of a binary tree to each character in $\mathcal{A}$. The assignments are carried out in descending order of probability. Each character α_i is placed at depth $\lceil -\log(P(X = \alpha_i)) \rceil$ immediately to the right of the previous character, so the leaves of the tree slant downward to the right.

We can analyze the behavior of this algorithm in terms of how the nodes that are assigned cover the "original" leaves; that is, the leaves of the starting uniform tree of depth h. *Note:*

- N_i covers a subtree with 2^{h-L_i} original leaves.

- Since L_i increases with i, for $j < i$, the number of original leaves pruned on the jth step, 2^{h-L_j} is an integer multiple of 2^{h-L_i}.

function ConstructCode(**in:** $\mathcal{A}$: Alphabet; P: probability distribution)
return: code for $\mathcal{A}$.

$\{\, k \leftarrow |\mathcal{A}|;$
  $\langle \alpha_1 \ldots \alpha_k \rangle \leftarrow \mathcal{A}$ sorted in descending order of $P(\alpha)$.
  $h = \lceil -\log(P(\alpha_k)) \rceil;$
  Construct a binary tree of depth h. Label each left arc '0' and each right arc '1'.
  for $(i \leftarrow 1 \ldots k)\ \{$
      $L_i \leftarrow \lceil -\log(P(\alpha_i)) \rceil;$
      $N \leftarrow$ the leftmost node of depth L_i that is neither marked or pruned;
      mark N as "in use";
      assign to α_i the code that is the sequence of arc labels from the root to N;
      prune the subtree under N;
      $\}$
  return the assignment of codes to the α_i;
$\}$

Algorithm 13.1. Construct code.

- Therefore, the total number of original leaves that have been labeled or pruned before the start of the ith iteration of the **for** loop is a multiple of 2^{h-L_i}.

- Therefore, in the course of executing the **for** loop, the original leaves are labeled or pruned in consecutive, left-to-right order.

- The fraction of original leaves that are covered on the ith step is 2^{-L_i}. Since $\sum_i 2^{-L_i} \le \sum_i p_i = 1$, we do not run out of original leaves.

Let Γ be the code corresponding to this labeling. Note that for every i, $|\Gamma(\alpha_i)| = \lceil -\log(p_i) \rceil < -\log(p_i) + 1$. Therefore,

$$\mathrm{BpC}(\Gamma, X) = \mathrm{Exp}(|\Gamma(X)|) = \sum_i p_i |\Gamma(\alpha_i)| < \sum_i p_i(-\log(p_i) + 1) = \mathrm{Ent}(X) + 1. \quad \square$$

Example 13.18. Suppose that $\mathcal{A} = \{a, b, c, d, e\}$, and $p_1 = 0.3$, $p_2 = 0.27$, $p_3 = 0.21$, $p_4 = 0.17$, and $p_5 = 0.05$. Then $L_1 = 2, L_2 = 2, L_3 = 3, L_4 = 3$, and $L_5 = 5$. The corresponding tree is shown in Figure 13.4. The code Γ_4 is

$$\Gamma_4(a) = 00,$$
$$\Gamma_4(b) = 01,$$
$$\Gamma_4(c) = 100,$$
$$\Gamma_4(d) = 101,$$
$$\Gamma_4(e) = 11000.$$

So, $\mathrm{BpC}(\Gamma_4, X) = \mathrm{Exp}(|\Gamma_4(X)|) = 2.53$ and $\mathrm{Ent}(X) = 2.155$.

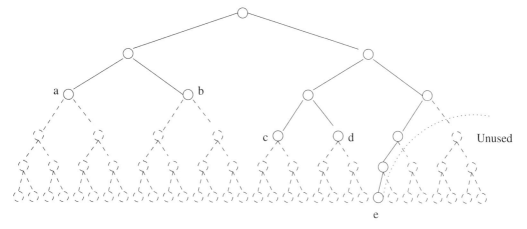

Figure 13.4. Tree for Example 13.18.

Clearly, this code isn't optimal, but it satisfies Theorem 13.17. We present an algorithm for generating the optimal code in Section 13.4.1.

Definition 13.19. Let $m \geq 1$. Let $\mathscr{A}^{\leq m}$ be the set of all strings of $\mathscr{A}$ of length at most m. An *m-block* code over alphabet $\mathscr{A}$ is an injection from $\mathscr{A}^{\leq m}$ to $\mathscr{B}^\omega$—that is, a code for every block of characters of size at most m. In particular, a character code is a 1-block code.

Code Γ is an *m-simple code* if

- code Γ restricted to $\mathscr{A}^{\leq m}$ is an m-block code;

- if s is a string of length greater than m, then $\Gamma(s)$ is obtained by breaking Γ into blocks of length m starting at the left, leaving a final block of length $\leq m$, and then concatenating the output of Γ on each of the blocks.

An m-simple code Γ is *prefix-free* if $\Gamma(\mathscr{A}^{\leq m})$ is a prefix-free set.

Theorem 13.20. *Let $\mathscr{A}$ be an alphabet and let X be a random variable over $\mathscr{A}$. Let $\epsilon > 0$, and let $m > 1/\epsilon$. There exists a prefix-free m-simple coding scheme Γ for $\mathscr{A}$ such that* $\mathrm{BpC}(\Gamma, X) \leq \mathrm{Ent}(X) + \epsilon$.

Proof: Consider $\mathscr{A}^m$ to be a "superalphabet" of character blocks of length m, and consider the distribution X^m over $\mathscr{A}$. Then, by Theorem 13.17, there exists a simple code Γ for $\mathscr{A}$ such that $\mathrm{BpC}(\Gamma, X^m) \leq \mathrm{Ent}(X^m) + 1$. (The encoding of the final blocks does not matter, since there is only one of these per string, and hence they add a negligible length as the length of the string goes to infinity.) However, since the "characters" of $\mathscr{A}^m$ are blocks of m characters of $\mathscr{A}$, we have $\mathrm{BpC}(\Gamma, X^m) = m \cdot \mathrm{BpC}(\Gamma, X)$. By Theorem 13.1(c), $\mathrm{Ent}(X^m) = m \cdot \mathrm{Ent}(X)$, so $m \cdot \mathrm{BpC}(\Gamma, X) \le m \cdot \mathrm{Ent}(X) + 1$. Dividing through by m, we get $\mathrm{BpC}(\Gamma, X) \le \mathrm{Ent}(X) + 1/m < \mathrm{Ent}(X) + \epsilon$. ☐

We will not give a rigorous proof of Theorem 13.22, that any coding scheme has a BpC of at least the entropy. Rather, we will give an approximate argument, and then wave our hands to indicate how the holes in the argument are fixed. First, we must prove Lemma 13.21.

Lemma 13.21. *Let D be a domain of size n, and let random variable X be uniformly distributed over D. Let Γ be an injection from D to $\mathcal{B}^\omega$. Then $\mathrm{Exp}(\Gamma(X)) \ge \log(n) - 3$.*

Note that we don't require that Γ is a prefix-free code, or that it will be an unambiguous code once we string characters and codes together. The only property we require of Γ is that it is one-to-one over D. Also, with a more careful analysis, we can improve the bound in this lemma, but it is sufficient as is for our purposes.

Proof: Clearly, the shortest average is attained when Γ uses all possible short bit strings. Note that there are two bit strings of length 1, four of length 2, eight of length 3, and so on. Therefore, the number of strings of length at most k is $2 + 4 + \ldots + 2^k = 2^{k+1} - 2$. To get n different strings, therefore, we must have $2^{k+1} - 2 \ge n$, so $k \ge \log(n+2) - 1 > \log(n) - 1$. Let us ignore the strings of length k. Considering only the strings of length less than k, their total length is $(\sum_{i=1}^{k-1} i \cdot 2^i) = (k-2) \cdot 2^k + 2$, so their average length is $((k-2) \cdot 2^k + 2)/(2^k - 2) > k - 2 > \log(n) - 3$. ☐

Theorem 13.22 (Shannon's theorem). *Let $\mathcal{A}$ be an alphabet and let X be a random variable over $\mathcal{A}$. Let Γ be a coding scheme for $\mathcal{A}$. Then $\mathrm{BpC}(\Gamma, X) \ge \mathrm{Ent}(X)$.*

Argument: Let $\alpha_1, \alpha_2, \ldots, \alpha_k$ be the characters of $\mathcal{A}$, and let $\langle p_1, p_2, \ldots, p_k \rangle$ be the distribution of X.

Let n be large, and consider the distribution X^n. With very high probability, a string of n characters chosen according to X^n has approximately $p_1 \cdot n$ occurrences of α_1, $p_2 \cdot n$ occurrences of α_2, ..., $p_k \cdot n$ occurrences of α_k. Assume, for simplicity, that $p_i \cdot n$ is an integer for all i.

Let us imagine, for the moment, that the only strings we have to worry about are those that have *exactly* $n \cdot p_i$ occurrences of α_i. That is, we let Y be a random variable that is uniformly distributed over those strings with *exactly* $n \cdot p_i$ occurrences of α_i. All of these strings are equally probable. According to the partition formula discussed in Section 8.3.4, the number of such strings Q is given by the formula

$$Q = C\left(\begin{array}{c} n \\ n \cdot p_1, n \cdot p_2, \ldots, n \cdot p_k \end{array} \right) = \frac{n!}{(n \cdot p_1)! \cdot (n \cdot p_2)! \ldots (n \cdot p_k)!}.$$

Therefore, by Lemma 13.21, for any coding scheme Γ, $\text{Exp}(|\Gamma(Y)|) > \log(Q) - 3$. Using a little algebra, we show that $\log(Q) \approx n \cdot \text{Ent}(p_1, \ldots, p_k)$. Each string of Y is, of course, n characters of X. Therefore, if X gave rise to Y, then we could say that $\text{BpC}(\Gamma(X)) \geq \lim_{n \to \infty} \text{Exp}(|\Gamma(Y)|)/n = \text{Ent}(p_1, \ldots, p_k)$, since the $3/n$ term becomes negligible.

Of course, the distribution over strings of length n that X actually generates is not Y; it is the multinomial distribution. However, the differences turn out not to matter in this calculation. To do the analysis it is convenient to divide strings into two categories; those that are "close to" Y (i.e., within some number of standard deviations) and those that are far from Y. We can then note the following gaps in our argument.

- The product $n \cdot p_i$ may not be an integer. That is purely of nuisance value; one can round to an integer with negligible effect on the computation.

- Some of the strings far from Y are, individually, much more probable than the strings in Y; an example is, the string that consists entirely of the most probable symbol repeated n times. But the total probability that X will generate *any* string far from Y is so small that they can be collectively ignored.

- Lots of strings close to Y don't exactly fit with Y; in fact, since we can vary by $\sqrt{n}$ independently in $k-1$ dimensions, there are on the order of $n^{(k-1)/2}$ times as many strings close to Y as there are in Y. But increasing the number of strings only makes it harder to keep the BpC small. It may seem surprising that in the opposite direction, multiplying the number of options by this large factor does not force an increase in the BpC, but the point is, taking the logarithm just amounts to adding a term of the form $(k-1)\log(n)/2$, which is negligible compared to n.

- Some of the strings close to Y are, in fact, more probable than the strings in Y (those that have more of the popular characters and fewer of the unpopular characters), and the total probabilities of these is nonnegligible. But they are actually not more probable enough to affect the calculation.

These arguments can be made rigorous, but doing so involves more calculation than is worthwhile here.

Let us return to the problem of computing $\log(Q)$:

$$\log(Q) = \log\left(\frac{n!}{(n \cdot p_1)! \cdot (n \cdot p_2)! \ldots (n \cdot p_k)!}\right)$$
$$= \log(n!) - [\log((n \cdot p_1)!) + \log((n \cdot p_2)!) + \log((n \cdot p_k)!)]. \qquad (13.2)$$

By Stirling's formula, $\ln(n!) = n\ln(n) - n + O(\ln(n))$, so $\log(n!) = n\log(n) - n/\ln(2) + O(\log(n))$. So Equation (13.2) becomes

$$\log(Q) = n\log n - n/\ln(2) - [(np_1\log(np_1) - np_1/\ln(2)) + \ldots$$
$$+ (np_k\log(np_k) - np_k/\ln(2)] + O(\log(n))$$
$$= n\log n - n/\ln(2) - np_1(\log n + \log p_1) - (n/\ln(2))p_1 + \ldots$$
$$+ np_k(\log n + \log p_k) - (n/\ln(2))p_k + O(\log(n).$$

By using the fact the p_i add up to 1, the $n\log n$ terms cancel, the $n/\ln(2)$ terms cancel, and what remains is

$$\log(Q) = -n(p_1\log(p_1) + \ldots + p_k\log(p_k)) + O(\log(n)) = n \cdot \text{Ent}(X) + O(\log(n)).$$

The key point about this argument is that the only assumption it makes about Γ is that Γ maps each string of characters to a unique bit string. Beyond that, it is just a counting argument; the argument is, essentially, that there are so many strings of length n with the right distribution that you will need $n \cdot \text{Ent}(p_1,\ldots,p_k)$ bits just to give each of these a different encoding.

Theorems 13.17, 13.20, and 13.22 support another argument in favor of measuring the information of E as $\log_2 P(E)$. If event E is one outcome of a random process P, and we need to communicate many long strings of outcomes of independent trials of P, then the best coding we can do will require, on average, $\log_2 P(E)$ bits for each occurrence of E. So it makes sense to say that E involves $\log_2 P(E)$ bits to be transmitted.

13.4.1   Huffman Coding

The Huffman coding algorithm computes the optimal simple code—that is, the simple code with the smallest possible BpC—given a distribution over an alphabet. The algorithm constructs a binary tree, comparable to Figure 13.2, in which the leaves are the alphabetic characters and the code for a letter corresponds to the path to the letter from the root.

The Huffman coding algorithm builds the tree bottom up. At each iteration of the main loop is a forest of disjoint trees. Initially, each character is in a tree by itself; labeled with a weight equal to its probability, at the end, there is a single tree. At each iteration, the two trees of minimum total weight become the two children of a new parent node; thus, the number of trees is reduced by one at each stage. The weight of the new tree is the sum of the weights of the children.

```
function T = Huffman(A: alphabet; P: distribution over A)
   for (each character u in A)
      create a leaf node for u labeled P(u);
   end
```

```
N4--->N3---> a {00}
     |       |
     |       |-> b {01}
     |
     |->N2---> c {10}
            |
            |->N1---> d {110}
                   |
                   |-> e {111}
```

Figure 13.5. Huffman code.

```
S = the set of all leaf nodes;
for I=1:size(A)-1
  [P,Q] = the two nodes in S with the smallest label
  create a new node N;
  make P and Q the two children of N;
  N.weight = P.weight + Q.weight;
  Add N to S and delete P and Q from S;
end
end
```

By applying this algorithm to the distribution in Example 13.6, we begin with five trees: characters, a, b, c, d, e, with weights 0.3, 0.27, 0.21, 0.17, 0.5, respectively. The algorithm proceeds through the following steps:

1. Make d and e children of $N1$ with weight 0.22.

2. Make c and $N1$ children of $N2$ with weight 0.43.

3. Make a and b children of $N3$ with weight 0.57.

4. Make $N2$ and $N3$ children of $N4$ with weight 1.

The resultant tree is shown in Figure 13.5. For the proof of the correctness of the Huffman coding algorithm, see Luenberger (2006).

13.5 Entropy of Numeric and Continuous Random Variables

The entropy of a discrete numeric random variable is defined in the same way in Section 13.3: $\text{Ent}(X) = \sum_v -P(X = v)\log(P(X = v))$. Note that this makes no use of the actual value v of X, just of the various values of the probabilities. As a consequence, it is a somewhat strange measure. For example, consider the following three random variables X, Y, Z, with probabilities given by

$$P(X = 0) = 1/2, \qquad\qquad P(X = 1) = 1/2,$$
$$P(Y = -0.01) = -1/4, \qquad P(Y = 0.01) = 1/4, \qquad P(Y = 1) = 1/2,$$
$$P(Z = 100) = 1/2, \qquad\qquad P(Z = 1000) = 1/2.$$

Intuitively, X seems much more similar to Y than to Z, and on most measures, such as expected value and standard deviation, it is. However, $\text{Ent}(X) = \text{Ent}(Z) = 1$, whereas $\text{Ent}(Y) = 3/2$.

As expected, the formula for the entropy of a continuous variable is obtained by replacing the probability by the probability density and summation by integration. If random variable X has a pdf $p(t)$, then

$$\text{Ent}(X) = \int_{-\infty}^{\infty} -p(t)\log(p(t))\,dt.$$

By working out the definite integrals, we can show that if X has the uniform distribution from l to $l + a$, then $\text{Ent}(X) = \log(a)$, which is very satisfying. We can also show that if X has the normal distribution $N_{\mu,\sigma}$, then $\text{Ent}(X) = \log(\sigma) + (1/2)\log(2\pi e)$, which is less elegant, although it is remarkable that it is a closed-form expression at all.

Note that this value can be negative; for instance, if $a < 1$ or $\sigma < 1/\sqrt{2\pi e}$. That seems odd; does finding out the value of this random variable constitute a *loss* of information? The answer is that we are not considering finding out the actual value, which would require infinite information. (Specifying all the digits of a real number is an infinite amount of information.) Rather, in the discrete case, we are considering moving from knowing that $P(X = v) = p$ to knowing that $P(X = v) = 1$; measuring the change in information; and averaging that information change over all v weighted by $P(X = v)$. In the continuous case, we are considering moving from knowing that $\tilde{P}(X = v) = f(v)$ to $\tilde{P}(X = v) = 1$; measuring the information change; and integrating that information change over all v weighted by $\tilde{P}(X = v)$. But if $\tilde{P}(X = v) > 1$, then indeed this constitutes a loss of information at v.

In general, if X is a continuous random variable and c is a constant, then $\text{Ent}(cX) = \log c + \text{Ent}(X)$. (The distribution of cX is c times wider and $1/c$ times the height of X, so the distribution is a lower probability density.)

See Luenberger (2006, Chapter 21) for further discussion of the entropy of continuous random variables.

13.6   The Principle of Maximum Entropy

The entropy of a random variable X is the expected gain in information from finding out the value of X. However, in many cases, it is more useful to think of it as a measure of *ignorance*;[1] it measures how much we *don't* know about the value of X if all we know is the distribution of X. This, in turn, suggests the following principle.

13.6.1 The Principle of Maximum Entropy

> If we do not know the distribution of a random variable X but we have some constraints on it, then we should assume that we are as ignorant as possible of the value of X, and therefore the true distribution is the one that maximizes the entropy, consistent with the constraints.

As we shall see, this principle has some pleasing consequences and some interesting applications to statistical inference. Carrying out the calculations, however, requires mathematical techniques that are beyond the scope of this book. The problem, in general, involves maximizing a nonlinear, multivariable function over a constraint space. The comparatively few cases where this can be done exactly require multivariable calculus; in most cases, it requires using techniques of numerical optimization. Therefore, we will state these consequences in Section 13.6.2 and we will describe some applications, but we will prove only one very simple case.

One very helpful feature of the entropy function, in either technique, is that entropy is a *strictly convex* function.[2] If the constraints are likewise convex, then (a) there is a unique local maximum, which is the true maximum; and (b) basic optimization techniques are guaranteed to converge to the true maximum.

13.6.2 Consequences of the Maximum Entropy Principle

Principle of indifference. If all we know about X is that it has k different values, then the uniform distribution $P(X = v) = 1/k$ is the maximum entropy distribution. By the maximum entropy principle, we should assume that $P(X = v) = 1/k$. This rule, of course, long predates the principle of maximum entropy; it is known as the *principle of indifference*.

We prove this here in the case $k = 2$; this is the only consequence of the maximum entropy principle that we actually prove. If p is the probability of one value, then $1 - p$ is the value of the other. Thus, we are looking for the value of p that maximizes $\text{Ent}(p, 1-p) = -(p\log(p) + (1-p)\log(1-p))$. By multiplying this expression through by $\ln(2)$, we get $-(p\ln(p) + (1-p)\ln(1-p))$, which clearly

---

[1] However, the stopped clock example of Section 13.2 illustrates that this view has to be taken with reservations. It would be strange to say that finding out that the clock is stopped makes you more ignorant of the time; it just makes you aware that you were already ignorant of the time.

[2] A function $f(\vec{v})$ is strictly convex if, for all $\vec{v}, \vec{u}$ and for all t such that $0 < t < 1$, $f(t\vec{v} + (1-t)\vec{u}) > tf(\vec{v}) + (1-t)f(\vec{u})$. A constraint $C(\vec{v})$ is convex if, for all $\vec{v}, \vec{u}$ satisfying C and for all t between 0 and 1, $t\vec{v} + (1-t)\vec{u}$ satisfies C.

has the same maximum but is easier to manipulate. Let us call this $f(p)$. The function f attains its maximum when $df/dp = 0$. Thus, $0 = df/dp = -(\ln(p) + 1 - \ln(1-p) - 1)$. Thus, $\ln(p) - \ln(1-p) = 0$, so $\ln(p) = \ln(1-p)$, or $p = 1 - p$, so $p = 1/2$.

Independent events. Suppose we know the probability distribution on X and the probability distribution on Y. Then the maximum entropy distribution on X, Y is the one for which X and Y are independent.

Specifically, suppose that we know $P(X = u) = p_u$ and $P(Y = v) = q_v$ for all values of v and u. Let $r_{u,v} = P(X = u, Y = v)$ be the unknown joint probability distribution over X, Y. Then we wish to maximize $\text{Ent}(\{r_{u,v}\})$ subject to the constraints $\sum_v r_{u,v} = p_u$ and $\sum_u r_{u,v} = q_v$. That maximum is attained with $r_{u,v} = p_u \cdot q_v$. (See Davis (1990 pp. 133–135) for the proof.)

This assumption can also be conditionalized: If $P(E \mid G)$, $P(F \mid G)$, $P(E \mid \neg G)$ and $P(F \mid \neg G)$ are specified, then the maximum likelihood assumption is that E and F are conditionally independent given G and $\neg G$. This is the assumption used in the analysis of independent evidence, Section 8.9.1.

Note, however, that the independence assumption is a consequence of the maximum entropy principle only in this particular state of knowledge; it does not follow in other states of knowledge. Suppose, for example, that X and Y are Boolean random variables, and that we know $P(X = t) = 1/10$ and $P(X = t, Y = t) = 1/50$. We might want to conclude, absent other information, that X can be taken to be independent of Y, and thus a random sample of Y, and that therefore $P(Y = t) = 1/5$. However, the maximum entropy principle does not at all support this. Rather, for the two unknown events, $X = f, Y = t$ and $X = f, Y = f$, it simply divides up the remaining probability evenly and decides that $P(X = f, Y = t) = P(X = f, Y = f) = 9/20$. Therefore, $P(Y = t) = 47/100$.

The problem is that the entropy function over X, Y treats the four events $\langle X = t, Y = t \rangle$; $\langle X = t, Y = f \rangle$; $\langle X = f, Y = t \rangle$; $\langle X = f, Y = f \rangle$ as just four separate, atomic values $\{tt, tf, ft, ff\}$ of the joint random variable $J = \langle X, Y \rangle$. The function loses track of the fact that these come from X and Y. Once it has done that, it has no reason to suspect that we are particularly interested in the event $Y = t$, which is $J = tt \cup J = ft$. We might just as well be interested in the event $Q = t$, defined as $J = tt \cup J = ff$. The maximum entropy calculation has no less reason to think that Q is independent of X than it does that Y is independent of X. But both statements cannot be true; in fact, it makes neither assumption.

Uniform distribution. If all we know about the continuous random variable X is that it is 0 outside the interval $[L, U]$, then the maximum entropy density function is the uniform distribution over $[L, U]$.

Normal distribution. If all we know about the continuous random variable X is that $\text{Exp}(X) = \mu$ and $\text{Std}(X) = \sigma$, then the maximum entropy density function is the normal distribution $N_{\mu,\sigma}$.

13.7   Statistical Inference

In the maximum entropy approach to statistical inference, we posit constraints that state that the actual probability of certain events is equal to their frequency in the data corpus; then we maximize the entropy relative to those constraints.

For example, the following entropy-based approach for an automated machine translation technique is discussed by Berger, Della Pietra, and Della Pietra (1996). We start with a corpus of parallel texts, such as the Canadian Hansard, which is the proceedings of the Canadian Parliament, published in English and French. We want to use the information in this corpus to translate a new text.

Let us suppose, for simplicity, that words in French can be matched one-to-one with words in English, and let us further suppose that this matching has been carried out. The problem still remains that the same word in one language is translated into the other in different ways, depending on its meaning, its syntactic function, its context, and so on. For example, the English word "run" is translated in the Hansards as *épuiser, manquer, écouler, accumular, aller, candidat, diriger,* and others. (The most obvious translation, *courir,* meaning the physical activity, does not occur much in the Hansards—the subject matter of Parliamentary debate is somewhat specialized.)

As clues to the correct translation for word W in sentence S here, we use the three words that precede W in S and the three words that follow W in S. Those six words give a large measure of context, which in most cases is sufficient to choose the correct translation. A human reader seeing only the seven word segment of the sentence can generally pick the right translation for the word.[3]

Thus, for each English word e, for each possible translation f, and for each set of six context words $c_1, c_2, c_3, c_4, c_5, c_6$, we wish to calculate the probability $P(Tr(e) = f \mid Context = \langle c_1, c_2, c_3, \cdot c_4, c_5, c_6 \rangle)$. For example, in translating the word "run" in the sentence, "Harold Albeck plans to run for comptroller of New York in 2011," we consider

$$P(Tr(\text{``}run\text{''}) = f \mid Context = \langle \text{``Albeck,'' ``plans,'' ``to''} \cdot \text{``for,'' ``comptroller'' ``of''} \rangle),$$

and then we choose the value of f for which this is maximal. Thus, for each English word e we have a probabilistic model of its translation in which the elementary events are tuples of the form $\langle f, c_1, \ldots, c_6 \rangle$.

However, of course, this particular sequence of seven words "Albeck plans to run for comptroller of" almost certainly never occurs in the Hansards, so we cannot calculate this probability directly. Instead, we take the following steps:

1. Identify a (large) number of patterns that are helpful in disambiguating "run," such as "either $c_4, c_5,$ or c_6 is the word 'for'," or "c_3 is 'to'."

---

[3]This has to be modified, of course, for the first three and last three words in the sentence, but the modification is straightforward. Essentially, we view every sentence as being separated from the next by three periods, and we take these periods as "words."

2. Conceptually, characterize each such pattern as the union of elementary events.

3. Impose the constraint that the conditional probability of the translation, given the pattern, is equal to the conditional frequency of the translation, given the pattern.

4. Find the maximum entropy solution for the elementary probabilities subject to the constraints in (3).

The wording of (4) suggests that we would have to precompute a probability for *every* possible translation of *every* word in *every* possible context of six words. Of course, such a huge distribution could not even be stored, let alone calculated. All we need is the relative conditional probabilities for the context that actually occurs in the sentence being translated, and this can be computed from the relevant patterns. However, the solution is the same as if we had carried out the immense calculation described above.

Exercises

Exercise 13.1. Compute the following quantities (use MATLAB)

(a) $\text{Ent}(1/3, 1/3, 1/3)$.

(b) $\text{Ent}(1/4, 1/4, 1/2)$.

(c) $\text{Ent}(1/5, 2/5, 2/5)$.

(d) $\text{Ent}(1/10, 1/10, 8/10)$.

Exercise 13.2. Let X be a random variable with values a, b, c, d, and let Y be a random variable with values p, q, r with the joint distribution shown in Table 13.1.

Using MATLAB, compute the following quantities: $\text{Ent}(X)$, $\text{Ent}(Y)$, $\text{Ent}(X, Y)$, $\text{Ent}(X \mid Y = p)$, $\text{CEnt}(X \mid Y)$, $\text{CEnt}(Y \mid X)$, $\text{MInf}(X, Y)$.

Exercise 13.3. Given the prefix-free code

'a' $\rightarrow$ 00   'b' $\rightarrow$ 01   'c' $\rightarrow$ 100   'd' $\rightarrow$ 101   'e' $\rightarrow$ 110   'f' $\rightarrow$ 111

decode the following string: 0110111001011101110001110.

Exercise 13.4. Suppose that $P(\text{'a'}) = 0.4$; $P(\text{'b'}) = 0.15$; $P(\text{'c'}) = 0.14$; $P(\text{'d'}) = 0.12$; $P(\text{'e'}) = 0.1$; $P(\text{'f'}) = 0.09$.

(a) What is the entropy of this distribution? (Use MATLAB.)

	a	b	c	d
p	0.2	0.05	0.02	0.03
q	0.01	0.04	0.08	0.07
r	0.1	0.3	0.04	0.06

Table 13.1. Values for Exercise 13.2.

(b) What is the BpC of the code in Exercise 13.3 relative to this distribution?

(c) Find the Huffman code for this distribution.

(d) What is the BpC of the Huffman code for this distribution?

Exercise 13.5. Section 13.6.2 claims that, given the distributions of two random variables X and Y, the maximum entropy for the joint distribution is achieved when the two variables are independent.

Suppose we have two three-valued random variables X and Y with the distributions

$$P(X = a) = 0.6, \qquad P(X = b) = 0.3, \qquad P(X = c) = 0.1,$$
$$P(Y = 1) = 0.8, \qquad P(Y = 2) = 0.1, \qquad P(Y = 3) = 0.1.$$

(a) Compute the entropy of the joint distribution of X, Y on the assumption that X and Y are independent (use MATLAB).

(b) Construct a different joint distribution for X, Y that is consistent with these constraints, and compute its entropy (use MATLAB).

Problem

Problem 13.1. What is the entropy of the binomial distribution $B_{n,p}(k)$? *Hint:* If you use Theorem 13.1, this is an easy problem. If you start to write down the terms of the binomial distribution, you are entirely on the wrong track.

Chapter 14

Maximum Likelihood Estimation

Maximum likelihood estimation (MLE) is one of the basic techniques of classical statistics. In this chapter, we discuss several problems that use MLE techniques. *Note:* Almost all the derivations of maximum likelihood estimates in this chapter require calculus, and most require multivariable calculus. The derivations appear at the end of each section. The results are more important than following the derivations.

Suppose that we have a collection of data, and that we have some reason to believe that these data were generated by a random process of some specific category, with some unknown parameters. Based on the data, we want to estimate the value of these parameters. Let's consider two examples.

1. The data are the result of randomly sampling k items out of a population of size n and testing each item for a specified property. The actual frequency of the property in the population is an unknown value p. We want to estimate p. This type of sampling problem is discussed in Chapter 11 and in Section 14.1 below.

2. The data are the result of independent samples of a normal distribution $N_{\mu,\sigma}$, where μ and σ are unknown. We want to estimate the values of μ and σ. This problem is discussed in Section 14.4.

We need to determine the parameters of the process from the data. This procedure can be cast in probabilistic terms as follows. Let D be the data. Let V be a random variable that ranges over the possible values of the parameter, in a given problem. In the first example, the domain of V is [0,1], the set of possible values of the parameter p. In the second example, the domain of V is $(-\infty,\infty) \times (0,\infty)$, the set of possible values of $\langle \mu, \sigma \rangle$.

We want to estimate the value of V given D. Casting this in probabilistic terms, we want to say something about the distribution of $P(V \mid D)$. By Bayes' law, for any value v, $P(V = v \mid D) = P(D \mid V = v) \cdot P(V = v)/P(D)$. The denominator $P(D)$ is a fixed normalization constant, and so does not affect the choice of v.

The term $P(V = v)$ is the prior probability distribution for V; that is, the evaluation of the likelihood of parameter values before you see any data. For example, Section 11.3 discussed using the uniform distribution over $[0, 1]$ for parameter p in the sampling problem of the first example.

In the second example, however, a reasonable choice for a prior distribution over $\langle \mu, \sigma \rangle$ is much less apparent. Since μ ranges over (∞, ∞) and σ ranges over $(0, \infty)$, there does not exist a uniform distribution; and it is not at all clear what alternative prior distribution would be "reasonable."

Lacking any information about $P(V = v)$, therefore, an approach that is often taken is to ignore this term altogether[1] and simply look for the value of v that maximizes $P(D \mid V = v)$ (or $\tilde{P}(D \mid V = v)$, for continuous models.)

In the remainder of this chapter, we discuss the calculation and application of MLE for a number of different kinds of random processes.

14.1 Sampling

Let $\vec{D}$ be a sequence of n bits containing m ones and $n - m$ zeros. Suppose that we conjecture that $\vec{D}$ was produced by n flips of a coin of some weight p. This is the situation analyzed at length in Chapter 11. If all we need is the maximum likelihood estimate, the analysis is much simpler; the maximum likelihood estimate of p is just m/n, which is what we would expect.

For example, suppose that $n = 7, k = 5$. Then the MLE for p is $5/7 = 0.714$, The probability $P(\vec{D}) = p^5(1-p)^2 = 0.0152$. By contrast, if $p = 1/2$, then $P(\vec{D}) = 0.0078$; and if $p = 3/4$ then $P(\vec{D}) = 0.0148$.

Derivation: The derivation of the MLE in this case is simple. We have $P(\vec{D}) = p^m \cdot (1-p)^{n-m}$. Let us write $f(p) = p^m \cdot (1-p)^{n-m}$. The function $f(p)$ reaches its maximum when

$$0 = \frac{df}{dp} = mp^{m-1}(1-p)^{n-m} - (n-m)p^m(1-p)^{n-m-1}.$$

Rearranging the terms and dividing through by $p^{m-1}(1-p)^{n-m-1}$, we get $(n-m)p = m(1-p)$, so $p = m/n$.

---

[1]This approach is generally considered inconsistent with the likelihood interpretation of probability, which would *require* instead that the data be combined with some prior distribution over V.

14.2   Uniform Distribution

Let $\vec{D}$ be a sequence of n real numbers. Suppose that we conjecture that $\vec{D}$ is the output of independent samples of a uniform distribution from L to U, where L and U are unknown. Then the MLE is $L = \min(\vec{D})$ and $U = \max(\vec{D})$.

For example, let $\vec{D} = \langle 0.31, 0.41, 0.59, 0.27, 0.18 \rangle$. Then the maximum likelihood estimate is $[0.18, 0.59]$. For that value, $\tilde{P}(\vec{D}) = 1/(0.59 - 0.18)^5 = 86.31$. By contrast, if we choose $L = 0, U = 1$, then $\tilde{P}(\vec{D}) = 1$.

Derivation: Let $X_{L,U}^n$ be the process that generates n independent samples of the uniform distribution over $[L, U]$. Then

$$\tilde{P}(X_{L,U}^n = \vec{D}) = \begin{cases} \frac{1}{(U-L)^n} & \text{if } \vec{D}[i] \in [L, U] \text{ for } i = 1, \ldots, n, \\ 0 & \text{otherwise.} \end{cases}$$

This is maximized when $L = \min(\vec{D})$ and $U = \max(\vec{D})$.

14.3   Gaussian Distribution: Known Variance

Let $\vec{D}$ be a sequence of n real numbers. Suppose that we conjecture that $\vec{D}$ is the output of independent samples of a Gaussian distribution $N_{\mu,\sigma}$, where σ is a fixed value but μ is unknown. The maximum likelihood estimate for μ is that it is the mean of $\vec{D}$, denoted $\text{Exp}(\vec{D})$: $\text{Exp}(\vec{D}) = \sum_{i=1}^n \vec{D}[i]/n$.

For example, again let $\vec{D} = \langle 0.31, 0.41, 0.59, 0.27, 0.18 \rangle$, and let $\sigma = 1$. Then the maximum likelihood estimate of μ is $\text{Exp}(\vec{D}) = (0.31 + 0.41 + 0.59 + 0.27 + 0.18)/5 = 0.352$. At that value, the probability density is

$$\tilde{P}(X_{\mu,\sigma}^n = D) = \Pi_{i=1}^n \frac{\exp(-(\vec{D}[i] - \mu)^2/2\sigma^2)}{\sqrt{2\pi}\sigma}$$

$$= \frac{\exp(-(0.31 - 0.352)^2/2)}{\sqrt{2\pi}} \cdot \ldots \cdot \frac{\exp(-(0.18 - 0.352)^2/2)}{\sqrt{2\pi}}$$

$$= 0.0096.$$

(To avoid squinting at complicated superscripts in small font, we write $\exp(x)$ for e^x. Do not confuse this with Exp, the expected value.)

By contrast, if we choose $\mu = 0.25$, then $\tilde{P}(X_{\mu,\sigma}^n = D) = 0.0094$. If we choose $\mu = 0.59$, then $\tilde{P}(X_{\mu,\sigma}^n = D) = 0.0084$.

Derivation: Let $X_{\mu,\sigma}^n$ be the process that generates n independent samples of the normal distribution. Then

$$\tilde{P}(X_{\mu,\sigma}^n = \vec{D}) = \Pi_{i=1}^n \frac{\exp(-(\vec{D}[i] - \mu)^2/2\sigma^2)}{\sqrt{2\pi}\sigma} = \frac{\exp(-\sum_{i=1}^n (\vec{D}[i] - \mu)^2/2\sigma^2)}{(\sqrt{2\pi}\sigma)^n}.$$

$$(14.1)$$

Obviously, for any fixed value of σ, Equation (14.1) is a decreasing function of $\sum_{i=1}^{n}(\vec{D}[i] - \mu)^2$. Call this $f(\mu)$. Thus, $\tilde{P}(X_{\mu,\sigma}^n = \vec{D})$ reaches its maximum when $f(\mu)$ reaches a minimum. The minimum is attained when the derivative $df/d\mu = 0$. Thus,

$$0 = df/d\mu = 2\sum_{i=1}^{n}(\vec{D}[i] - \mu) = 2n\mu - 2\sum_{i=1}^{n}\vec{D}[i].$$

So $\mu = \sum_{i=1}^{n}\vec{D}[i]/n = \text{Exp}(\vec{D})$.

All of the remaining MLE computations we discuss in this chapter are based on Gaussian distributions of one kind or another, so they all involve minimizing sums of squares of some kind. Such minimizations are called *least squares* techniques. In fact, one of the major motivations that leads scientists and statisticians to assume Gaussian distributions is precisely that the assumption leads to these elegant least squares problems.

14.4   Gaussian Distribution: Unknown Variance

As in Section 14.3, let $\vec{D}$ be a sequence of n real numbers, and let us conjecture that $\vec{D}$ is the output of independent samples of a normal distribution $N_{\mu,\sigma}$. However, this time let us suppose that both μ and σ are unknown. The maximum likelihood estimate of μ is again $\text{Exp}(\vec{D})$. The maximum likelihood estimate of σ is calculated as follows. We define the *variance* and the *standard deviation* of $\vec{D}$ as $\text{Var}(\vec{D}) = \sum_{i=1}^{n}(\vec{D} - \mu)^2/n$, and $\text{Std}(\vec{D}) = \sqrt{\text{Var}(D)}$, respectively. The maximum likelihood estimate of σ is $\text{Std}(\vec{D})$.

The mean, variance, and standard deviation of a *data collection* are closely related to the expected value, variance, and standard deviation of a *random variable* discussed in Chapter 9, and we use the same symbols in both cases, but the distinction should be kept in mind.

For example, we again let $\vec{D} = \langle 0.31, 0.41, 0.59, 0.27, 0.18 \rangle$. The MLE for μ is 0.352 and the MLE for σ is 0.1401. For these values, $\tilde{P}(X_{\mu,\sigma}^n = \vec{D}) = 15.39$. This may be contrasted with the value obtained in Section 14.3 of $\tilde{P}(X_{\mu,\sigma}^n = \vec{D}) = 0.0096$ for $\mu = 0.352, \sigma = 1$.

It is meaningful also to compare this with the pdf of 86.31 computed in Section 14.2 for this same dataset with a uniform distribution over $[0.18, 0.59]$. The significance of the comparison is that if we consider a heterogeneous space of models containing both normal and uniform distributions, then the MLE is the uniform distribution over $[0.18, 0.59]$.

For some other statistical purposes, not discussed in this book,[2] the variance of a data collection is calculated as $\sum_{i=1}^{n}(\vec{D}[i] - \mu)^2/(n-1)$. In particular,

---

[2]This alternative calculation gives an *unbiased* estimate.

in MATLAB, the function calls std(D) and var(D) use the denominator $n-1$.
To get the variance and standard deviation computed with denominator n, we
have to use the function calls std(D,1) and var(D,1). We have to be careful
with this.

Derivation: The derivation of the MLE for μ is the same as that in Section 14.3.
The derivation of σ takes a little more work. We again start with the expression

$$\tilde{P}(X_{\mu,\sigma}^n = \vec{D}) = \Pi_{i=1}^n \frac{\exp(-(\vec{D}[i]-\mu)^2/2\sigma^2)}{\sqrt{2\pi}\sigma}.$$

Note that if σ is close to 0, the pdf $\tilde{P}(X = \vec{D})$ is very small because the exponent $-\sum_{i=1}^n (\vec{D}[i]-\mu)^2/2\sigma^2)$ is very large and negative; and if σ is very large,
then again $\tilde{P}(X = \vec{D})$ is small because the exponent is about 0, the exponential is therefore about 1, so the overall expression has size proportional to $1/\sigma$.
Thus, the maximum is reached at some value in between.

Let

$$g(\sigma) = \ln(\tilde{P}(X_{\mu,\sigma}^n = \vec{D}))$$

$$= \left[\sum_{i=1}^n -(\vec{D}[i]-\mu)^2/2\sigma^2 - \ln(\sigma) \right] - \ln(\sqrt{2\pi})$$

$$= \left[\sum_{i=1}^n -(\vec{D}_i - \mu)^2/2\sigma^2 \right] - n\ln(\sigma) - \ln(\sqrt{2\pi}).$$

The maximum is attained when $0 = \partial g/\partial\sigma = [\sum_{i=1}^n (\vec{D}[i]-\mu)^2/\sigma^3] - n/\sigma$, so
$\sigma^2 = \sum_{i=1}^n (\vec{D}[i]-\mu)^2/n = \text{Var}(\vec{D})$, and $\sigma = \text{Std}(\vec{D})$.

Note that in doing this MLE calculation, it is not correct to ignore the normalization factor $1/\sqrt{2\pi}\sigma$ because it changes from one distribution to another.
Normalization factors can be ignored in comparing two probabilities or probability densities from the *same* distribution, but not, generally, in comparing
probabilities or densities from two *different* distributions.

14.5   Least Squares Estimates

Let $D = \{\langle x_1, y_1 \rangle, \ldots, \langle x_n, y_n \rangle\}$ be a collection of n two-dimensional vectors. Let
us conjecture that the y-coordinate is a linear function of the x-coordinate
$f(x) = ax + b$ plus some noise that follows a normal distribution around $f(x)$
with some fixed standard deviation σ. We take the x-coordinates to be given;
we are not concerned with how they have been generated. We wish to find the

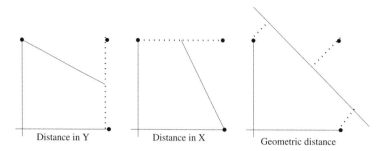

Figure 14.1. Least squares approximations.

best values of a and b. Let $X_{a,b}^n$ be the associated distribution. Then

$$\tilde{P}(X_{a,b}^n = D) = \Pi_{i=1}^n \frac{\exp(-(y_i - (ax_i + b))^2/2\sigma^2}{\sqrt{2\pi}\sigma}$$
$$= \frac{\exp(-(\sum_{i=1}^n (y_i - (ax_i + b))^2/2\sigma^2}{(\sqrt{2\pi}\sigma)^n}. \qquad (14.2)$$

For fixed σ, Equation (14.2) is maximized for the values of a and b where $\sum_{i=1}^n (y_i - (ax_i + b))^2$ is minimized. The line $y = ax + b$ is the *least squares approximation* of the data. This is the line L that comes "closest" to the points in D, where distance is measured in terms of the sum of the squares of the difference in y-coordinates. That is, if we draw a vertical line from each point in D to L, and you compute the sum of the squares of the lengths of the lines, we find the line L that minimizes that sum.

Note that the line L does *not* minimize the sum of the squares of the geometric distance from the points to the line; that would correspond to normal noise in both the x- and y-coordinates. (Section 14.6 shows how to solve that problem.) Nor does it minimize the sum of the squares of the distance from the points to the line along the x-coordinate, which would correspond to viewing x as a function of y with noise. Each of these possibilities gives a different line (Figure 14.1).

For example, consider the dataset $D = \{\langle 0,1 \rangle, \langle 1,1 \rangle, \langle 1,0 \rangle\}$. Viewing y as a function of x, the least squares line is $y(x) = 1 - 0.5x$. The mean square error is $[(1-y(0))^2 + (1-y(1))^2 + (0-y(1))^2]/3 = 1/6$. Viewing x as a function of y, the least squares line is $y(x) = 1 - 0.5x$. The mean square error is again $1/6$ (the equality of the two mean square errors is an artifact of the symmetry of the point set and does not hold in general). The geometrically closest fit is the line $x + y = 4/3$. The mean squared distance from the points to the line is $1/9$.

A drawback of the least squares approximation is that it tends to do badly in the presence of *outliers*, rare erroneous or anomalous cases that lie far away from the general pattern. Since the least squares technique "charges" a cost that is quadratic in the distance, it tends to work hard to try to reduce the distance to the outliers; if the outliers are anomalies or garbage, that is probably misdirected effort. A better estimation function in these cases may be the median approximation, which minimizes the cost function $\sum_i |y_i - (ax_i + b)|$. This corresponds to the maximum likelihood estimate for the random process $Y = aX + b + N$, where the distribution of the noise is given as $P(N = d) = e^{-|d|}/2$. However, the calculation of the MLE for this case is more difficult.

Derivation: Let $f(a,b) = \sum_{i=1}^{n} (y_i - (ax_i + b))^2$. We wish to compute the values of a and b where f is minimal. These can be computed by finding the values of a and b where the partial derivatives of $f(a,b)$ are 0:

$$0 = \frac{\partial f}{\partial a} = \sum_{i=1}^{n} 2x_i \cdot (y_i - (a \cdot x_i + b)),$$

$$0 = \frac{\partial f}{\partial b} = \sum_{i=1}^{n} -2(y_i - (a \cdot x_i + b)).$$

Collecting common terms in a and b, we get the pair of equations, linear in a and b,

$$\left(\sum_{i=1}^{n} x_i^2 \right) a + \left(\sum_{i=1}^{n} x_i \right) b = \sum_{i=1}^{n} x_i y_i,$$

$$\left(\sum_{i=1}^{n} x_i \right) a + nb = \sum_{i=1}^{n} y_i.$$

For example, with the sample set $D = \{\langle 0,1 \rangle, \langle 1,1 \rangle, \langle 1,0 \rangle\}$, we get the system of equations

$$2a + 2b = 1,$$
$$2a + 3b = 2.$$

The solution is $a = -0.5, b = 1$.

The least squares approximation for a function of an n-dimensional vector. is similar. Let D be a sequence of pairs $\langle \vec{D}_i, y_i \rangle$, where $\vec{D}$ is an n-dimensional vector. Suppose we conjecture that y is a linear function of $\vec{D}$ plus normal noise; that is, $y_i = \vec{a} \bullet \vec{D}_i + b + n_i$, where n_i is normally distributed. We wish to find the maximum likelihood estimate of $\vec{a}$. The analysis is exactly analogous to the analysis above of the case where y is a function of the single variable x. The MLE for $\vec{a}$ is the value that minimizes $\sum_{i=1}^{n} (y_i - \vec{a} \bullet \vec{D}_i - b)^2$; setting each of the partial derivatives to zero gives a set of linear equations satisfied by $\vec{a}, b$.

14.5.1   Least Squares in MATLAB

As discussed in Section 5.4, the MATLAB back-slash operator returns the least-squares solution to an overconstrained system of equations. Let Q be an $m \times n$ matrix such that Rank(A) = $n < m$, and let $\vec{c}$ be an m-dimensional vector. The system of equations $Q\vec{x} = \vec{c}$ has m equations in n unknowns; in general, it is overconstrained and has no solutions. The MATLAB expression Q \ c returns the m-dimensional vector x such that $|\vec{c} - Q \cdot \vec{x}|$ is minimal.

We can use this operator to solve the least-squares problem as follows. Let $\{\vec{D}_1, \ldots, \vec{D}_m\}$ be a set of m n-dimensional vectors, and let $\vec{Y} = \langle y_1, \ldots, y_m \rangle$ be a vector of values. We are looking for the values of $\vec{a}$ and b such that $\sum_i (y_i - (\vec{a} \bullet \vec{D}_i + b))^2$ is minimal. Therefore, we construct the $m \times (n+1)$ matrix Q whose ith row is D followed by 1, and call Q \ Y. The result is an $(n+1)$-dimensional vector $\vec{x}$ such that $\vec{x}[1, \ldots, n] = \vec{a}$ and $\vec{x}[n+1] = b$.

For example, the least squares calculation in Figure 14.1 is carried out as follows:

```
>> Q = [0,1;1,1;1,1];
>> Y = [1;1;0];
>> X=Q\Y;
>> A=X(1)
A =
    -0.5000

>> B=X(2)
B =
    1.0000
```

As another example, consider the following points:

$$\vec{D}_1 = \langle 0,0,0 \rangle, y_1 = 1;$$
$$\vec{D}_2 = \langle 0,1,1 \rangle, y_2 = 6;$$
$$\vec{D}_3 = \langle 1,2,2 \rangle, y_3 = 12.05;$$
$$\vec{D}_4 = \langle 2,1,0 \rangle, y_4 = 5.03;$$
$$\vec{D}_5 = \langle 2,2,2 \rangle, y_5 = 13.08.$$

(These points are generated by the function $y = 1 + \vec{D}[1] + 2\vec{D}[2] + 3\vec{D}[3] + 0.01\vec{D}[1]^2 - 0.01\vec{D}[2]^2 + 0.01\vec{D}[3]^3$.) Then we can find the least squares approximation as follows:

```
>> Q = [0,0,0,1; 0,1,1,1; 1,2,2,1; 2,1,0,1; 2,2,2,1];
>> Y = [1; 6; 12.05; 5.03; 13.08];
>> X=Q\Y;
>> A = X(1:3)
```

```
A  =
     1.0360
     1.9600
     3.0460

>> B=X(4)
B  =
   0.9980
```

As can be seen, the linear part of the function is recovered quite accurately. The accuracy of the approximation can be quantified by computing $\vec{C} = D \cdot \vec{a} + b$ and evaluating $|\vec{C} - \vec{Y}|$:

```
>> C= Q(:,1:3)*A +B
C  =
     0.9980
     6.0040
    12.0460
     5.0300
    13.0820

>> norm(C-Y)
ans  =
     0.0063
```

14.6 Principal Component Analysis

Principal component analysis (PCA) is a further variant of the least squares estimate described in Section 14.5. Let $\mathscr{D} = \{\vec{D}_1, \ldots, \vec{D}_n\}$ be a collection of m-dimensional vectors. We conjecture that the vectors are generated by taking points in some k-dimensional affine space $\mathscr{S}$ and adding a noise vector $\vec{e}$. The direction of the noise $\vec{e}$ is uniformly distributed over the space of vectors orthogonal to $\mathscr{S}$; its length is normally distributed, with variance σ^2. Which subspace $\mathscr{S}$ is the maximum likelihood estimate?

This problem can be solved by using the theory of the singular value decomposition (SVD) discussed in Section 7.7. Let $\vec{\mu}$ be the mean of the vectors in $\mathscr{D}$. Let M be the matrix such that $M[i,:] = \vec{D}_i - \vec{\mu}$. Let $\hat{u}_1, \ldots, \hat{u}_k$ be the first k right singular vectors of M. Then the space $\{\vec{\mu} + \sum_{i=1}^{k} t_i \hat{u}_i \mid t_i \in \mathbb{R}\}$ is the maximum likelihood estimate.

The right singular vectors $\hat{u}_1, \ldots, \hat{u}_m$ are called the *principal components* of the data set $\mathscr{D}$, and the process of computing them is *principal component analysis*. Historically, singular value decomposition was developed for use in differential geometry by Beltrami and Jordan, independently, in the 1870s, and PCA was developed for use in statistics by Pearson in 1901.

Another way of viewing PCA is as follows. Define the mean and variance of the collection $\mathscr{D}$ analogously to the mean and variance of a set of numbers: that is, the mean of $\mathscr{D}$ is $\vec{\mu} = \mathrm{Exp}(\mathscr{D}) = \sum_{i=1}^{n} \vec{D}/n$, and the variance of $\mathscr{D}$ is $\mathrm{Var}(\mathscr{D}) = \sum_{i=1}^{n} |\vec{D}_i - \vec{\mu}|^2/n$. (Note that the mean is an m-dimensional vector and the variance is a scalar quantity whose dimension is distance squared.)

Lemma 14.1 is easily proven (Problem 14.1).

Lemma 14.1. *Let D be a collection of m-dimensional vectors. Let $\mathscr{U}$ and V be orthogonal complements in $\mathbb{R}^m$. Then $\mathrm{Var}(D) = \mathrm{Var}(\mathrm{Proj}(D, \mathscr{U})) + \mathrm{Var}(\mathrm{Proj}(D, V))$.*

Therefore, the k dimensional subspace $\mathscr{U}$ that maximizes the value of $\mathrm{Var}(\mathrm{Proj}(D, \mathscr{U}))$ is the orthogonal complement of the $(m - k)$-dimensional subspace V that minimizes the value of $\mathrm{Var}(\mathrm{Proj}(D, V))$. We say that the subspace $\mathscr{U}$ *accounts for a fraction f of the variance of D* where $f = \mathrm{Var}(\mathrm{Proj}(D, \mathscr{U}))/\mathrm{Var}(D)$. The k-dimensional subspace that accounts for the maximal fraction of the variance is thus the subspace spanned by the k first singular vectors; the $m - k$ dimensional subspace that accounts for the minimal fraction of the variance is the subspace spanned by the $m - k$ last singular vectors.

For instance, consider the previous dataset: $D = \{\mathbf{a}, \mathbf{b}, \mathbf{c}\}$, where $\mathbf{a} = \langle 1, 0 \rangle$, and $\mathbf{b} = \langle 1, 1 \rangle$, and $\mathbf{c} = \langle 0, 1 \rangle$. The mean of this set is $\mathbf{m} = \langle 2/3, 2/3 \rangle$. The total variance is $\mathrm{Var}(D) = (d(\mathbf{a}, \mathbf{m})^2 + d(\mathbf{b}, \mathbf{m})^2 + d(\mathbf{c}, \mathbf{m})^2)/3 = (5/9 + 2/9 + 5/9)/3 = 4/9$.

The first principal component is $\hat{u}_1 = \langle \sqrt{2}/2, -\sqrt{2}/2 \rangle$ and the second principal component is $\hat{u}_2 = \langle \sqrt{2}/2, \sqrt{2}/2 \rangle$. The principal axes L_1, L_2 are the lines parallel to these through $\mathbf{m} = \langle 2/3, 2/3 \rangle$. L_1 is the line $x + y = 4/3$ and L_2 is the line $x = y$. Let $\mathbf{a}_1 = \mathrm{Proj}(\mathbf{a}, L_1) = \langle 1/6, 7/6 \rangle$, $\mathbf{b}_1 = \mathrm{Proj}(\mathbf{b}, L_1) = \langle 2/3, 2/3 \rangle$, and $\mathbf{c}_1 = \mathrm{Proj}(\mathbf{c}, L_1) = \langle 7/6, 1/6 \rangle$. Therefore, $\mathrm{Var}(\mathrm{Proj}(D, L_1)) = (d(\mathbf{a}_1, \mathbf{m})^2 + d(\mathbf{b}_1, \mathbf{m})^2 + d(\mathbf{c}_1, \mathbf{m})^2)/3 = (1/2 + 0 + 1/2)/3 = 1/3$.

Likewise, let $\mathbf{a}_2 = \mathrm{Proj}(\mathbf{a}, L_2) = \langle 1/2, 1/2 \rangle$, $\mathbf{b}_2 = \mathrm{Proj}(\mathbf{b}, L_2) = \langle 1, 1 \rangle$, and $\mathbf{c}_2 = \mathrm{Proj}(\mathbf{c}, L_2) = \langle 1/2, 1/2 \rangle$. Therefore, $\mathrm{Var}(\mathrm{Proj}(D, L_2)) = (d(\mathbf{a}_2, \mathbf{m})^2 + d(\mathbf{b}_2, \mathbf{m})^2 + d(\mathbf{c}_2, \mathbf{m})^2)/3 = (1/18 + 2/9 + 1/18)/3 = 1/9$.

Thus, L_1 accounts for $(1/3)/(4/9) = 3/4$ of the variance and L_2 accounts for $1/4$ of the variance. Note that, as stated in Lemma 14.1, $\mathrm{Var}(D) = \mathrm{Var}(\mathrm{Proj}(D), L_1) + \mathrm{Var}(\mathrm{Proj}(D), L_2)$

Derivation: We wish to find the MLE for the following family of models: There exists a k-dimensional affine space $\mathscr{S} \subset R^m$ and n points $\vec{p}_1, \ldots, \vec{p}_n$ in $\mathscr{S}$. The data point $\vec{D}_i = \vec{p}_i + q_i \hat{v}_i$, where the distance q_i follows the normal distribution $N_{0,1}$ and $\hat{v}_i$ is uniformly distributed over the space of unit vectors orthogonal to $\mathscr{S}$. The problem is to find the MLE for $\mathscr{S}$, given $\vec{D}_1, \ldots, \vec{D}_n$.

There are two parts to this derivation. First, we need to show that the MLE is the one that minimizes the sum of the squares of the distances from the points to the subspace $\mathscr{S}$. This is exactly analogous to the argument for the least squares estimate in Section 14.5.

Second, we need to show that the sum of the squares of the distances is minimized when $\mathscr{S}$ is the affine space through the mean parallel to the first k singular vectors. Let $\vec{p}$ be a point in $\mathscr{S}$. Let $\mathscr{U}$ be the k-dimensional subspace of $\mathbb{R}^m$ parallel to $\mathscr{S}$; that is, $\mathscr{U} = \{\vec{s} - \vec{p} | \vec{s} \in \mathscr{S}\}$. Let V be the orthogonal complement of $\mathscr{U}$, and let $\hat{v}_1, \ldots, \hat{v}_{m-k}$ be an orthogonal basis for V.

For any data point $\vec{D}$, $d^2(\vec{D}, \mathscr{S}) = \sum_{j=1}^{m-k}((\vec{D} - \vec{p}) \bullet \hat{v}_j)^2$. Therefore,

$$\sum_{i=1}^n d^2(\vec{D}_i, \mathscr{S}) = \sum_{i=1}^n \sum_{j=1}^{m-k}((\vec{D}_i - \vec{p}) \bullet \hat{v}_j)^2 = \sum_{j=1}^{m-k} \sum_{i=1}^n ((\vec{D}_i - \vec{p}) \bullet \hat{v}_j)^2.$$

For any fixed $\hat{v}_j$ and $\vec{D}_i$, the value of $\sum_{i=1}^n((\vec{D}_i - \vec{p}) \bullet \hat{v}_j)^2$ is minimized when the coordinate of $\vec{p}$ in the $\hat{v}_j$ direction is equal to the mean of the coordinates of $\vec{D}_i$ in the $\hat{v}_j$ direction. Therefore, for any fixed choice of $\mathscr{U}$ and V, the minimum is achieved when $\mathscr{S}$ contains the mean $\mathrm{Exp}(D)$.

We can therefore recast the problem as follows: Let $\vec{m} = \mathrm{Exp}(D)$. For $i = 1, \ldots, n$, and let $\vec{F}_i = \vec{D}_i - \vec{m}$. We wish to find an orthonormal set of vectors $\hat{v}_1, \ldots, \hat{v}_{m-k}$ for which $\sum_{i=1}^n \sum_{j=1}^{m-k}(\vec{F}_i \bullet \hat{v}_j)^2$ is minimal. Let F be the $n \times m$ matrix whose rows are the $\vec{F}_i$. For any unit vector $\hat{x}$, $F \cdot \hat{x}$ is the column vector $\langle \vec{F}_1 \bullet \hat{x} \ldots \vec{F}_n \bullet \hat{x} \rangle$ so $|F \cdot \hat{x}|^2$ is equal to $(\vec{F}_1 \bullet \hat{x})^2 + \ldots + (\vec{F}_n \bullet \hat{x})^2$. Thus, we are looking for a set of orthonormal vectors $\hat{v}_1, \ldots, \hat{v}_{m-k}$ that minimizes $\sum_{i=1}^{m-k}|F \bullet \hat{v}_i|^2$. But, as we have stated in Theorem 7.13, these are the $m - k$ smallest left singular vectors of F.

14.7 Applications of Principal Component Analysis

We discuss here five applications of PCA: visualization, data analysis, to find bounding boxes for graphics and other geometric applications, to find surface normals, and to cluster related words by their appearance in documents.

14.7.1 Visualization

Suppose that you have a collection of n-dimensional data points that you want to display in a picture that reveals as much as possible of the structure. That is you want to find the best possible two-dimensional mapping of the data. Assuming that the mapping is a projection onto a plane in $\mathbb{R}^n$, then the "best" plane to choose, under the least-squares measure of "bestness," is the plane whose basis is the first two principal components.

14.7.2 Data Analysis

Suppose that we have a data collection D of vectors in $\mathbb{R}^n$. We conjecture that the points in D fundamentally lie on an affine space $\mathscr{S}$ within $\mathbb{R}^n$, and that they are slightly perturbed in the dimensions orthogonal to $\mathscr{S}$ by some small noise. The noise may be due to small random processes, or to errors in the process of measurement, or even to floating-point roundoff in the processing of the points of D. Assume that the signal-to-noise ratio is large; specifically, assume that the variance in each of the dimensions within $\mathscr{S}$ is much larger than the variance in any of the noise dimensions. Then $\mathscr{S}$ can be recovered by using the following procedure:

1. Let $\vec{\mu} = \mathrm{Exp}(D)$.

2. Let M be the matrix whose rows are $\vec{d_i} - \vec{\mu}$, where $\vec{d_i} \in D$.

3. Let $\langle \sigma_1, \ldots, \sigma_m \rangle$ be the singular values of M, and let $\langle \hat{u}_1, \ldots, \hat{u}_n \rangle$ be the right singular vectors of M.

4. Look for a sudden dropoff in the sequence of singular values—that is, an index q for which σ_{q+1} is much less than σ_q.

5. Conjecture that $\mathscr{S}$ is the q-dimensional space $\mathscr{S} = \{\vec{\mu} + t_1 \hat{u}_1 + \ldots + t_q \hat{u}_q\}$.

Once $\mathscr{S}$ is recovered, it can be used for a variety of purposes:

Prediction. Given $q - 1$ coordinates of a vector $\vec{v}$, predict that the remaining coordinates are those that will place it in the space $\mathscr{S}$.

Denoising. If the dimensions orthogonal to $\mathscr{S}$ are indeed noise, then projecting the data onto $\mathscr{S}$ may improve the quality of the data.

Data compression. Approximate each vector $\vec{d_i} \in D$ by its projection in $\mathscr{S}$. This is a generalization of the lossy data compression technique discussed in Section 7.9.3.

Another use of the principal components is to posit that they correspond to separate causal factors that combine to generate the data values. The first principal component is the most important determinant of the value, the second is the second most important, and so on. The fact that these principal components are orthogonal guarantees that these causes are independent. In fact, PCA was first invented for this kind of analysis of the results of intelligence tests. This kind of causal inference, however, lies on shaky ground; for an extensive discussion and critique, see Stephen Jay Gould's *The Mismeasure of Man*, (Gould, 1981).

14.7.3   Bounding Box

In three-dimensional geometric applications, such as graphics, robotics, and computer-aided design, detailed models of three-dimensional objects are generally given in terms of surface models with thousands or tens of thousands of

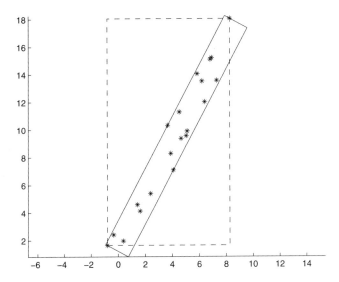

Figure 14.2. Bounding box.

surface points. A critical operation with these models is to determine whether two objects intersect. The complete algorithm that gives an accurate answer when the objects are close together is very time-consuming for complex models. Since most pairs of objects are nowhere near intersecting, it is important to have a quick way to determine that two objects do not intersect when they are actually not close to intersecting. A standard approach is to compute a *bounding box* for each object—that is, a rectangular box that contains the entire object. If the bounding boxes do not intersect, then the objects certainly do not intersect. It then becomes a problem to choose the directions of the axes of the bounding box. Ideally, we would want to choose the axes that give the box of minimal volume; but that is a difficult problem. A good approximation for this is to use the principal components as the axes.

Figure 14.2 shows a collection of points together with the bounding boxes along the x, y-axes and along the principal component axes.

14.7.4 Surface Normals

Another problem that arises with three-dimensional surface models is to find the normal at a point. This can be done as follows. Given a surface point $\mathbf{q}_0$ on the surface, find a collection of nearby surface points $\mathbf{q}_1, \ldots, \mathbf{q}_k$. Compute the PCA of the points $\langle \mathbf{q}_0, \ldots, \mathbf{q}_k \rangle$. Since these points all lie close to the tangent plane at $\mathbf{q}_0$, the tangent plane is approximately given by the first two right singular vectors, so the normal is given by the third right singular vector.

14.7.5   Latent Semantic Analysis

The following technique, called *latent semantic analysis*, is used for automatically grouping together words of related meaning by using their co-occurrence in documents.

Suppose we have an online library of documents, and we want to study the occurrence of words in the documents. Section 2.2 considered the vector model of documents, in which a document was considered a vector in a vector space whose dimensions corresponded to words. Here we take the dual approach. Let us consider a high-dimensional vector space, where each dimension corresponds to one document. A word can then be considered as a vector in this space, where the component of word $\vec{w}$ in document d is some measure of the importance of word $\vec{w}$. A cluster of words in document space is thus a collection of words that tend to appear in the same documents and therefore are presumably related. We can find these clusters as follows: (a) normalize all the word vectors to unit length; (b) carry out a PCA of the word vectors; (c) take the top n right singular vectors; (d) for each right singular vector $\hat{u}_i$, form a group G_i of all the words that are nearly parallel to $\hat{u}_i$. (Principal component analysis is not an effective way to do clustering in general; it works in this case because the sets of documents corresponding to two different word clusters are substantially disjoint and therefore orthogonal.)

Exercises

Use MATLAB for all of these exercises.

Exercise 14.1. Suppose that you have sampled 100 items from a large population and found that 75 have a specified property. The formula in Section 14.1 states that the MLE for the true fraction in the population is $75/100 = 3/4$. What is the probability of this outcome if the true probability is the MLE 0.75? What is it if the true fraction is 0.7? 0.5? 0.2?

Exercise 14.2.

(a) Using the MATLAB function `randn`, generate a vector $\vec{D}$ of ten numbers following the normal distribution $N_{0,1}$. What is the probability density $\tilde{P}(\vec{D})$ of $\vec{D}$ given the distribution $N_{0,1}$?

(b) Compute the MLE for the mean μ on the assumption that $\vec{D}$ follows the distribution $N_{\mu,1}$. What is the probability density $\tilde{P}(\vec{D})$ under this distribution?

(c) Compute the MLE for the mean μ and standard deviation σ on the assumption that $\vec{D}$ follows the distribution $N_{\mu,\sigma}$. What is the probability density $\tilde{P}(\vec{D})$ under this distribution?

(d) Compute the MLE, assuming that $\vec{D}$ was generated by a uniform distribution over the interval $[L, U]$. What is the probability density $\tilde{P}(\vec{D})$ for this estimate?

Exercise 14.3.

(a) Using the MATLAB function rand, generate a vector $\vec{D}$ of ten numbers following the uniform distribution over $[0,1]$. What is the probability density $\tilde{P}(D)$ for this distribution?

(b)–(d) Repeat Exercise 14.2, parts (b)–(d) for this new dataset.

Exercise 14.4. Consider the following set of three-dimensional points:

$$\{\langle 0, 2, 1 \rangle, \langle 0, 4, 3 \rangle, \langle 1, 4, 5 \rangle, \langle 1, 8, 6 \rangle, \langle 1, 8, 10 \rangle, \langle 4, 8, 14 \rangle, \langle 5, 9, 13 \rangle\}.$$

(a) Find the least squares estimate for z taken as a function of x and y.

(b) Find the least squares estimate for x taken as a function of y and z.

(c) Find the best-fit plane, in terms of principal component analysis.

Problems

Problem 14.1. Prove Lemma 14.1.

Problem 14.2. Prove that the bounding box for a set of points along the principal component axes is not always equal to the bounding box of minimal volume. *Hint:* Consider what happens to the two rectangles "the bounding box along the principal component axes" and "the bounding box of minimal volume" if you move interior points around.

Programming Assignments

Assignment 14.1. Write a function InvPowerLawMLE(R,EPS) that computes the MLE among inverse power laws for a frequency distribution R, to accuracy EPS > 0. Specifically, R is an n-dimensional vector, where R[I] is the number of occurrences of the Ith most common element in a sample. Therefore, R[I] is a nonincreasing function of I. Assume that this sample is generated by an inverse power law $P(X_I) = \gamma / I^\alpha$ for $I = 1, \ldots, N$, where $\alpha > 1$ and γ are constants. Here γ is a normalization factor, whose value depends on α. The function InvPowerLawMLE(R) uses a numerical binary search to compute the MLE of the parameter α, given a dataset R.

You should proceed as follows:

- Write a function InvPowerNormalize(ALPHA,N), which returns the correct normalization factor for given values of α and n.

- Write a function InvPowerLawProb(ALPHA,F), which computes the probability of F for a given value of α.

- Write the function InvPowerLawMLE(F) by doing a binary search on values of α. The binary search algorithm for finding a maximum of a continuous function $f(x)$ works as follows. In the first stage, we expand outward from a starting value until we find three values a, b, c such that $f(a) < f(b) > f(c)$. At that point, we can be sure that f has a local maximum between a and c. In the second stage, we work our way inward, testing points either between a and b or between b and c to narrow the range of search for the maximum.

In particular, since the maximum for α is certainly not less than 0, you can execute the following pseudocode.

```
function m = FindMax(f,eps)
if (f(0) > f(1))
    b = 1/2
    while (f(b) < f(0) && b > eps)
        b = b/2
        end
    c = 2*b
  else
    b = 1
    c = 2
    while (f(b) < f(c))
        b = c;
        c = 2*c;
      end
end
m = BinarySearchForMax(f,eps,a,b,c)
end

function m = BinarySearchForMax(f,eps,a,b,c)
while (c-a > eps)
  if (b-a > c-b)
      d = (a+b)/2
      if (f(d) > f(b))
          c = b
          b = d
        else
          a = d
        end
    else
      d=(b+c)/2
      if (f(d) > f(b))
          a = b
```

```
            b = d
        else
            c = d
        end
    end
m = b
end
```

Note: In many applications, it would be more reasonable to consider that the distribution holds for $I = 1,\ldots,\infty$ rather than $I = 1,\ldots,N$. The latter is valid only when you can assume that the sample includes all possible values. The computation is more difficult, however, particularly in efficiently computing an accurate value for γ when α is only slightly greater than 1.

Assignment 14.2. This assignment experiments with latent semantic analysis and dimensionality reduction for data visualization.

Step 1. Choose five or six related topics, such as { math, physics, chemistry, biology, electrical engineering, astronomy } or { presidents, actors, scientists, writers, composers, athletes }. For each topic, download 20 Wikipedia articles.

Step 2. Do data cleaning: delete stop words (see Assignment 2.1); delete HTML markup, delete words specific to Wikipedia, delete any words that occur in only one article.

Step 3. For each remaining word W, the associated *document vector $\vec{w}$* is the vector whose ith component is the number of occurrences of w in the ith document. Write a program to construct the matrix whose rows are the document vectors for all the remaining words in the corpus.

Step 4. Carry out a principal component analysis of the document vectors.

Step 5. Project the word vectors onto the plane of the first two principal components. Plot the points (or as many of them as you can put in a readable plot) onto an image.

Does this process put closely related words together? Can you attribute any semantic meaning to the two principal components (are these two directions recognizably "about" anything in particular)?

References

Adam Berger, Stephen Della Pietra, and Vincent Della Pietra. 1996. "A Maximum Entropy Approach to Natural Language Processing." *Computational Linguistics.* 22, 39–71.

D. Coppersmith and S. Winograd. 1990. "Matrix Multiplication via Arithmetic Progressions." *J. Symbolic Computation.* 9, 251–280.

Thomas H. Cormen. Charles E. Leiserson, Ronald L. Rivest, and Clifford Stein. 2009. *Introduction to Algorithms*, Third Edition. Cambridge, MA: MIT Press.

Ernest Davis. 1990. *Representations of Commonsense Knowledge.* San Mateo, CA: Morgan Kaufmann.

Philip J. Davis. 1960. *The Lore of Large Numbers.* New York: Random House.

Philip J. Davis and William G. Chinn. 1985. *3.1416 and All That,* second edition. Boston: Birkhauser.

Tobin A. Driscoll. 2009. *Learning MATLAB* . Philadelphia: Society for Industrial and Applied Mathematics.

Terence Fine. 1973. *Theories of Probability: An Examination of Foundations.* New York: Academic Press.

George Fishman. 1996. *Monte Carlo: Concepts, Algorithms, and Applications.* New York: Springer.

James Foley, Andries van Dam, Steven Feiner, and John F. Hughes. 1990. *Computer Graphics: Principles and Practice,* second edition. Reading, MA: Addison Wesley.

Malik Ghallab, Dana Nau, and Paolo Traverso. 2004. *Automated Planning: Theory and Practice,* San Mateo, CA: Morgan Kaufmann.

Amos Gilat. 2008. *MATLAB: An Introduction with Applications.* New York: Wiley.

Matthew L. Ginsberg. 1999. "GIB: Steps toward an Expert-Level Bridge-Playing Program." *Sixteenth International Joint Conference on Artificial Intelligence.* 228–233.

Stephen Jay Gould. 1981. *The Mismeasure of Man.* New York: Norton.

Stuart G. Hoggar. 2006. *The Mathematics of Digital Images: Creation, Compression, Restoration, Recognition.* Cambridge, UK: Cambridge University Press.

John P. A. Ioannidis. 2005. "Why Most Published Research Findings Are False." *PLoS Medicine.* 2:8.

Richard Karp, Michael Luby, and Neal Madras. 1989. "Monte-Carlo Approximation Algorithms for Enumeration Problems." *Journal of Algorithms.* 10, 439–448.

Adam Kilgarriff. 2010. "BNC Database and Word Frequency Lists." Available at http://www.kilgarriff.co.uk/bnc-readme.html.

Amy N. Langville and Carl D. Meyer. 2006. *Google's PageRank and Beyond: The Science of Search Engine Rankings.* Princeton, NJ: Princeton University Press.

Steven Leon. 2009. *Linear Algebra with Applications,* Eighth Edition. New York: Prentice Hall.

David Luenberger. 2006. *Information Science.* Princeton, NJ: Princeton University Press.

Robert J. Marks II. 2009. *Handbook of Fourier Analysis and Its Applications.* Oxford, UK: Oxford University Press.

Michael Overton. 2001. *Numerical Computing with IEEE Floating Point Arithmetic.* Philadelphia: Society for Industrial and Applied Mathematics.

Lawrence Page, Sergey Brin, Rajeev Motwani, and Terry Winograd. 1998. "The PageRank Citation Ranking: Bringing Order to the Web." Stanford Digital Library Technologies Project. http://ilpubs.stanford.edu:8090/422/

Mark Petersen. 2004. "Musical Analysis and Synthesis in MATLAB." *The College Mathematics Journal.* 35:5, 396–401.

Jeffrey Rouder and Richard Morey. 2011. "A Bayes Factor Meta-Analysis of Bem's ESP Claim." *Psychonomic Bulletin and Review.* 18:4, 682–689.

Stuart Russell and Peter Norvig. 2009. *Artificial Intelligence: A Modern Approach.* New York: Prentice Hall.

George Salton (ed.). 1971. *The SMART Retrieval System: Experiments in Automatic Document Processing.* New York: Prentice Hall.

Dennis Shasha and Manda Wilson. 2008. *Statistics is Easy!* San Rafael, CA: Morgan and Claypool.

S. Sinha and R. K. Pan. 2006. "How a 'Hit' is Born: The Emergence of Popularity from the Dynamics of Collective Choice," in *Econophysics and Sociophysics: Trends and Perspectives,* edited by B. K. Chakarabarti, A. Chakraborti, and A. Chatterjee. Berlin: Wiley.

Lloyd Trefethen and David Bau. 1997. *Numerical Linear Algebra.* Philadelphia: Society for Industrial and Applied Mathematics.

Chee Yap. 2000. *Fundamental Problems in Algorithmic Algebra.* Oxford, UK:Oxford University Press.

Notation

Mathematical Notation

The page references here are to the page where the notation is described or defined; therefore standard mathematical notations that are used in the text without explanation have no page reference.

Many of the basic operators are polymorphic; that is, multiplication can be an operator on two scalars, on a scalar and a vector, and so on. This list does not enumerate the different possibilities or provide page references for them. Some notations that are used in only a single section are omitted.

Special Notations

Notation	Definition	Page
$\vec{v}$	Vector	17
$\hat{v}$	Unit vector	
$\vec{e}^i$	Unit vector in the ith dimension	17
$\vec{0}$	Zero vector	17
$\vec{1}$	Vector $\langle 1,1,\ldots,1 \rangle$	17
$\overset{\Rightarrow}{x}$	Arrow	143
$\mathcal{V}$	Set of vectors	
$\mathbb{R}$	The real line	
$\mathbb{R}^n$	Space of n-dimensional vectors	
$\langle x_1,\ldots,x_n \rangle$	n-dimensional vector with components $x_1,\ldots,x_n$	
$\{\ldots\}$	Set	
$e = 2.71828\ldots$	Constant	
$\pi = 3.14159\ldots$	Constant	
$|x|$	Absolute value of x	
$|\vec{v}|$	Length of vector $\vec{v}$	

Notation	Definition	Page
$\begin{bmatrix} a & b & c \\ d & e & f \end{bmatrix}$	Matrix	47
I_n	$n \times n$ identity matrix	64
$\emptyset$	Empty set	
$\top$	True	
$\perp$	False	
Ω	Sample space	

Operators

Notation	Definition	Page
$x + y$	Addition	
$x - y$	Subtraction	
$x \cdot y, \ x * y$	Multiplication	
x / y	Division	
$x = y, \ x \neq y$	Equality, inequality	
$x < y, x > y,$	Order relations: less than, greater than, less than or	
$x \leq y, x \geq y$	equal to, more than or equal to	
$x \approx y$	x is approximately equal to y	
x^y	Exponentiation	
$\sqrt{x}$	Square root	
$n!$	Factorial	226
$[l, u]$	Closed interval from l to u	
$\vec{u}[i]$	ith component of vector $\vec{v}$	17
$\vec{u} \bullet \vec{v}$	Dot product	23
$M[i, j]$	Element i, j of matrix M	47
$M[i, :]$	ith row of matrix M	47
$M[:, j]$	jth column of matrix M	47
M^{-1}	Matrix inverse	93
M^T	Matrix transpose	49
$m \times n$	m by n (matrix dimension)	
$f \circ g$	Composition of functions f and g.	
	$(f \circ g)(x) = f(g(x))$	
f^{-1}	Inverse of function f	
$x \in U$	x is an element of set U	
$U \cup V$	Set union	
$U \setminus V$	Set difference	
$U \cap V$	Set intersection	
$\mathcal{U} \oplus \mathcal{V}$	Direct sum of vector spaces	88
$\neg p$	Not p	
$p \wedge q$	p and q	
$p \vee q$	p or q	

Notation	Definition	Page
$p \Rightarrow q$	p implies q	
$p \Leftrightarrow q$	p if and only if q	
E, F	Joint event of E and F	
X, Y	Joint random variable of X and Y	
$\#_T(E)$	The number of elements in T satisfying E	
$\sum$	Summation	
$\prod$	Product	
$\int$	Integral	
dx/dt	Derivative	
$\partial x/\partial t$	Partial derivative	
$\lim$	Limit	
$O(f(n))$	Order of magnitude growth	
$\lfloor x \rfloor$	Round x to lower integer	
$\lceil x \rceil$	Round x to upper integer	

Functions

Notation	Definition	Page
$B_{n,p}(k)$	Binomial distribution with parameters n, p	273
$\text{BpC}(\Gamma, X)$	Bits per character of coding scheme Γ. X is a probability distribution over an alphabet.	365
$C(n, k)$	Number of combinations of k out of n	227
$\text{CEnt}(X \mid Y)$	Conditional entropy of random variable X given Y	362
$\text{Coords}(\vec{v}, \mathcal{B})$	Coordinates of vector $\vec{v}$ relative to basis $\mathcal{B}$	84
$\text{Coords}(\mathbf{p}, \mathcal{C})$	Coordinates of point $\mathbf{p}$ relative to coordinate system $\mathcal{C}$	145
$\cos(\theta)$	Trigonometric cosine	
$d(\mathbf{p}, \mathbf{q})$	Distance between points $\mathbf{p}$ and $\mathbf{q}$	24
$\text{Det}(M)$	Determinant of matrix M	159, 172
$\text{Dim}(\mathcal{V})$	Dimension of vector space $\mathcal{V}$	86
$\text{Dir}(\vec{v})$	Direction of vector $\vec{v}$	147
$\text{Ent}(X)$	Entropy of random variable X	361
$\exp(x), e^x$	Exponent	
$\text{Exp}(X)$	Expected value of random variable X	260
$\text{Exp}(D)$	Mean of dataset D	386
$\text{Freq}_T(E)$	Frequency of event E in T	
$\text{Hc}(\mathbf{p}, \mathcal{C})$, $\text{Hc}(\vec{x}, \mathcal{C})$	Homogeneous coordinates of point $\mathbf{p}$ or arrow $\vec{x}$ in coordinate system $\mathcal{C}$	163
$\text{Inf}(E)$	Information of event E	359
$\text{Image}(M)$, $\text{Image}(\Gamma)$	Image space of matrix M or linear transformation Γ	89

Notation	Definition	Page
$\log(x)$	Logarithm of x to base 2	
$\ln(x)$	Natural logarithm of x (base e)	
$\max(S)$	Maximal element of set S	
$\min(S)$	Minimal element of set S	
$\mathrm{MInf}(X, Y)$	Mutual information of variables X and Y	364
$N_{\mu,\sigma}(x)$	Normal distribution with mean μ and standard deviation σ	287
$\mathrm{Null}(M), \mathrm{Null}(\Gamma)$	Null space of matrix M or linear transformation Γ	89
$\mathrm{Odds}(E)$	Odds on event $E = P(E)/(1 - P(E))$	241
$P(E)$	Probability of event E	224
$P(E \mid F)$	Conditional probability of event E given event F	231
$\tilde{P}(X = x)$	Probability density of random variable X at value x	282
$\mathrm{Proj}(\mathbf{a}, X)$	Projection of point $\mathbf{a}$ onto affine space X	154
$\mathrm{Rank}(M)$, $\mathrm{Rank}(\Gamma)$	Rank of matrix M or linear transformation Γ	90
$\sin(\theta)$	Trigonometric sine	
$\mathrm{Span}(\mathcal{V})$	Span of the set of vectors $\mathcal{V}$	82
$\mathrm{Std}(X)$	Standard deviation of random variable X	260
$\mathrm{Std}(D)$	Standard deviation of dataset D	386
$\mathrm{Var}(X)$	Variance of random variable X	260
$\mathrm{Var}(D)$	Variance of dataset D	386

MATLAB Notation

The list of MATLAB functions and operators includes just those that are mentioned in the text plus a few additional standard functions. Many of these have several uses other than those mentioned in this index or in the text. The plotting functions are omitted. References are given to the page where the notation is defined or discussed.

MATLAB Operators

Notation	Definition	Page
x=y	Assignment	1
x+y, x-y, x*y, x/y	Arithmetic	
x^y	Exponentiation	
x==y	Equality	
x~=y	Inequality	
x < y, x > y, x <= y, x >= y	Comparators	

Notation	Definition	Page
`x & y,x	y,` `~x`	Boolean operators	2
`Inf`	Infinity	3
`NaN`	Indeterminate numerical value (not a number)	3
`v(i)`	Index into vector `v`	31
`[x,y,z]`	Construct the vector $\langle x, y, z \rangle$	31
`[a,b,c;d,e,f]`	Construct the matrix $\begin{bmatrix} a & b & c \\ d & e & f \end{bmatrix}$	68
`I:J`	Construct the arithmetic sequence from `I` to `J`	31
`I:J:K`	Construct the arithmetic sequence from `I` to `J` with step size `K`	31
`{a, b, c}`	Construct the cellular array with elements `a`, `b`, `c`	75
`c{i}`	`i`th element of celluar array `c`	75
`m\c`	Find the solution to the system of equations `mx=c`	127

MATLAB Functions

Notation	Definition	Page
`asin(x),acos(x),` `atan(x)`	Inverse trigonometric functions	
`det(m)`	Determinant of matrix `m`	173
`diag(v)`	Construct the diagonal matrix with the elements in vector `v`	68
`dot(u,v)`	Dot product of vectors `u`,`v`	34
`erf(x)`	Error function	292
`erfinv(x)`	Inverse error function	292
`eye(n)`	`n` × `n` identity matrix	68
`factorial(n)`	$n!$	226
`full(m)`	Convert sparse matrix `m` into a full array	36, 73
`length(u)`	Length of vector `u`	34
`lu(m)`	LU factorization of matrix `m`	128
`max(u)`	Maximum of vector `u`	34
`mean(u)`	Mean of vector `u`	34
`median(u)`	Median of vector `u`	34
`min(u)`	Minimum of vector `u`	34
`nchoosek(n,k)`	Number of combinations of `k` items out of `n`; `n/(k! (n-k)!)`!	291
`norm(u)`	Euclidean length of vector `u`	34
`null(m)`	Null space of matrix `m`	93
`quad(f,a,b)`	Integral of `f` from `a` to `b`	93
`rand`	Random number uniformly distributed between 0 and 1	352
`rand(m,n)`	`m`×`n` matrix of random numbers	35

Notation	Definition	Page
randn	Random number distributed according to the standard Gaussian $N_{0,1}$	
randperm(n)	Random permutation of the numbers 1 to n	35
RandStream	Create a reusable stream of random numbers	352
rank(m)	Rank of matrix m	93
sin(x), cos(x), tan(x)	Trigonometric functions	
sort(u)	Vector u in sorted order	34
sparse(m,n)	Create an empty m×n sparse array	36, 73
sparse(m)	Convert full matrix M into a sparse array	36, 73
sqrt(x)	Square root	
sum(u)	Sum of elements of vector u	34
svd(m)	Singular value decomposition of matrix m	213

Other MATLAB Command and Control Structures

Notation	Definition	Page
ans	Last value computed	
else	Condition branch	4
for	For loop	4
format compact	Eliminates line space	1
format long	Display 15 digits of precision	1
format short	Display 4 digits of precision	1
format rat	Display in rational format	1
function	Function definition	7
if	Conditional	4
while	While loop	4

Index